TMS ICME

MANUFACTURING

DESIGN MATERIALS

Proceedings of the
2nd World Congress
on Integrated Computational Materials Engineering (ICME)

Sponsored by

TMS (The Minerals, Metals & Materials Society)

Held

July 7-11, 2013 at Salt Lake Marriott Downtown at City Creek,
Salt Lake City, Utah

Edited by

**Mei Li, Carelyn Campbell, Katsuyo Thornton,
Elizabeth Holm, and Peter Gumbsch**

Editors
Mei Li
Carelyn Campbell
Katsuyo Thornton
Elizabeth Holm
Peter Gumbsch

ISBN 978-3-319-48585-0
ISBN 978-3-319-48194-4 (eBook)
DOI 10.1007/978-3-319-48194-4

Chemistry and Materials Science: Professional

Published by Springer International Publishers, Switzerland, 2016
Reprint of the original edition published by John Wiley & Sons, Inc., 2013, 978-1-11876-689-7

Printed on acid-free paper

This Springer imprint is published by Springer Nature
The registered company is Springer International Publishing AG
The registered company address is: Gewerbestrasse 11, 6330 Cham, Switzerland

TABLE OF CONTENTS
2nd World Congress on Integrated Computational Materials Engineering

ICME Success Stories and Applications

Process Optimization

Materials Data for ICME

Building Blocks for ICME

ICME Challenges and Education

Preface

This is a collection of manuscripts presented at the 2nd World Congress on Integrated Computational Materials Engineering, a specialty conference organized by The Minerals, Metals & Materials Society (TMS) and the five conference organizers, and held in Salt Lake City, Utah, USA, on July 7-11, 2013.

Integrated Computational Materials Engineering (ICME) has received international attention as it has been proven to shorten product and process development time, while lowering cost and improving outcome. Building on the great success of the 1st Congress on Integrated Computational Materials Engineering in 2011 and the motivation of the Materials Genome Initiative (MGI) announced in June 2011, the 2nd World Congress on ICME convened researchers, educators, and engineers to assess the state-of-the-art ICME and determine paths to further the global advancement of ICME. Over 200 authors and attendees from all over the world contributed to this conference in the form of presentations, lively discussions, and papers presented in this volume. The international advisory committee members representing 14 different countries actively participated and promoted the conference.

The specific topics highlighted during this conference included ICME Success Stories and Applications with separate sessions on lightweighting, composites, ferrous, and non-ferrous applications; Process Optimization; Materials Data for ICME; Building Blocks for ICME with separate sessions on experimental tools, first principles and atomistic tools, computational thermodynamic and kinetics, process and performance modeling; and ICME Challenges and Education. The conference consisted of both all-conference single sessions and parallel sessions and integrated 10 keynote presentations from international experts, 2 panel discussions, 83 contributed presentations, and 70 poster presentations. From the two evening posters sessions, outstanding posters were selected for awards, which were presented to the authors at the conference dinner. The first panel discussion highlighted the need for the materials data infrastructure and initial efforts to develop such an infrastructure. The panel consisted of materials data experts from industry, academia, and government. The conference ended with a closing panel of experts focusing the discussion on the needed next steps forward to help ensure a broader and more global implementation of ICME in the future.

The 45 papers presented in these proceedings are divided into five sections: (1) ICME Success Stories and Applications; (2) Process Optimization; (3) Materials Data for ICME; (4) Building Blocks of ICME; and (5) ICME Challenges and Education. These manuscripts represent a cross section of the presentations and discussions from this conference. It is our hope that the 2nd World Congress on ICME and these proceedings will further the global implementation of ICME, broaden the variety of applications to which ICME is applied, and ultimately help industry design and produce new materials more efficiently and effectively.

Acknowledgements

The organizers/editors would like to acknowledge the contributions of a number of people without whom this 2nd World Congress, and the proceedings, would not have been possible.

First, we would like to offer many thanks to the TMS staff who worked tirelessly to make this an outstanding conference and excellent proceedings.

Second, we want to thank the international advisory committee for their input in the planning of the conference, the promotion of the conference, and their participation in the conference.

This international committee included:

John Ågren, KTH - Royal Institute of Technology, Sweden
John Allison, University of Michigan, USA
Dipankar Banerjee, Indian Institute of Technology, Bangalore, India
Yves Brechet, Institute National Polytechnic de Grenoble, France
Dennis Dimiduk, USAF Research Lab, USA
Masato Enomoto, Ibaraki University, Japan
Juergen Hirsch, Hydro Aluminum, Germany
Dorte Juul Jensen, Risoe National Laboratory, Denmark
Nack Kim, Pohang University of Science and Technology, Korea
Milo Kral, University of Canterbury, New Zealand
Peter Lee, Imperial College, United Kingdom
Baicheng Liu, Tsinghua University, China
Jiangfeng Nie, Monash University, Australia
Tresa Pollock, UC Santa Barbara, USA
Gary Purdy, McMaster University, Canada
Antonio J. Ramirez, Brazilian Synchrotron Light Laboratory, Brazil
K.K. Sankaran, Boeing Company, USA
James Warren, National Institute of Standards and Technology, USA
Deb Whitis, General Electric, USA

Finally, we would especially like to acknowledge the financial support of all our sponsors, especially ESI and UES, Inc. We are also grateful for the participation and contributions of all of the attendees.

Conference Editors/Organizers

Mei Li is a Technical Expert and Group Leader of Light Metals Research and ICME at Ford Research and Advanced Engineering Laboratory. She is one of the key developers of Ford Virtual Aluminum Casting (VAC), a successful example of ICME applications in the automotive industry. The knowledge gained in metallurgy, physics, mechanics, and the computational models developed for microstructural evolution and property predictions in VAC have been extended to other materials and processes. This includes the development of computational tools for gear steels during the heat treatment process, high-pressure die casting of aluminum alloys for additional powertrain and body applications and high-pressure die casting of magnesium alloys for body applications. Her group is also developing advanced aluminum and heat-resistant alloys through ICME guidance and experimental validation as Ford continues to seek the competitive advantage in manufacturing lightweight, durable, and energy-efficient vehicles.

She received her Ph.D. degree from the Department of Materials and Metallurgy at McGill University in 2000. She has been actively involved in practicing, advancing, and promoting ICME and has organized numerous symposia, served as Technical Advisor for the ICME Committee within TMS, is a member of the Board of Review for the TMS journal *Integrating Materials and Manufacturing Innovation* (IMMI).

Carelyn Campbell is the acting group leader for the Thermodynamics and Kinetics group in the Materials Science and Engineering Division in the Material Measurement Laboratory at the National Institute of Standards and Technology (NIST). Her research is focused on diffusion in multicomponent multiphase systems and the development of data and informatics tools for phase-based data. For the past ten years, she has sponsored the NIST Diffusion Workshop series, which brings together experimentalists and theorists to improve the development of diffusion mobility databases and the prediction of diffusion controlled microstructure evolution in multicomponent multiphase systems. She received both her B.S. and Ph.D. in Materials Science and Engineering from Northwestern University. She began her tenure at NIST in 1997 as a National Research Council Postdoctoral Fellow. In 2010, she received a Bronze Medal from the Department of Commerce for superior federal service in leading the NIST Diffusion Workshop series.

Katsuyo Thornton is an associate professor of materials science and engineering at the University of Michigan, Ann Arbor. Her research focuses on computational studies of the evolution of microstructures and their effects in a wide range of materials, including metals, semiconductors, oxides, and biomaterials. She received her Ph.D. from the Department of Astronomy and Astrophysics at the University of Chicago in 1997. She was a Postdoctoral Fellow at Northwestern University and a Visiting Lecturer and Scientist at MIT, both in Materials Science and Engineering, followed by three years as a research assistant professor in Materials Science and Engineering at Northwestern University. She has organized numerous symposiums, and she was the primary guest editor of the *MRS Bulletin*, June 2008 issue on three-dimensional materials science. She also has served in a leadership role as the inaugural Chair of the Integrated Computational Materials Engineering (ICME) Committee within TMS. Transforming undergraduate and graduate education in materials science and engineering to include computational approaches is a passion of hers, which has led her to establish the Annual Summer School for Integrated Computational Materials Education with a grant from the National Science Foundation. To date, over 70 participants who are interested in incorporating computation into materials science and engineering curricula have attended the Summer School and associated short course. She is the recipient of several prestigious awards, including the TMS Early Career Faculty Fellow Award, NSF Faculty Early Career Development (CAREER) Award, and Carl Sagan Excellence in Teaching Award.

Elizabeth Holm is a professor of Materials Science and Engineering at Carnegie Mellon University. Prior to joining CMU in 2012, she spent 20 years as a computational materials scientist at Sandia National Laboratories, working on simulations to improve processes for lighting manufacture, microcircuit aging and reliability, and the processing and welding of advanced materials. Her research areas include the theory and modeling of microstructural evolution in complex polycrystals, the physical and mechanical response of microstructures, atomic-scale properties of internal interfaces, and the wetting and spreading of liquid metals. Dr. Holm obtained her B.S.E. in Materials Science and Engineering from the University of Michigan, S.M. in Ceramics from MIT, and dual Ph.D. in Materials Science and Engineering and Scientific Computing from the University of Michigan. Active in professional societies, Dr. Holm has received several honors and awards, is a Fellow of ASM International, is President of The Minerals, Metals & Materials Society for 2013–2014, is an organizer of several international conferences, and has been a member of the National Materials Advisory Board. Dr. Holm has authored or co-authored over 110 publications.

Peter Gumbsch received the diploma degree in physics in 1988 and his Ph.D. degree in 1991 from the University of Stuttgart. After extended visits at the Sandia National Laboratories in Livermore, postdoctoral work at the Imperial College, London and the University of Oxford he returned to the Max-Planck-Institute in Stuttgart as a group leader and established the group Modeling and Simulation of Thin Film Phenomena. In 2001 he took the chair for Mechanics of Materials at Karlsruhe Institute of Technology KIT and the position as head of Fraunhofer Institute for Mechanics of Materials IWM in Freiburg and Halle.

His research activities focus on modeling and simulation of materials, in particular multiscale modeling approaches. The intention is to make materials and components safer, more reliable, and more durable. He and his team are pioneering the integration of material modeling in the product development process. His current interest is in the investigation of friction and wear processes, where complex interactions of mechanics, physics, and chemistry are important. The aim is to improve material and energy efficiency in technical systems.

Among other recognitions he was awarded the Masing Memorial Award (1998) and the Gottfried Wilhelm Leibniz Prize (2007). He is chairman of the section Engineering Sciences of the German Academy of Sciences Leopoldina. In 2009 he was nominated as Hector Fellow and in 2011 he was appointed as Advisory Professor of Shanghai Jiao Tong University, China.

2nd World Congress on Integrated Computational Materials Engineering (ICME)

ICME Success Stories and Applications

2nd World Congress on Integrated Computational Materials Engineering
Edited by: Mei Li, Carelyn Campbell, Katsuyo Thornton, Elizabeth Holm, and Peter Gumbsch
TMS (The Minerals, Metals & Materials Society), 2013

APPLICATION OF COMPUTATIONAL THERMODYNAMICS AND CALPHAD IN MAGNESIUM ALLOY DEVELOPMENT

Alan A. Luo

General Motors Global Research and Development
30500 Mound Road, Warren, MI 48090, U.S.A.

Keywords: Magnesium alloy, CALPHAD, computational thermodynamics, alloy development

Abstract

This paper presents two examples of using computational thermodynamics and CALPHAD modeling in the selection and development of new magnesium alloys. The Scheil model simulation of the solidification microstructure suggests that calcium is more effective than rare earth elements (such as cerium) in suppressing the formation of $Mg_{17}Al_{12}$ phase in binary Mg-Al alloys. Calcium additions also introduce more thermally stable phases in the ternary alloys, thus improving their creep resistance and strength at elevated temperatures.

Introduction

Application of lightweight materials is a cornerstone for all major automakers to address the challenging fuel economy targets recently set for the industry. Magnesium, the lightest structural metal, will thus see increased use in a wide range of structural and functional applications for energy generation, energy storage, propulsion, and transportation [1]. Current applications of magnesium alloys are limited to non-structural or semi-structural components in the transportation industry, due to limited mechanical properties of conventional Mg-Al-based alloys such as AZ91 (Mg-9Al-1Zn[1]) and AM60 (Mg-6Al-0.3Mn). New magnesium alloys are being developed with higher strength, ductility and creep resistance at room and elevated temperatures [2-9].

While most alloy development has occurred through meticulous experiments and microstructure-composition relationships, recent approaches rely on computational methods based on phase diagrams for materials design and process optimization. These methods allow for efficient manipulation of alloy composition and process parameters to achieve the desired microstructure and properties. Originated from the early work of Kaufman and Bernstein [10], the CALPHAD (CALculation of PHAse Diagrams) technique, based on computational thermodynamics of alloy systems, has matured over the past few decades; and many commercial software packages, such as ThermoCalc [11], FactSage [12] and Pandat [13], have become important ICME (integrated computational materials engineering) tools used in the development of new materials and products [14]. This paper demonstrates two examples of applying computational thermodynamics and CALPHAD modeling in the development of new creep-resistant magnesium alloys using Pandat8.1 code and its PanMg2012 database.

Computational Thermodynamics

The CALPHAD approach is based on the thermodynamic description of an alloy system, which denotes a set of thermodynamic parameters for all phases in the system. In a ternary alloy

[1] All compositions in wt.% except otherwise stated.

system, the phases of interest are solid, liquid and intermetallic phases. The Gibbs energy per mole of a liquid or a substitutional solid solution is [16],

$$G_m^{\phi} = \sum_i x_i {}^{\circ}G_i^{\phi} + RT\sum_i x_i \ln x_i + \Delta^{xs}G_m^{\phi} \tag{1}$$

The first term on the right-hand side (RHS) of Eq. (1) is the Gibbs energy of the component elements in the reference state at a constant temperature (T) and a pressure (P) of 1 bar, the second is the ideal Gibbs energy of mixing, and the third the excess Gibbs energy. The last term is described by the Redlich-Kister equation as given below. The number of parameters is limited to three at a constant T and P of 1 bar.

$$\Delta^{xs}G_m^{\phi}(x_i, T) = x_i x_j \sum_{\nu=0}^{\nu=2} {}^{(\nu)}\lambda(x_i - xj_j)^{\nu} \tag{2}$$

The symbols G_m, R, λ, x_i, x_j are, respectively, the molar Gibbs energy, universal gas constant, model parameters, mole fraction of component i and that of component j. For the intermetallic phases with more than one sublattice, the compound energy formalism is used and given below:

$$G_m = \sum y_p^{(i)} y_q^{(j)} G_{(p:q)} + RT\sum f_i y_p^{(i)} \ln y_p^{(i)} + \sum y_p^{(i)} y_q^{(i)} y_r^{(j)} \sum_k L_{(p,q:r)}^{(k)} (y_p^{(i)} - y_q^{(i)})^k \tag{3}$$

where G_m is the Gibbs energy expressed as a function of the concentrations of the sublattice species. The first term on the RHS of Eq. (3) is the reference term, the second ideal Gibbs energy of mixing on the sublattices, and the last term the excess Gibbs energy on the sublattice. The y's are the mole fractions of the species on a specific sublattice, f_i the fraction of a specific sublattice within the crystal, and $L_{(p,q:r)}$'s are the model parameters. The superscripts (i), (j) specify the sublattice in question and the subscripts p and q the species on the sublattices.

The above descriptions of alloy systems were coded in the Pandat [14] software for phase equilibria calculations in this paper. The latest thermodynamic database PanMg8 was used in the calculations.

Magnesium Alloy Development

Mg-Al System

Aluminum is the most widely used alloying addition in magnesium for strengthening and castability. Fig. 1 shows the calculated Mg-Al phase diagram. There are two eutectic reactions that are important to the phase constitution of Mg-Al binary alloys:

1) At 450°C L → $Al + Mg_2Al_3$
2) At 436°C L → $Mg + Mg_{17}Al_{12}$

Commercial cast and wrought magnesium alloys (AZ91, AM60 and AZ31) contain less than 10% Al, and the microstructure of these Mg-Al based alloys is generally characterized by the formation of $Mg_{17}Al_{12}$ phase. The low eutectic temperature (436°C) of $Mg_{17}Al_{12}$ phase limits the application of Mg-Al alloys to temperatures below 125°C, above which the discontinuous precipitation the $Mg_{17}Al_{12}$ phase leads to substantial creep deformation [2]. Therefore, possible approaches for improving creep resistance in Mg-Al based alloys include: 1) suppressing the formation of the $Mg_{17}Al_{12}$ phase; 2) pinning grain boundary sliding; and 3) slowing solute diffusion in the magnesium matrix.

Mg-Al-Ce System

Earlier experimental work [7, 17] has shown that additions of RE in the form of mischmetal can improve the creep resistance of Mg–Al based alloys, especially when the aluminium content was low (less than 4%). This led to the development of AE series alloys, AE42 (Mg-4Al-2RE) and AE44 (Mg-4Al-2RE) where the mischmetal RE generally contains more than 60%Ce (balance La, Nd and Pr). Fig. 2 shows the calculated liquidus projection of the Mg-Al-Ce system in the Mg-rich corner. Generally, the liquidus temperature decreases with Al addition (up to at least about 10%) and Ce (up to at least about 10%), with the following two type II invariant reactions marked at 871K (598°C) and 835K (562°C), respectively:

1) At 598°C $L + (Al,Mg)_{12}Ce \rightarrow Mg + Mg_{12}Ce$
2) At 562°C $L + (Al,Mg)_{12}Ce \rightarrow Mg + Al_{11}Ce_3$

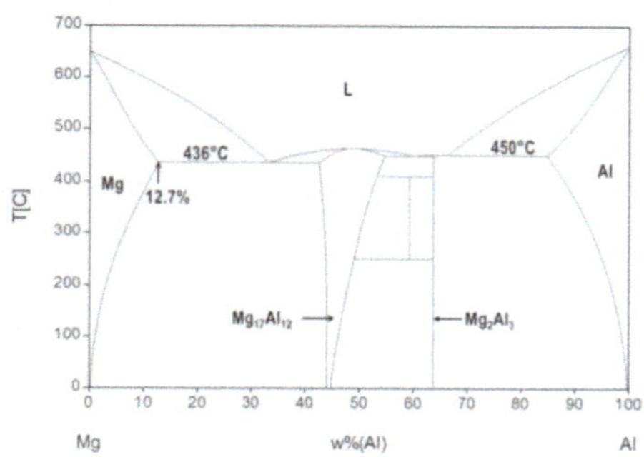

Fig. 1. Calculated Mg-Al phase diagram.

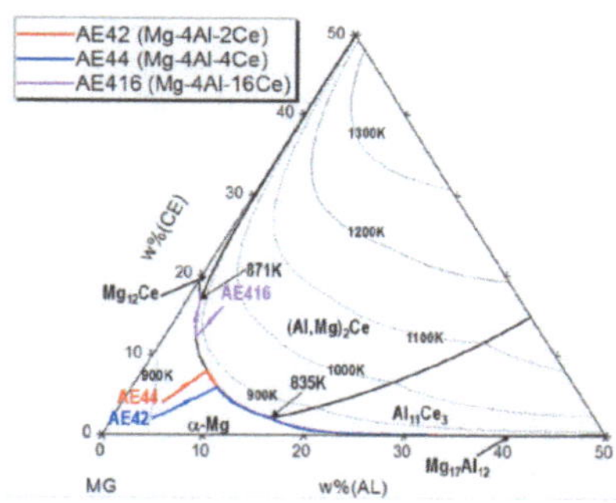

Fig. 2. Calculated Mg-Al-Ce liquidus projection and solidification paths of experimental Mg-Al-Ce alloys.

The calculated solidification paths of AE42 and AE44 alloys using the Scheil model, based on the assumption of complete mixing in the liquid but no diffusion in the solid, are superimposed in the phase diagram shown in Fig. 2. Based on the simulation results, the solidification sequence for both alloys is as follows:

1) Nucleation of primary magnesium: $L \rightarrow L + Mg$
2) Binary eutectic reaction: $L \rightarrow L + Mg + (Al,Mg)_{12}Ce$
3) Type II invariant reaction: $L + (Al,Mg)_{12}Ce \rightarrow L + Mg + Al_{11}Ce_3$
4) Ternary eutectic reaction: $L \rightarrow Mg + Al_{11}Ce_3 + Mg_{17}Al_{12}$

It is clear that the additions of 2-4% Ce to Mg-Al alloys have resulted in the formation of $Al_{11}Ce_3$ in addition to the $Mg_{17}Al_{12}$ phase in the Mg-Al binary system. Fig. 3(a) shows the effect of Ce addition to Mg-4Al alloy on the fraction of $Mg_{17}Al_{12}$ phase in the ternary alloy as calculated using the Scheil model. This indicates that it takes about 15% Ce to completely suppress the formation of $Mg_{17}Al_{12}$ phase in the Mg-4Al alloy. Similar calculations were made for Mg-Al-Ce alloys with various Al and Ce contents, Fig. 3(b), which can be used as guidance to design the ternary alloy avoiding $Mg_{17}Al_{12}$ phase formation for elevated temperature applications. Fig. 2 also shows the solidification sequence of AE416 (Mg-4Al-16Ce) alloy as follows:

1) Nucleation of $(Al,Mg)_{12}Ce$ phase: $L \rightarrow L + (Al,Mg)_{12}Ce$
2) Binary eutectic reaction: $L \rightarrow L + Mg + (Al,Mg)_{12}Ce$
3) Type II invariant reaction: $L + (Al,Mg)_{12}Ce \rightarrow L + Mg + Mg_{12}Ce$
4) Binary eutectic reaction: $L \rightarrow Mg + Mg_{12}Ce$

The eutectic temperatures for $Al_{11}Ce_3$, $(Al,Mg)_{12}Ce$ and $Mg_{12}Ce$ phases are calculated as 560°C, 622°C and 867°C, respectively; which are all significantly higher than of the $Mg_{17}Al_{12}$ phase (436°C). The Scheil model was also used to calculate the fraction of phases formed in the three AE alloys according to the above solidification paths. The results of these calculations are summarized in Table 1 and compared with commercial AM50 (Mg-5Al-0.3Mn) alloy. In AE alloys, 4-5%Al is generally needed for die castability, while it is very expensive to use 16%Ce (e.g., AE416 alloy) to suppress formation of $Mg_{17}Al_{12}$. On the other hand, AE44 alloy has significantly lower fraction of $Mg_{17}Al_{12}$, and, thus, much better high-temperature strength compared with AE42 or AM50 alloy [18]. Therefore, AE44 alloy was selected for the Corvette engine cradle application where the operating temperature would approach 150°C [19].

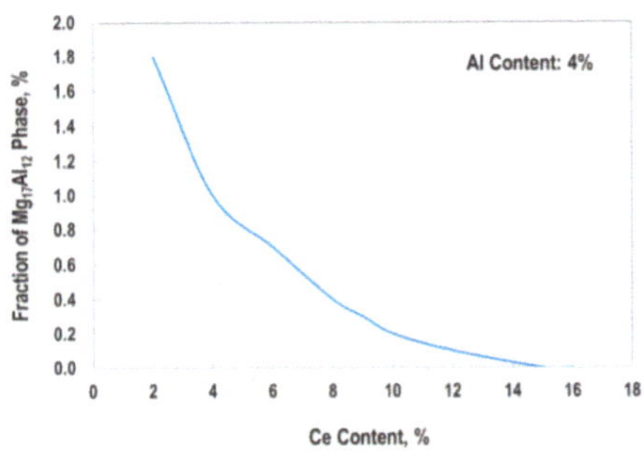

Fig. 3(a). Effect of Ce content on the fraction of $Mg_{17}Al_{12}$ phase in Mg-4Al-Ce alloys following solidification based on the Scheil simulation.

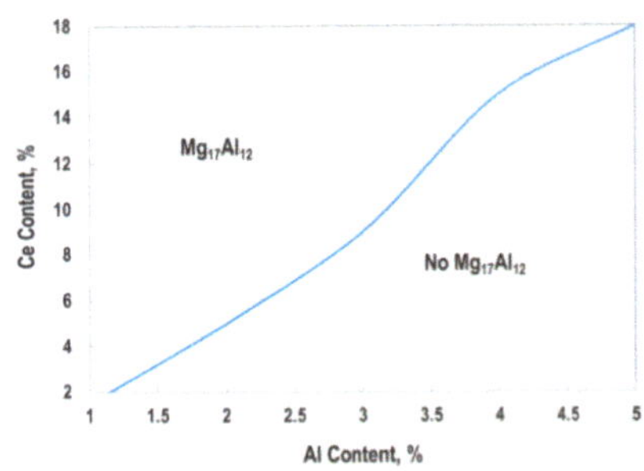

Fig. 3(b). Effect of Ce and Al content on the formation of $Mg_{17}Al_{12}$ phase in Mg-Al-Ce alloys following solidification based on the Scheil simulation.

Table 1. Scheil simulation (vol.%) of Mg-Al-Ce alloys (Baseline: AM50 alloy)

Alloy	$(Al,Mg)_2Ce$	$Al_{11}Ce_3$	$Mg_{17}Al_{12}$	$Mg_{12}Ce$
AM50	-	-	4.3	
AE42	0.9	0.2	1.8	
AE44	2.0	0.1	1.0	
AE416	9.5	0	0	0.7

<u>Mg-Al-Ca System</u>

Mg-Al-Ca system was investigated to replace the more expensive AE alloys. Fig. 4 shows the calculated liquidus projection of the Mg-Al-Ca system, superimposed by the solidification paths of three Mg-Al-Ca alloys; AX51 (Mg-5Al-1Ca), AX52 (Mg-5Al-2Ca) and AX53 (Mg-5Al-3Ca), calculated using the Scheil model. Based on the simulation results, the solidification sequence for AX51 and AX52 alloys are as follows:

1) Nucleation of primary magnesium: $L \rightarrow L + Mg$

2) Binary eutectic reaction: $L \rightarrow L + Mg + (Mg,Al)_2Ca$
3) Type II invariant reaction: $L + (Mg,Al)_2Ca \rightarrow L + Mg + Mg_{17}Al_{12}$
4) Binary eutectic reaction: $L \rightarrow Mg + Mg_{17}Al_{12}$

AX53 alloy has a different ternary eutectic reaction where Mg_2Ca is formed instead of $Mg_{17}Al_{12}$, resulting in a slightly different solidification path:

1) Nucleation of primary magnesium:
   $L \rightarrow L + Mg$
2) Binary eutectic reaction:
   $L \rightarrow L + Mg + (Mg,Al)_2Ca$
3) Ternary eutectic reaction :
   $L \rightarrow Mg + (Mg,Al)_2Ca + Mg_2Ca$

Fig. 5(a) shows the effect of Ca on the fraction of $Mg_{17}Al_{12}$ phase formed in these alloys during solidification as determined by the Scheil simulation. It is evident that the Ca content has to be greater than 2.8% in order to completely suppress the formation of $Mg_{17}Al_{12}$ in the Mg-5Al alloy. Furthermore, the critical Ca contents were calculated for Mg-Al-Ca ternary alloys containing 3-9% Al, and plotted in Fig. 5(b), which is an important composition map for optimizing creep-resistant alloys in this system.

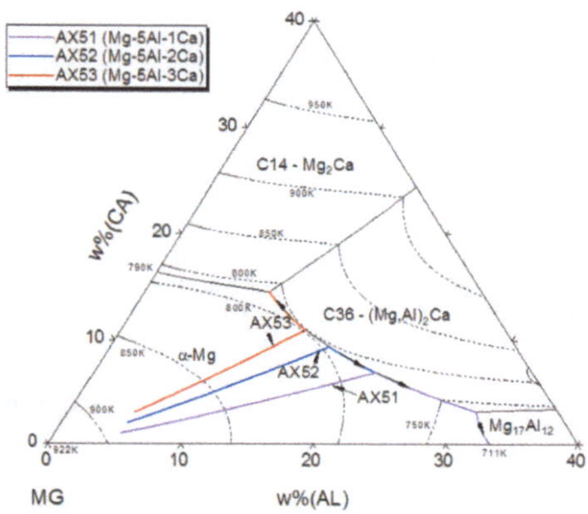

Fig. 4. Calculated Mg-Al-Ca liquidus projection and the solidification paths of the experimental Mg-Al-Ca alloys.

The presence of $(Mg,Al)_2Ca$ phase in the die-casting microstructure of Mg-Al-Ca alloys [20] and Mg_2Ca eutectic phase in AX53 (die casting) have been confirmed in previous experimental results [20, 21]. While C14 is a complete hcp (hexagonal close packed) structure with 100% hexagonality, C36 is an intermediate structure between hcp and fcc (face centered cubic) with 50% hexagonality [2]. The role of C14 and C36 phases in creep resistance of Mg-Al-Ca-based alloys has been discussed in previous investigations [20, 21]. The calculated Mg-Ca binary diagram suggests that the C14 phase has a high eutectic temperature (517°C) and melting point (710°C) and can thus exhibit better thermal stability. The C36 phase is more stable than C14 and $Mg_{17}Al_{12}$ phases in terms of the relative change in the eutectic structure during annealing. This implies an advantage of the C36 phase as a strengthener at grain boundaries in creep-resistant alloys. This computational alloy design approach confirms the 2-3% Ca present in these alloys to have significant fractions of C14 and C36 phases for creep resistance, and the AX53 alloy to have improved castability due to its reduced freeze range compared with AX51 and AX52 alloys [22]. AX53 alloy is presently being developed by GM for automotive powertrain applications.

Summary

Computational thermodynamics and CALPHAD modeling, when combined with critical experimental validation, can be used to guide the selection and development of new magnesium alloys. Ca is more effective than RE elements (such as Ce) in suppressing the formation of $Mg_{17}Al_{12}$ phase in binary Mg-Al alloys and introducing more thermally stable phases in the ternary alloys; thus, improving their creep resistance and strength at elevated temperatures. AE44 alloy is used in engine cradle applications and AX53 alloy is being developed by GM for automotive powertrain applications.

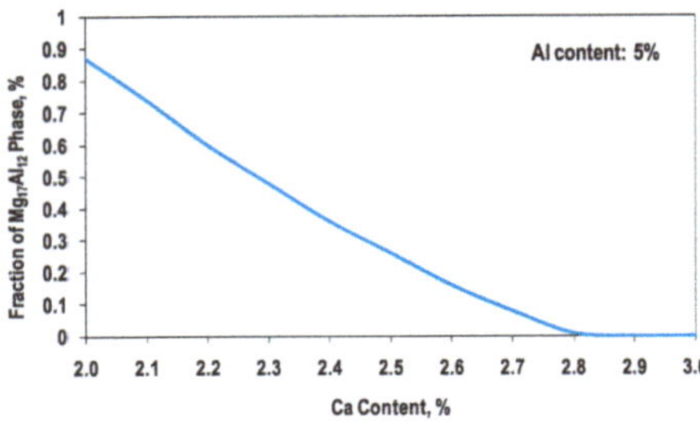

Fig. 5(a). Effect of Ca content on the fraction of $Mg_{17}Al_{12}$ phase in Mg-4Al-Ca alloys following solidification based on the Scheil simulation.

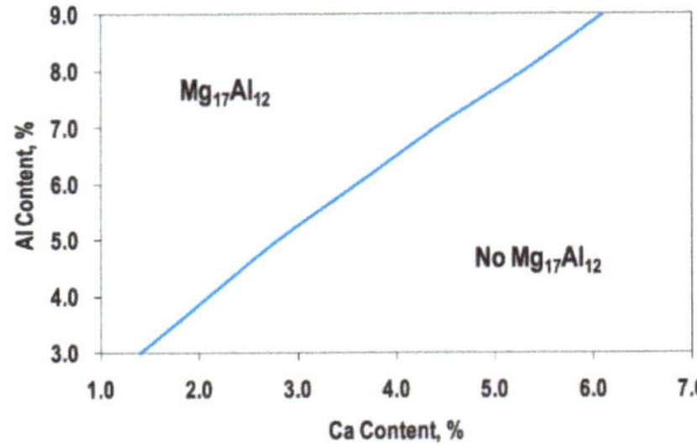

Fig. 5(b). Effect of Ca and Al content on the formation of $Mg_{17}Al_{12}$ phase in Mg-Al-Ca alloys following solidification based on the Scheil simulation.

References

1. T.M. Pollock, Science, 2010, 328, 986.
2. A.A. Luo, International Materials Reviews, 2004, 49, (1), 13.
3. P. Bakke, K. Pettersen, D. Albright, in Magnesium Technology 2004, TMS, 289.
4. S. Cohen, et al G.R. Goren-Muginstein, S. Avraham, G. Dehm, M. Bamberger, in Magnesium Technology 2004, TMS, 301.
5. A.L. Bowles, C. Blawert, N. Hort, K.U. Kainer, in Magnesium Technology 2004, TMS, 307.
6. A.A. Luo and A.K. Sachdev, in Magnesium Technology 2009, TMS, 437.
7. P. Bakke and H. Westengen, in Magnesium Technology 2005, TMS, 291.
8. P. Lyon, T. Wilks, I. Syed, in Magnesium Technology 2005, TMS, 203.
9. M.S. Dargusch, K. Pettersen, K. Nogita, M.D. Nave, G.L. Dunlop, Met. Trans., 47, 2006, 977.
10. L. Kaufman, H. Bernstein, Computer calculation of phase diagrams, New York, Academic Press, 1970.
11. http://www.thermocalc.com/
12. http://www.factsage.com/
13. http://www.computherm.com/
14. Y.A. Chang, S. Chen, F. Zhang, X. Yan, F. Xie, R. Schmid-Fetzer, W.A. Oates, Prog. Mater. Sci., 2004, 49, 313.
15. A.A. Luo, R.K. Mishra, B.R. Powell, A.K. Sachdev, Materials Science Forum, 2012, 706-709, 69.
16. Pandat 8.0 - Phase Diagram Calculation Software for Multi-component Systems, CompuTherm LLC, Madison, WI, USA, 2008.
17. W.E. Mercer II, SAE Technical Paper 900788, Society of Automotive Engineers, Warrendale, PA, USA, 1990.
18. A.A. Luo, keynote talk, Magnesium Technology 2012 Symposium, TMS Annual Meeting, Orlando, FL, USA, March 11-15, 2012.
19. J. Aragones, K. Goundan, S. Kolp, R. Osborne, L. Ouimet, W. Pinch, SAE Technical Paper No. 2005-01-0340, SAE International, Warrendale, PA, USA, 2005.
20. A.A. Luo, M.P. Balogh, B.R. Powell, Metall Mater Trans A, 2002, 33, 567.
21. A. Suzuki, N.D. Saddock, J.W. Jones, T.M. Pollock: Acta Materialia, 2005, 53, 2823.
22. A.A. Luo, B.R. Powell, A.K. Sachdev, Intermetallics, 2012, 24, 22.

2nd World Congress on Integrated Computational Materials Engineering
Edited by: Mei Li, Carelyn Campbell, Katsuyo Thornton, Elizabeth Holm, and Peter Gumbsch
TMS (The Minerals, Metals & Materials Society), 2013

MODELLING PRECIPITATION KINETICS DURING AGING OF AL-MG-SI ALLOYS

Qiang Du[1], Jepser Friis[1]

[1]SINTEF Materials and Chemistry, Trondheim, Norway

Keywords: precipitation kinetics modeling, Al-Mg-Si alloys, KWN model, morphology change

Abstract

A classical Kaufmann-Wagner numerical model is employed to predict the evolution of precipitate size distribution during the aging treatment of Al-Mg-Si alloys. One feature of the model is its fully coupling with CALPHAD database, and with the input of interfacial energy from ab-initial calculation, it is able to capture the morphological change of the precipitates. The simulation results will be compared with the experimental measurements.

Introduction

The Al-Mg-Si (6xxx) aluminium alloys, with 1-2wt% Mg and Si added to pure Al, have been widely used in construction, automobile, and marine industries due to their excellent properties (high strength/weight ratio, good formability and weldability, and excellent corrosion resistance). They are age harden-able, and the increase in strength results from the precipitation of different types of metastable phases. Experimental investigations [1] have shown that the hardening process in very complex and difficult to optimize as many parameters including alloy composition, heating rate, aging temperature and time, solution treatment temperature and cooling rate from solution, storing time at room temperature prior to aging are involved. To optimize the hardening process, great efforts have been made to develop a predictive integrative through process model [2], which consists of solidification microstructure microsegregation model, meso-scale precipitate kinetics model, solute and precipitation hardening model, etc.

The focus of this paper is to improve one important chain of the integrative model, i.e., the meso-scale precipitation model. The one that have been used is based on the numerical framework developed by Kampmann and Wagner [3] (referred to as KWN model). The essence of this model as well as its later variants is that the continuous Particle Size Distribution (PSD) curve is subdivided into size classes, each of which is associated with a precipitate number. The temporal evolution of size distribution is then tracked by following the size evolution of each discrete size class. The KWN framework has been used in various alloys systems with modifications related to the treatment of the Gibbs Thompson effect and the numerical solution procedures [4,5, 6,7,8,9, 10]. However, one assumption adopted in the model, i.e., precipitates being of spherical shape, has restricted its usage to the alloy systems such as AA6xxx alloys where precipitates are of needle shape. The aim of this paper is to remove this restriction and apply the extended model to simulate precipitation of needle-shape precipitates.

In the ealier work due to Frank S. Ham [11], the needle-shape precipitate were approximated as prolate spheroid, and a simple equation was derived which describes the diffusional flux that brings in solute to a growing precipitate by employing a variational method. Ham's work has been confirmed very recently with the computational experiments by phase field method for systems with small anisotropy and small super-saturations in [12]. However Ham did not consider Gibbs-Thomson effect and his model was only for binary alloys. Phase field model is also capable of modeling the growth and morphological evolution of several non-spherical precipitates, but it is still beyond the computational power of a personal computer for phase filed model to handle concurrent nucleation, growth and coarsening of multi-component alloys. In this paper, we will firstly extend Ham's treatment to take into account Gibbs-Thomson effect of multi-component alloys. Then the extended model is combined with empirical morphological evolution law to simulate precipitation kinetics of needle-shape precipitates during aging treatment of AA6xxx alloys.

Model Description

Experimentally it has been observed that precipitates during aging treatment of AA6xxx alloys are of needle-shape. Follow Ham's treatment, a needle-shape precipitate is approximated by a prolate spheroid with length of L and radius of r_0:

$$\frac{x^2}{b^2}+\frac{y^2}{b^2}+\frac{z^2}{a^2}=1 \tag{1}$$

Where $a=\frac{L}{2}, b=r_0$. The aspect ratio, α, eccentricity e, volume, V, and surface area, S, of the prolate precipitate are given by:

$$\alpha=\frac{L/2}{r_0} \tag{2.a}$$

$$e=\sqrt{1-\frac{4{r_0}^2}{L^2}} \tag{2.b}$$

$$V=\frac{4\pi a^3}{3}(1-e^2) \tag{2.c}$$

$$S=2\pi{r_0}^2(1+\frac{a}{r_0 e}sin^{-1}e) \tag{2.d}$$

In general α is much larger than 1, and e is close to unity.

It is useful to find the radius, R, of an equivalent sphere, whose volume is identical to the prolate:

$$R=a\sqrt[3]{1-e^2} \tag{3}$$

As to be shown later, the precipitation kinetics of the prolate and its equivalent sphere are closely related.

As in the case of the growth of spherical precipitate, it is assumed that the compositional profile of solute i in the front of the migrating precipitate-matrix interface satisfies the steady state diffusion equation:

$$\nabla^2 c_i(x)=0 \tag{4}$$

By introducing of spheroidal coordinate, the compositional profile surrounding a growing prolate could be described [11]:

$$c_i(x)=c_i^m+\Delta c_i\left(1-\frac{\ln\frac{(x+1)(x_0-1)}{(x-1)(x_0+1)}}{\ln\frac{(x_s+1)(x_0-1)}{(x_s-1)(x_0+1)}}\right) \tag{5}$$

Where $c_i(x)=c_i^m+\Delta c_i$ on the prolate surface $x=x_0$, and $c_i(x)=c_i^m$ for large $x\sim x_s$

Eq. (5) should be contrasted with the compositional profile surrounding the equivalent spherical precipitate:

$$c_i^{Spherical}(x) = c_i^m + \Delta c_i \frac{L}{2} \frac{\sqrt[3]{1-e^2}}{x} \tag{6}$$

With the variational method, Ham [11] derived the following equation describing the flux that brings in solute to the growing particle:

$$J_i^m = \frac{4\pi D_i \Delta c_i L e}{\ln\frac{1+e}{1-e}} \tag{7}$$

Eq. (7) should be contrasted with its counter-part for the equivalent spherical precipitate:

$$J_i^{m\,Spherical} = 4\pi D_i \Delta c_i \frac{L}{2} \sqrt[3]{1-e^2} \tag{8}$$

With the solute conservation law, the rate of the volume change of the prolate could be derived from Eq.(7):

$$J_i^m dt = (c_i^m + \Delta c_i - c_i^p) dV \tag{9.a}$$

$$\dot{V} = \frac{4\pi D_i \Delta c_i L e}{\ln\frac{1+e}{1-e}(c_i^m + \Delta c_i - c_i^p)} \tag{9.b}$$

Eq.(9) allows for the tracking of the evolution of precipitate volume providing Δc_i is known. It is trivial to calculate $\Delta c_i$for binary alloys in the absence of Gibbs-Thomson effect. However, as pointed out by Mou and Howe in [13], Gibbs-Thomson effect would modify Δc_i along the precipitate interface due to the local curvature:

$$\gamma = \frac{\frac{L}{2}(2r_0{}^2 + \left(\frac{L^2}{4} - r_0{}^2\right)\cos^2\beta)}{2r_0(r_0{}^2 + \left(\frac{L^2}{4} - r_0{}^2\right)\cos^2\beta)^{3/2}}, -\frac{\pi}{2} < \beta < \frac{\pi}{2} \tag{10}$$

Obviously the local curvature depends on β, so-called parametric latitude. To simplify the mathematical treatment, the local variation is treated in an average way, i.e., we assume that the modification to the interfacial matrix composition on the whole prolate surface is represented by the mean curvature of the prolate surface. This assumption may be justified by short-circuit effect originated from fast diffusion along the migrating surface. More rigorous treatment with the variational method is under development. Therefore we have

$$\gamma_{mean} = \frac{1}{\pi} \int_{-\frac{\pi}{2}}^{\frac{\pi}{2}} \frac{\frac{L}{2}\left(2r_0{}^2 + \left(\frac{L^2}{4} - r_0{}^2\right)\cos^2\beta\right)}{2r_0\left(r_0{}^2 + \left(\frac{L^2}{4} - r_0{}^2\right)\cos^2\beta\right)^{\frac{3}{2}}} d\beta \tag{11}$$

Fig. 1 plots the dependence of the normalized mean curvature, $\hat{\gamma}_{mean}$, on aspect ratio where $\hat{\gamma}_{mean} = \frac{\gamma_{mean}}{\frac{1}{R}}$. Clearly the curvature effect is more significant than in the equivalent spherical precipitate. With this assumption it is possible to obtain Δc_i together with kinetic constrains and local equilibrium assumption.

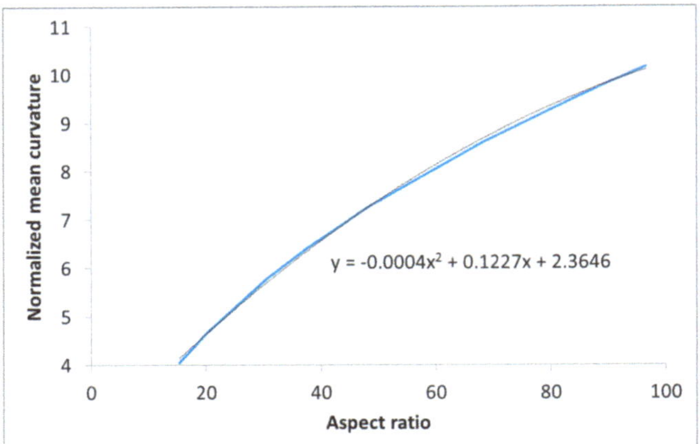

Fig. 1 the dependence of the normalized mean curvature on aspect ratio

Theoretically the evolution of aspect ratio could be due to anisotropy in interfacial energy such as in the case of the development of dendrite during solidification. Other factors might be contributing too. For example one could speculate that preferential growth occurs at the needle tip while thickening of the cross section is slow. It might be possible for the entire surface of the particle serves to capture solute atom from solution, and there might be a mechanism to transport them to the active region of the surface [14]. It is a very interesting topic to be investigated with phase field model while our treatment of the evolution of aspect ratio is of empirical nature. Experimentally it is known [15] that aspect ratio is not constant but scale with the size parameter of the cross section of the precipitate, *d*, i.e.,

$$\alpha = 14\, nm^{-1}\, d \tag{12}$$

Finally some words on the phase diagram used in our simulation. The coupling with CALPHAD method has been implemented to get access to multi-component phase diagram. Our final goal is to simulate the precipitation of β" phase. However the simulations to be reported in the next section are performed for Mg_2Si phase due to the unavailability of the relevant metastable phase diagram. The construction of the metastable phase diagram for β" is in progress.

Results and Discussions

The input parameters use in the simulations are listed in Table 1.

Table 1. The input parameters used in the simulations

Surface energy (J/m^2)	Molar volume (m^3)	Thermodynamic database	aging temperature	Mg and Si Diffusivity within matrix phase (m^2/s) [16]	
0.24	1.0×10^{-5}	TTAL6*	250 °C	1.9×10^{-19}	3.8×10^{-19}

* TTAL6 is the Al-based alloys database developed by Thermotech, Surrey, UK.

The first simulation is for the growth of single small prolate precipitate from the center of a finite volume of 0.0072 μm^3 (corresponding a sphere with a diameter of 240 nm). Initially the precipitate has a r_0 of 0.5 nm and aspect ratio of 7.2. It grows for about 30 minutes to r_0 of 6.2 nm and aspect ratio of 86. The predicted precipitate fraction evolution curve is plotted in Fig. 2 together with the one for a spherical precipitate (i.e., with constant aspect ratio equal to 1) growing in the identical conditions. Both of the two curves approaches to the final equilibrium fraction (1.06%), but the prolate precipitate grows much faster than the spherical one.

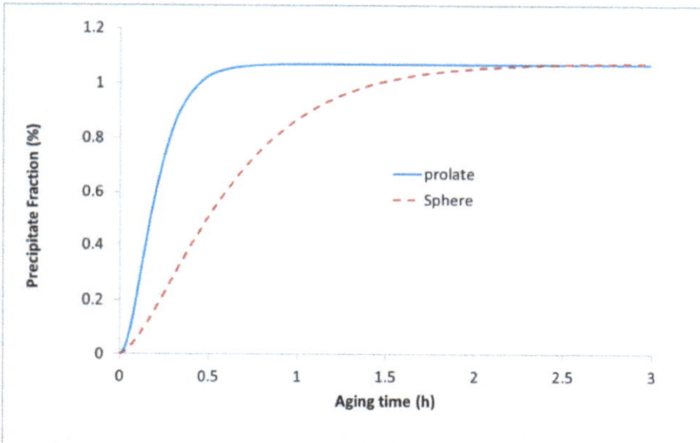

Fig. 2 The comparison of two predicted precipitate fraction evolution curves obtained with different approximations of precipitate shape (prolate and sphere) for the growth of single Mg2Si precipitate at 250 °C from Al-0.65wt%Mg-0.59wt%Si solid solution.

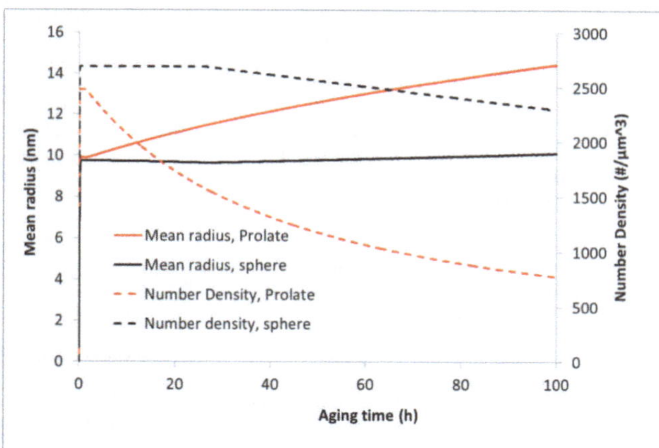

Fig. 3 The comparison of the predicted mean radius and precipitate number density evolution curves obtained with different approximations of precipitate shape (prolate and sphere) for the aging treatment of Al-0.65wt%Mg-0.59wt%Si at 250 °C.

The second group of simulations is for the aging treatment of Al-0.65wt%Mg-0.59wt%Si at 250 °C. Concurrent precipitate nucleation, growth, and coarsening have been taken into account. The predicted mean radius and number density are plotted in Fig. 3. Clearly after 100 hours aging treatment significant coarsening occurs in the simulation with prolate shape precipitate while negligible coarsening in the one with spherical precipitates. It implies better agreement could be obtained with the experimental results reported in [1] where significant coarsening was observed after only 3 hours of aging heat treatment.

Conclusions

The KWN model has been extended to take into account the effect of precipitate shape on precipitation kinetics during aging treatment of Al-Mg-Si alloys. In the proposed model the growth rate of a prolate shape precipitate, in comparison to a spherical precipitate with identical volume, is magnified by a factor of $\frac{\frac{2}{\sqrt[3]{1-e^2}}e}{\ln\frac{1+e}{1-e}}$, where e is eccentricity. Its curvature effect is magnified by a factor of $-00044\alpha^2 + 0.122\alpha + 2.36$, where α is aspect ratio.

The model has been applied to simulate the aging experiments reported in the literature, and reasonable agreements have been obtained. Further improvement is required to handle shape coarsening.

Acknowledgement

The financial support from the Research Council of Norway under the eVITA programme with contract number of NFR Project 205353/V30 "Multiscale Modelling of hardening Precipitate Interfaces in Alloy Design" is gratefully acknowledged.

[1] C. D. Marioara, H. Nordmark, S. J. Andersen, R. Holmestad, "Post-β" phases and their influence on microstructure and hardness in 6xxx Al-Mg-Si alloys", J. Mater. Sci 41 (2006) 471-478.
[2] Myhr, Ole Runar; Grong, Oystein; Pedersen, Ketill Olav, "A Combined Precipitation, Yield Strength, and Work Hardening Model for Al-Mg-Si Alloys", METALLURGICAL AND MATERIALS TRANSACTIONS A-PHYSICAL METALLURGY AND MATERIALS SCIENCE Volume: 41A Issue: 9 Pages: 2276-2289.
[3] Wagner R, Kampmann R. Homogeneous second phase precipitation, in: Cahn RW (Eds.). Materials Science and technology: a comprehensive treatment, Weinheim: John Wiley & Sons Inc.; 1991.
[4] Maugis P, Goune M. Acta Materialia 2005; 53: 3359.
[5] Myhr OR, Grong O. Acta Materialia 2000; 48: 1605.
[6] Robson JD. Acta Materialia 2004; 52: 4669.
[7] Gandin CA, Jacot A. Acta Materialia 2007; 55: 2539.
[8] Perez M, Dumont M, Acevedo-Reyes D. Acta Materialia 2008; 56: 2119.
[9] Kamp N, Sullivan A, Tomasi R, Robson JD. Acta Materialia 2006; 54: 2003.
[10] Q. Du, W. Poole, M. Wells, "A mathematical model coupled to CALPHAD to predict precipitation kinetics for multicomponent aluminum alloys", Acta Materialia 60 (2012) 3830-3839.
[11] Frank S. Ham, "Theory of diffusion-limited precipitation", J. Phys. Chem. Solids, 1958, Vol.6. pp335-351.
[12] Rajdip Mukherjee, T.A. Abinandanan, M. P. Gururajan, "Phase field models as computer experiments: Growth kinetics of anisotropic precipitates", Materials Science Forum, Volume: Advances in Materials Development, 2013.
[13] Y. Mou, J. M. Howe, "Diffusion fields associated with prolate spheroids in size and shape coarsening", Acta Mater. Vol. 45, No. 2, pp.823-835.
[14] Frank S. Ham, "Diffusion-limited growth of precipitate particles", Journal of applied physics, Volume 30, number 10, October 1959. 1518-1525.
[15] J. Friis, internal report, 2012.
[16] Y. Du et al, "Diffusion coefficients of some solutes in fcc and liquid Al: critical evaluation and correlation", Materials Science and Engineering A363 (2003) 140-151.

2nd World Congress on Integrated Computational Materials Engineering
Edited by: Mei Li, Carelyn Campbell, Katsuyo Thornton, Elizabeth Holm, and Peter Gumbsch
TMS (The Minerals, Metals & Materials Society), 2013

MODELING PROCESSING-PROPERTY RELATIONSHIPS TO PREDICT FINAL ALUMINUM COIL QUALITY

Kai F. Karhausen, Galyna Laptyeva, Stefan Neumann

Hydro Aluminium Rolled Products GmbH
R&D, Georg-von-Boeselager-Str. 21
53117 Bonn, Germany

Keywords: Aluminum, Rolling, Process Simulation, Property Simulation

Abstract

In the definition of the term ICME the integration of multiple length-scale models is the key requisite to obtain information required to design products. While product properties are a measure on a macroscopic length scale, they are controlled by the microstructure on a microscopic or even atomistic length-scale. They are generated on production facilities on a very large scale.

In translation of the definition of ICME to the production of rolled Aluminum semi-fabricated products alloys and processing routes must be combined to deliver improved final customer properties to enable the development of new designs and products. Since the relationships between processing conditions, microstructural evolution and derived properties are highly non-linear, this task can only be achieved by computer aided methods.

Critical properties of Aluminum coils and sheet are on the one hand the physical and chemical properties of the metal, such as strength, elongation, anisotropy, formability, corrosion resistance etc.. But on the other hand, geometrical tolerances and surface quality are equally important and often result from metallurgical events during processing. In some cases they can be derived from integrated process and microstructure simulation methods as well.

This paper describes the state of the art in through process modeling of Aluminum coils and strip with a view of tracing the microstructural development and deriving property information. Especially new approaches to critical parameters such as coil stability and sheet flatness, resulting from microstructural mechanisms are treated by sample computations of industrial processing chains.

Introduction

The properties of aluminum wrought alloys and their products strongly depend on the process conditions imposed during almost each fabrication step (i.e. casting, homogenization, hot and cold rolling and annealing). It is specific to aluminum as compared to other materials, that already very early process stages take effect on the final properties. Therefore, the control and optimization of properties requires sound knowledge of the underlying mechanisms and the sometimes complex interactions between metallurgical and processing parameters. These are well developed in plant experience, usually based on trial and error. But the implementation of new alloys and processing routes involves extensive testing, which is very costly and time consuming. For these reasons, research in the area of what is termed today Integrated Computational Materials Engineering (ICME) has been conducted in the Aluminum industry since the late 1990s (e.g.[1]) at that time termed Through Process Modeling (TPM). In 2000, three coordinated projects, funded by the European Union as VIR* were initiated to develop and apply new simulation tools and methods to integrate physically based models to the main wrought alloys and their fabrication routes. The result was a "TPM" exercise that proved the validity of the different material and microstructure models, integrated into process models of industrial practices [2].

Meanwhile, models are well advanced and increased computation power allows for applications in many areas. Thus TPM is often not only used for designing new alloys or novel production technologies or routings but it serves also the analysis of running processes in order to achieve more stable process windows, to allow for tighter tolerances in final properties or to avoid production problems.

Modeling of the Rolling Process Chain

The models being applied in this study are briefly described in the following. RoseRoll [3] (ROSE=Rolling Simulation Environment of Hydro Aluminium) represents a thermo-mechanical code, employed to simulate rolling. It is a dedicated 2D code for fast application on multi-pass hot and cold rolling including inter-pass times. A number of material models are incorporated in a fully coupled mode. These are Strucsim [4] for the modeling of separate structures as they occur in partial recrystallisation situations and 3IVM [5-7] for work hardening and dynamic and static recovery, based on dislocation statistics.

The combined models Strucsim and 3IVM in their original versions accounted for the state of solute and precipitates (microchemistry) by the use of metallo-physical parameters, which were adjusted for a specific alloy. In the extension as described below, they now take as input the amount of solutes as well as the amount and size of particles being formed in previous operations. These must be determined either by measurement or by a microchemical simulation. The numerical model used for this purpose is ClaNG (**C**lassical **N**ucleation and **G**rowth) [8, 9], which considers the processing and the chemistry effects on particle precipitation and dissolution. ClaNG is based on the theory of nucleation and growth of secondary phases and multi-component thermodynamics and is linked to thermodynamic databases, which provide the phase diagram information [10]. The main input to the model is the time/temperature history and the alloy composition, whereby Si, Fe, Mn, Mg, Cu, Cr, Ti can be selected for modeling. For an accurate prediction of the nucleation sites for precipitation, the dislocation density accumulated during the preceding operations, has to be provided.

Furthermore, RoseTem is a FDM based thermal model to simulate coil heating and cooling processes in combination with RoseStat to determine static recovery and recrystallisation during these processes. Furthermore a module RoseWind [11] is available to calculate the mechanics of winding operations of strip to coils. These models run in a decoupled mode. A successive coupled mode is accomplished between all models via standardized interfaces, which enable the exchange of pre-treated materials between the models and allows for a comparative application of different model combinations on the same processes or material state.

<u>Material models for prediction of hardening and softening behavior</u>

The 3IVM+ model (3-Internal-Variable-Model) predicts the flow stress evolution of cell building metals, such as Aluminium and Aluminium-alloys.

The three internal state variables of the model are the mobile dislocation density ρ_m, the immobile dislocation density in cell interiors ρ_i and in cell walls ρ_w. These variables represent the microstructure of the modeled alloy.

In the course of a forming process, dislocations move and a number of interactions with other dislocations, e.g. annihilation, dipole formation, lock formation take place. The corresponding evolution of each type of dislocation density is described by the structure evolution equation:

$$\dot{\rho}_{m,i,w} = \dot{\rho}^{+}_{m,i,w} - \dot{\rho}^{-}_{m,i,w} \quad (1)$$

where $\dot{\rho}^+$ and $\dot{\rho}^-$ are the dislocation production and reduction rates, which are influenced by temperature, strain rate and the microstructure itself.

In order to calculate the macroscopic flow stress k_f, which is required to achieve the imposed strain rate $\dot{\varepsilon}$ for a given temperature T and microstructure, the kinetic equation of state (a partial derivative of the Orowan equation) is applied:

$$\dot{\gamma} = \dot{\varepsilon} \cdot \overline{M} = \rho_m \cdot b \cdot \upsilon(\tau_{eff}, T) \tag{2}$$

where $\dot{\gamma}$ is the shear rate, $\overline{M}$ the average Taylor factor, b the magnitude of the Burgers vector; υ is the average dislocation velocity, which is a function of effective stress τ_{eff} and temperature T.

The shear stress in the cell interiors and cell walls, which enables dislocation slip, is the sum of the passing stress of dislocations and the increment of shear stress coming from alloy chemistry τ_{chem}:

$$\tau_{i,w} = \alpha \cdot G \cdot b \cdot \sqrt{\rho_{i,w} + \rho_m} + \tau_{chem} \tag{3}$$

where α is the geometric constant, G the shear modulus, $\tau_{chem} = \tau_p + \tau_{sol}$; τ_p is the increment of shear stress from particles [12] and τ_{sol} the increment of shear stress from solutes [7]. The macroscopic flow stress can be then calculated using the following equation:

$$k_f = \overline{M}(f_i \cdot \tau_i + f_w \cdot \tau_w) \tag{4}$$

where f_i and f_w are the volume fractions of the cell interiors and cell walls, respectively.

To model the static recrystallisation, a dislocation density based recrystallisation model is combined with 3IVM. This approach enables an accurate consideration of the recovery effect on recrystallisation nucleation and nuclei growth. The driving force for static recrystallisation arises from the total dislocation density, which is composed of the dislocation density fractions in the cell interiors and cell walls, being predicted by 3IVM.

The velocity of the growth (=grain boundary velocity υ_{GB}) is calculated using the following simplified expression:

$$\upsilon_{GB} = m \cdot p \tag{5}$$

where m is the grain boundary mobility and p is the driving force for grain boundary motion. In the presence of dispersoids and solutes the driving force p is strongly affected by the Zener drag p_{Zener} and the solute drag p_{Sol}:

$$p = p_{RX} - p_{Zener} - p_{sol} \tag{6}$$

In the model, Zener drag is considered for the two most common particle types in the alloy. Solute drag is calculated using the approach, proposed by Lücke and Detert [13].

Finally, the recrystallized fraction f_{RX} can be calculated using the "up-dated" grain boundary velocity values, by integrating the function for the recrystallisation rate:

$$\frac{df_{RX}}{dt} = (1 - f_{RX}) \cdot 4 \cdot \pi \cdot N_0 \cdot r^2 \cdot \upsilon_{GB} \tag{7}$$

where $(1\text{-}f_{RX})$ is the fraction of the unrecrystallized microstructure, N_0 the density of nuclei and r the recrystallized grain radius.

Simulation of creep

A special attention in this study is devoted to the effect of creep on the coil quality, whereby the focus has been put on the sheet flatness. For creep modelling a phenomenological description of creep rate $\dot{\varepsilon}$ as a function of the internal stress σ and temperature T is commonly used:

$$\dot{\varepsilon} = A \cdot (\sinh(B \cdot \sigma))^n \cdot \exp\left(-\frac{Q}{RT}\right) \quad (8)$$

where Q is the activation energy for creep, R is the gas constant and A, B and n are material constants. A can be formulated as function of k_f assuming that the dislocation reactions during static recovery are proportional to the reactions during creep. Thus, the creep strain rate is a function of the materials pre-treatment.

The material constants and the activation energy were determined using stress relaxation measurements, performed in a temperature range of RT-110°C and a load range of 10-60 MPa. Depending on the alloy chemistry and precipitation conditions, different initial internal stresses can be generated during rolling. After the subsequent winding operation and during storage, these internal stresses are driving the creep deformation [15].

Performance of the material models

To obtain the required material parameters for the flow behavior, a set of uniaxial compression tests was conducted on a computer-controlled servo-hydraulic hot forming facility at the Institute of Metal Forming of the RWTH Aachen. A constant strain rate during compression is achieved by setting the tool velocity via displacement control. Throughout the measurement the specimen, the tools and the surrounding atmosphere are held at constant homogeneous temperature in a furnace. Hence, uniform conditions within the samples are ensured for the whole series of tests. The test range covers a spectrum of temperatures from 25 to 550°C at constant strain rates of 0.1, 1, 10 and 100 s^{-1}. The adiabatic heating during the test was corrected by an iterative procedure [14]. Fig.1 shows selected measured flow curves and their representation by the 3IVM model over a large span of conditions.

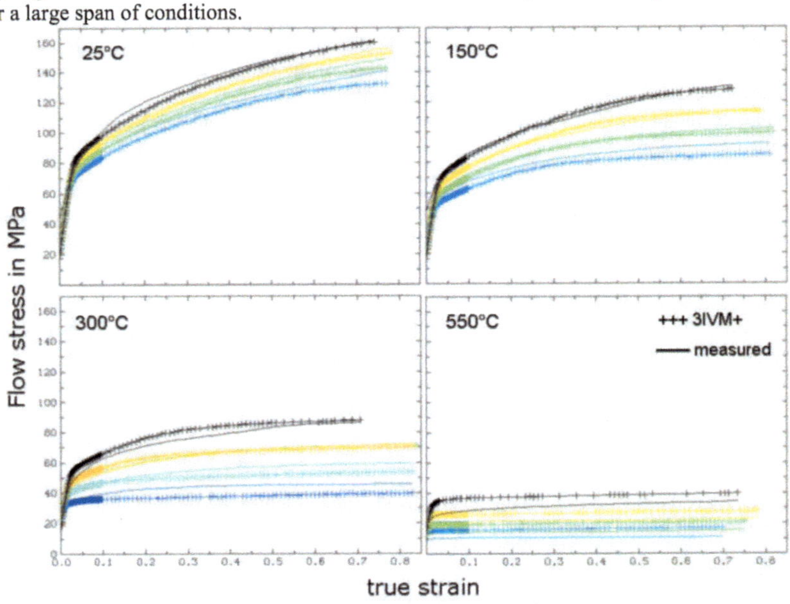

Figure 1: Measured and calculated flow curves of alloy AA8079 for strain rates of 0.1 (blue), 1 (light blue), 10 (yellow) and 100 (black) s^{-1}

In the alloy AA8079 the main elements that contribute to changes in the microchemical composition are Fe and Si. These form particles from which Al_3Fe and α-AlFeSi are the most prominent. The particles form in the cast slab during pre-heating before hot rolling, while the state of elements in solid solution changes accordingly. A ClaNG simulation of the thermal history of the samples tested in hot compression (see above) revealed a concentration of Fe=1.06e-2 atm.% and Si=9.61e-4 atm.% in solid solution and particle concentrations of Al_3Fe=1.4 vol.% and α-AlFeSi=0.03 vol.%. Fig.2 shows experimental recrystallisation data and simulations with the above chemistry. Furthermore, two extreme cases were simulated, in which the lowest and highest states of solid solution in this alloy were assumed and the particle concentrations were modified accordingly. It is obvious that the microchemistry has a considerable effect on recrystallisation kinetics within the bounds of realistic temperature pre-treatments. It must be further noticed, that the RX kinetics do not only directly depend on the microchemical state during the holding period, but are furthermore determined by the different work hardening kinetics of prior deformations, which also depend on the state of microchemistry.

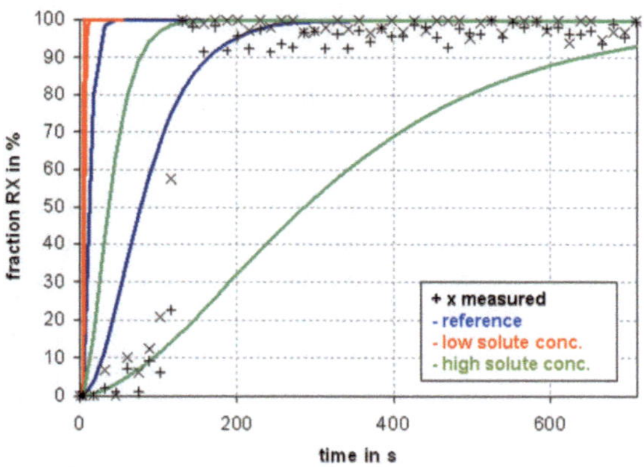

Figure 2: Measured and simulated RX-kinetics for three microchemistry states in AA8079 at two temperatures 400°C and 500°C.

Sample case: Hot Rolling

A first case is calculated with the TPM on hot rolling of alloy AA8079. In industrial practice it was found, that this alloy sometimes was difficult to wind to a coil after the final hot rolling pass. The winding takes place on an expandable mandrel, which is taken off after the coils has been transported from the mill to an intermediate storage position. It was sometimes observed that after some time of storage, while cooling down from rolling temperature, the coil slowly deformed into an oval shape. In such case the coil cannot be further processed and must be scrapped. However, this problem occurred in single cases only and did not correlate to obvious changes in the processing.

A proposed explanation of this effect is, that the hot strip may be in an unstable process window with respect to recrystallisation directly after the last pass. E.g. fluctuations in pre-heating time, exit temperature or mill speed could in some critical combinations cause recrystallisation to take place directly after winding, causing the coil to loose stability, while in most cases RX would be retarded.

To check this theory, a TPM simulation of the process chain was run from slab pre-heating, hot rolling on a roughing reversing mill, hot rolling on a 4-stand tandem mill, coil winding and coil storage for cooling. Variations in pre-heating were performed with varying times in a temperature area between 530 and 600°C leading to different initial conditions for hot rolling in terms of the microchemistry of the slab. Variations were also calculated for the rolling speed in the tandem mill by assuming ±50% variations in the entry velocity. Fig.3 presents some simulated process data of the finishing tandem mill. Microchemistry and microstructure alone do not affect the process very much although a slight variation in flow stress can be observed leading also to slight temperature variations. The change of the rolling speed however has a considerable effect. Shorter inter-stand times and increased work hardening at higher strain rates lead to an increase in exit temperature of 78°C. It is also interesting to notice, that the flow stress of the variant with lower solute content but higher amount of particles is initially higher but due to accelerated softening this difference diminishes slightly.

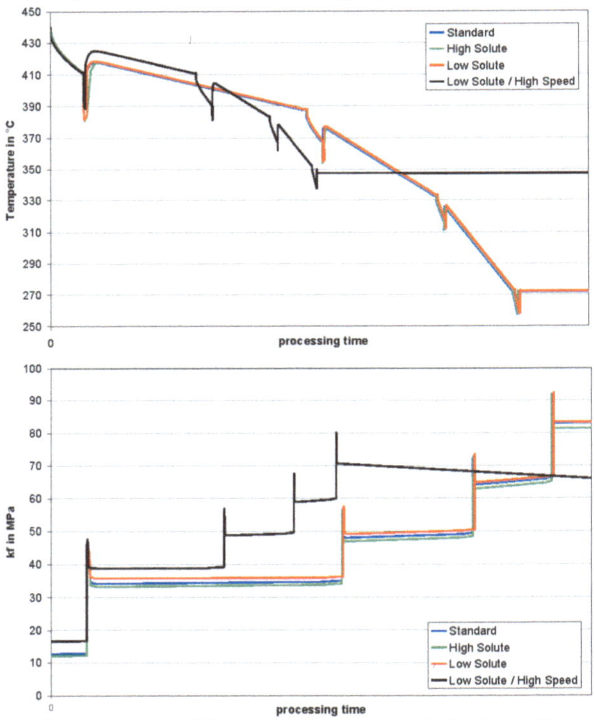

Figure 3: Computed temperature and flow stress development in the 4-stand hot tandem mill.

As a result of the computations, the recrystallisation kinetics in the coil after rolling are dramatically changed as depictured in fig.4. A decreasing solute content and the corresponding increasing particle fraction accelerate the recrystallization under constant processing conditions. Process variations, e.g. increased speed, can further accelerate the recrystallisation by providing higher driving forces. Altogether this results in a large flow stress spectrum of the material exiting the mill and being wound to a coil.

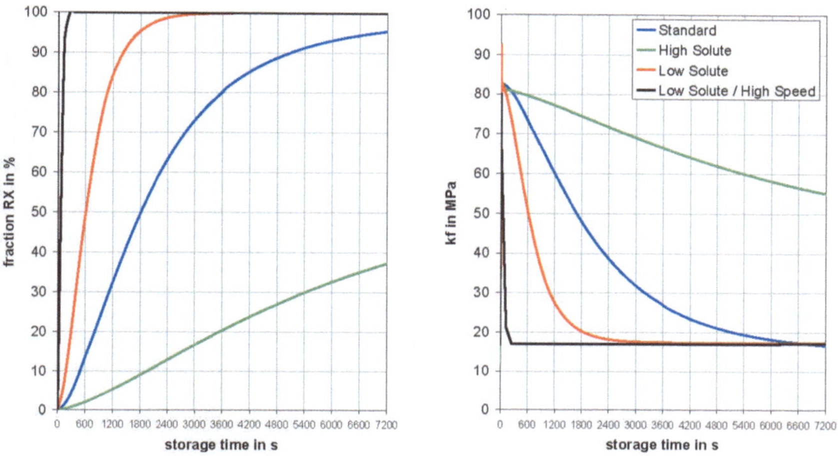

Figure 4: Computation of the material condition after exiting the hot mill and being wound and stored: a) recrystallised volume fraction and b) flow stress

The process of winding a coil after rolling must be controlled within a certain process window. A too loose winding can cause the coil to open in a spiral mode, a too tight winding will introduce high stresses and may cause the coil to deform. In the further computation of the TPM the coil winding is simulated thermo-mechanically as described in [11].

Applying a winding stress of 18 MPa, a stress distribution is obtained as depicted in Fig.5a directly after winding when the expandable mandrel is still in place. After removal of the mandrel a stress re-organization takes place, as depicted in fig.5b. We now observe high stresses above 60 MPa at the coil eye, which is not supported anymore. If the stresses exceed the flow stress, the coil will deform. It is now interesting to compare these internal stresses with the computed flow stress of fig.4b. In the material with low solute content, processed at increased speed, recrystallisation will complete within a few minutes after rolling. If the mandrel is removed, the coil will not be able to withstand the wound in stresses. The situation becomes safer, for a higher content of elements in solid solution.

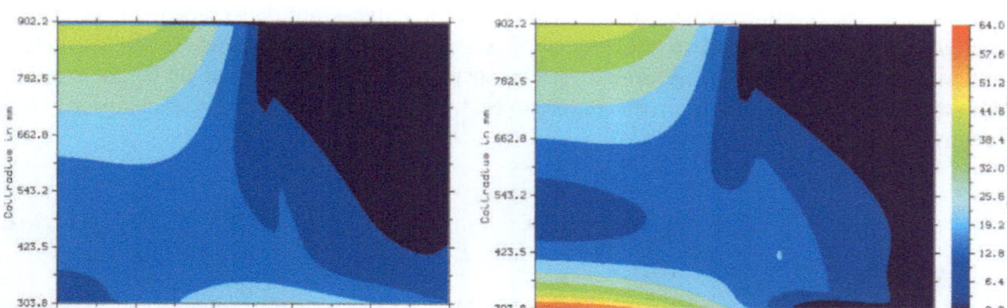

[redacted] [redacted]

Figure 5: v.Mises stress distribution in a coil a) directly after winding, b) after removal of mandrel.

Sample case: Cold Rolling

The microstructural and microchemical state takes effect on material properties further down the production route. In multi pass cold rolling on a single stand non-reversing mill, the material work hardens and subsequently recovers/recrystallizes during annealing stages or during storage times. A dominant feature of a coil as delivered to a final customer is the flatness of the strip upon unwinding. Thus it is important, that the stress state in the coil does induce strains, which may cause the strip to form buckles or waves. Stresses are introduced by the winding process itself while the straining is essentially a material characteristic, which depends on the creep properties of the alloy and the accumulative work hardening prior to the last winding operation.
In the following sample of an industrial case, the winding of a 0.25 mm thick strip is calculated, as it leaves a leveling line. Thus, it can be assumed that there is no initial unflatness in the material prior to winding. The winding of a profiled strip even with a rather low tension of 8 MPa again introduces internal stresses in the coil which are shown in fig.6a. Since aluminum is a material prone to creep deformation even at room temperature, this material deforms during the subsequent storage time. In fig.6b the strains are shown, which have formed after one week of storage of a low alloyed material. Alloys of the 1xxx or 8xxx class can gain in creep resistance, if the main alloying elements Fe and Si can be kept in solid solution during the preceding processing stages. Thus, a suitable pre-heating before hot rolling or intermediate annealing in-between the cold rolling passes offer means to minimize this effect.

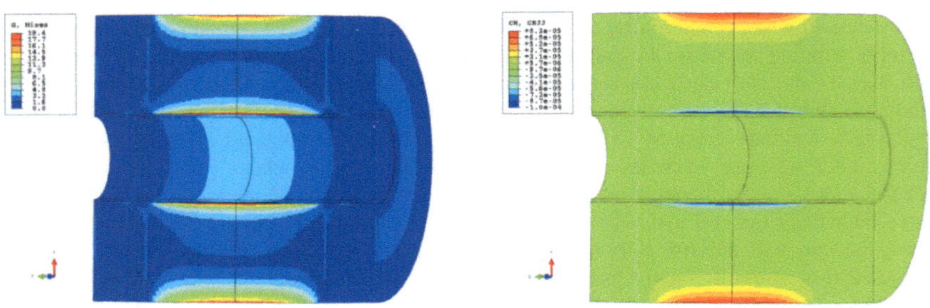

Figure 6: Calculated v.Mises stress distribution in a coil after winding and creep strains after one week of storage.

As a result of the creep deformation, the strip can show flatness problems when it is unwound, although it has been leveled in the preceding stage. The effect is shown in fig.7 where a simulation of the flatness was performed for two strip positions in the coil of fig.6. The material from the coil core shows characteristic edge-waves, while on the outside a tendency to formation of centre waves can be observed.

Figure 7: Calculated flatness of a strip cut out of the coil and laid on a table a) from coil core b) from coil outer layer

Conclusions

ICME methods are well advanced in the production of aluminum strip. While it is not realistic to expect quantitative prediction of final alloy properties after a long processing chain like Al-rolling, it is shown in this article, that predictions are used in industrial practice to design process routes and alloy compositions under different aspects. Most commonly standardized material properties for a potential customer are predicted by ICME tools such as strength, elongation or r-value. In this paper, however, the possibilities of using ICME to achieve a higher process stability and reliability are documented. Especially for aluminum and its alloys, very early process stages, such as the pre-heating before hot rolling already take severe influence on processability and final properties of the strip.

Two industrial cases were calculated. In hot rolling it was shown, that the condition of the alloy combined with non-optimal processing parameters can lead to failures of coils after winding. The use of a through process model showed, how the material can be pre-conditioned to increase process stability to avoid such occurrences. In cold rolling one of the major issues of strip quality is the strip flatness after unwinding a coil. In this paper it was demonstrated by computational methods, how waves or buckles are formed in a coil and how the material properties take effect.

References

[1] J.Hirsch, K.Karhausen, R.Kopp: “Microstructure simulation during hot rolling of Al-Mg alloys“ Proc. 4th Int. Conf. on Aluminium Alloys (ICAA4), Atlanta GA, Sept. 11-16 (1994)

[2] “Virtual Fabrication of Aluminium Products” Microstructural Modeling in Industrial Aluminum Production ; Wiley-VCH Verlag, Weinheim 2006 (ISBN: 3-527-31363-X)

[3] K.F.Karhausen, R.Kopp, Proc. 1st Int. Conf. on Modelling of Metal Rolling Processes, London, Sept. 21-23 (1993) p.66

[4] K.F.Karhausen, Integrierte Prozeß- und Gefügesimulation bei der Warmumformung, Umformtechnische Schriften Bd.52, Verlag Stahleisen mbH, 1995

[5] M. Goerdeler, G. Gottstein, Mat. Sci. and Eng. A. 309 (2001), 377

[6] F. Roters, D. Raabe, G. Gottstein, Acta Mater. 48 (2000), 4181

[7] V. Mohles, X. Li, C. Heering, G. Hirt, S. Bhaumik, G. Gottstein, *in Proc. of 11th Int. Esaform Conf. on Material Forming*, INSA de Lyon, Springer, (2008)

[8] M. Schneider, G. Gottstein, L. Löchte, J. Hirsch. Mat. Sci. Forum 396-402 (2002), 637

[9] L. Löchte, M. Schneider, G. Gottstein. Aluminium 80 (2004), 685

[10] M.Schneider (PhD thesis, 2006, Aachen)

[11] M. Wimmer, A. Cremer, S. Neumann, K. Karhausen, "Optimisation the winding of Al-strip", Light Metals, Proc. TMS 2008, New Orleans, LA, USA

[12] G. Gottstein, Physikalische Grundlagen der Metallkunde, Springer Heidelberg (1998)

[13] K. Lücke, K. Detert, *Acta Metall.* 5 (1957)

[14] K. Karhausen, M. Tschirnich, R. Luce, R. Kopp, Determination of constitutive equations for the hot rolling of Al-alloys by combinatoric optimisation methods, Proc 3rd int. conf. on Modelling of Metal Rolling Processes, London, Dec. 13-15, 1999, IOM, 300-309

[15] S. Neumann, K.F. Karhausen, "Through process effects on final Al-sheet flatness", Light Metals, Proc. TMS 2011 , San Diego, CA, USA

2nd World Congress on Integrated Computational Materials Engineering
Edited by: Mei Li, Carelyn Campbell, Katsuyo Thornton, Elizabeth Holm, and Peter Gumbsch
TMS (The Minerals, Metals & Materials Society), 2013

RESIDUAL STRESS MODELING IN ALUMINUM WROUGHT ALLOYS

Bowang Xiao, Qigui Wang, Cherng-Chi Chang, Josie E Rewald

GM Global Powertrain Headquarter, 823 Joslyn Ave, Pontiac, MI 48326, USA

Keywords: Residual stress, Aluminum Wrought Alloys, quenching, stretching, simulation

Abstract

Aluminum wrought alloys are usually subjected to heat treatment which includes quenching after solution treatment to improve aging responses and mechanical properties. Rapid quenching can lead to high residual stress and severe distortion which significantly affect dimension stability, functionality and particularly performance of the product. Following quenching, a mechanical stretching is usually applied to reduce as-quench residual stress and straighten the products. To model residual stress and distortion during heat treatment and mechanical stretching of aluminum wrought alloys for robust product design and durability assurance, a finite element based approach was developed by coupling a nodal-based transient heat transfer algorithm with material thermo-viscoplastic constitutive model. The integrated residual stress model has demonstrated its robustness in predicting residual stresses and optimizing heat treatment of aluminum wrought alloys.

Introduction

Aluminum wrought alloys are widely used in aerospace and automotive industry to reduce weight and improve fuel efficiency because of their high strength to weight ratio. Many aluminum wrought alloys are heat-treated through solution, quenching and ageing to improve mechanical properties like tensile strengths and hardness. However, high cooling rates in water quenching can produce significant amount of residual stresses [1, 2] that can result in severe distortion [3], even cracking. Tensile residual stress produced during water quenching is also detrimental to fatigue properties of the material [2, 4]. Therefore, product geometry design and heat treatment processing must be optimized so that the mechanical properties can be improved and meanwhile residual stress and distortion are minimized. As a result, stretching process is usually applied to aluminum wrought alloys after quench to relieve high internal residual stresses. The stretching process is helpful to relive the residual stresses, but it comes with the question how much stretching is enough and some extra manufacturing cost. How to optimize the product design, heat treating processes and pre-stretching processes therefore becomes important. This can be achieved with the development of a numerical approach to accurate prediction of residual stress and distortion in the quenched components.

The numerical modeling of quenching process is full of challenges, due to the complexity of part geometry and boundary conditions and in particular the variation of material behaviors under different thermal and stress conditions [5-6]. In this article, the numerical modeling of quenching, stretching and simplified machining processes of aluminum wrought alloys is demonstrated. In the transient thermal analysis, experimentally calibrated heat transfer coefficient (HTC) data has been applied to accurately predict temperature distribution at different time stages.

After thermal analysis, the subsequent structural analysis is performed to predict quenching results such as residual stress and distortion based on the temperature-profile of quenched aluminum wrought alloys and material constitutive model. The stretching process is studied on how much residual stress is relieved. Finally, a simplified machining process is simulated to evaluate the re-balanced and re-distributed residual stresses after some materials are machined off.

Transient Thermal Analysis

During quenching, heat is mainly transferred from aluminum wrought alloy component to the surrounding quench media by convection as expressed in Equation 1. Heat flux and HTC data vary from location to location and play very important roles in quenching results. Temperature-time profile of the quenched part at different time stages can be predicted by FEA tools with experimentally calibrated HTC.

$$\dot{Q} = h \cdot A \cdot (T_{surface} - T_{water}) \tag{1}$$

where

$\dot{Q}$ =heat rate, J/s

h =heat transfer coefficient, $w/m^2\,{}^oC$

A =surface area, m^2

$T_{surface}$ =work piece surface temperature, ${}^oC$

T_{water} =water temperature, ${}^oC$

An aluminum alloy picture frame-shape casting has been used in quench experiments to measure the temperature responses to various quench conditions at different locations during quenching. The temperature-time curves have been used to optimize HTC iteratively as shown in Figure 1 [7-8].

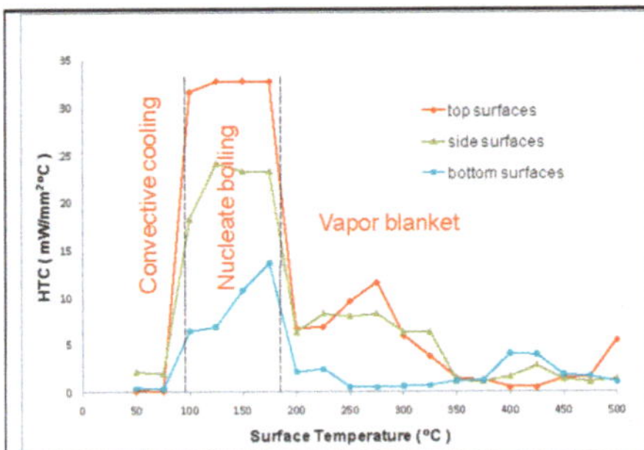

Fig. 1 Iteratively determined HTC from aluminum casting quenching experiments[7]

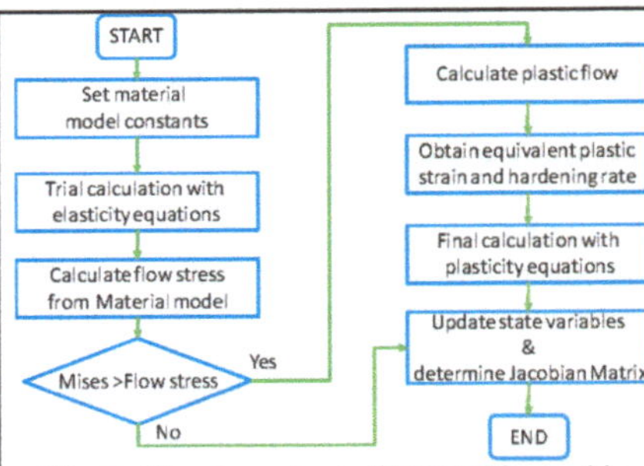

Fig. 2 The flow chart of UMAT is used in FEA analysis.

Structural Analysis

In the structural simulation, the temperature-time profile of the quenched work piece from the thermal simulation is read in. The finite element code then calculates local (node to node) displacements based on the temperature distribution and the relevant materials properties. During quenching, the part shrinks because of the temperature drop. The non-uniform thermal shrinkage of the part is constrained by the geometric structure and material properties that vary with temperatures and strain rates [9-11]. Thus an accurate representation of the material behavior at different temperatures and strain rates is extremely important to the simulation results. To accurately predict the residual stress and distortion during quenching, the temperature- and strain rate-dependent material constitutive behavior needs to be modeled. Following Newman *et al*, a material constitutive model called Mechanical Threshold Stress (MTS), was adapted in this work for water quenching process [9].

A. MATERIAL CONSTITUTIVE MODEL

During quenching, the total thermal strain can be simply divided into elastic and plastic strains. The plastic strain can be described by time-independent plastic model. As shown in Eqn, 2, the flow stress is the sum of two components: an elastic stress modeling yielding, a plastic stress evolving with strain hardening.

$$\frac{\sigma}{\mu(T)} = C_e(\dot{\varepsilon},T)\frac{\hat{\sigma}_e}{\mu_0} + C_p(\dot{\varepsilon},T)\frac{\hat{\sigma}_p}{\mu_0} \tag{2}$$

where $C_e(\dot{\varepsilon},T)$, and $C_p(\dot{\varepsilon},T)$ are referred as scaled temperature-strain rate parameters; T is measured in Kelvin; μ_0 (=28.815GPa) is the reference modulus value at 0 K and $\dot{\varepsilon} = 10^7 s^{-1}$ for cast aluminum; and $\mu(T)$ is the temperature-dependent shear modulus, given as

$$\mu(T) = \mu_0 - \frac{3440}{\exp(\frac{215}{T}) - 1} \tag{3}$$

•Before yield, the stress-strain curve is treated as fully elastic, depending only on the Young's modulus E and yield stress σ_y. E is determined from the stress-strain curves of tensile tests at different temperatures and strain rates using a second-order polynomial.

•At yield, $\hat{\sigma}_p$ =0, and the yielding stress σ_y depends only on the intrinsic strength $\hat{\sigma}_e$, scaled by $C_e(\dot{\varepsilon},T)$.

•After yield, the flow stress is modeled through the evolution of $\hat{\sigma}_p$. Following Voce, a liner form is used in this model for strain hardening.

$$\hat{\sigma}_p = \hat{\sigma}'_p + \frac{\mu(T)}{\mu_0}\theta_0\left[1 - \frac{\hat{\sigma}'_p}{\hat{\sigma}_{os}}\right]d\varepsilon \tag{4}$$

where θ_0 represents the slope of the stress-strain curve at yield in the reference state (0 K, $\dot{\varepsilon} = 10^7 s^{-1}$) and $\hat{\sigma}_{os}$ is a material parameter.

B. COUPLING MATERIAL CONSTITUTIVE MODEL IN FEA

The material constitutive model is coupled in FEA software ABAQUS via UMAT. Fig. 2 shows the flow chart of the user materials subroutine (UMAT). In residual stress analysis, FEA code (for instance ABAQUS) chooses a proper time increment for each step and calls UMAT for calculating thermal strains and stresses at each integration point.

Application of Residual Stress Modeling

In aluminum wrought alloys, stretching process is commonly used to reduce as-quenched residual stress, i.e. the quenched billet is stretched to certain level of plastic deformation and then released. Although it is generally accepted that plastic deformation in the stretching process can help relieve some residual stresses, stretching increases process complexity and manufacturing cost. In practice, there are two questions that are usually asked with regard to stretching process for an individual component: 1) is the stretching process necessary? 2) If it is needed, how much should the extrusion billet be stretched for minimal residual stress and manufacturing cost?

To answer these two questions with regard to the stretching process, an example of a compressor wheel machined from the extrusion billet is given to evaluate the residual stress changes with respect to different heat treatment and stretching processes. It is assumed that the extrusion billet is typically subject to T651 heat treatment and stretching process. To study the effects of stretching and the amount of stretching on the residual stress in the final finished component, the developed residual stress models have been used.

Figure 3 shows the mesh modes for the extrusion billet and the compressor wheel and the locations where the wheel is machined from the billet. Figure 4 shows the predicted residual stress distributions in the extrusion billet, and the stress distribution after the compressor wheel is machined. It is clearly shown that a big amount of residual stresses are relieved because some materials are machined off. Figure 5 shows the predicted residual stress distributions in the extrusion billet, the stress distribution in the billet after 1% stretching, and the stress distribution in compressor wheel after machining. Similarly, figure 6 shows the predicted residual stress distributions in the extrusion billet, the stress distribution in the billet after 2% stretching, and the stress distribution in compressor wheel after machining. As expected, proper stretching can reduce residual stress significantly in the extrusion billet. It is possible the stretching process before machining can reduce the residual stresses to a near zero stress distribution for this small part.

A significant reduction of residual stress in the given component is seen after machining. The maximum residual stress in the final machined wheel is less than 13MPa even without stretching process. After 2% stretching of the extrusion billet, the maximum residual stress in final machined part is near zero.

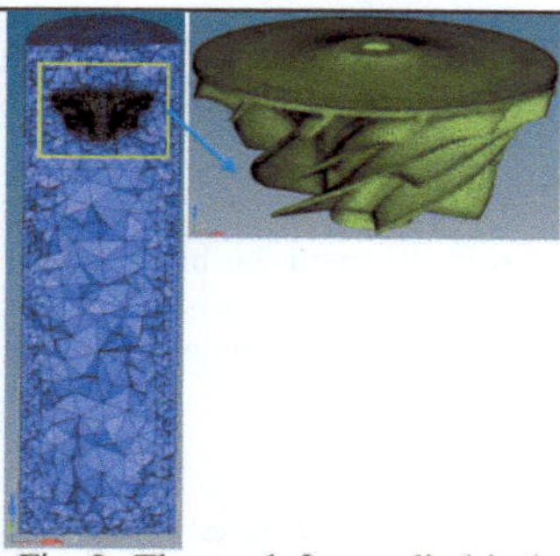

Fig. 3 The mesh for a cylindrical extrusion billet and a final machined compressor wheel

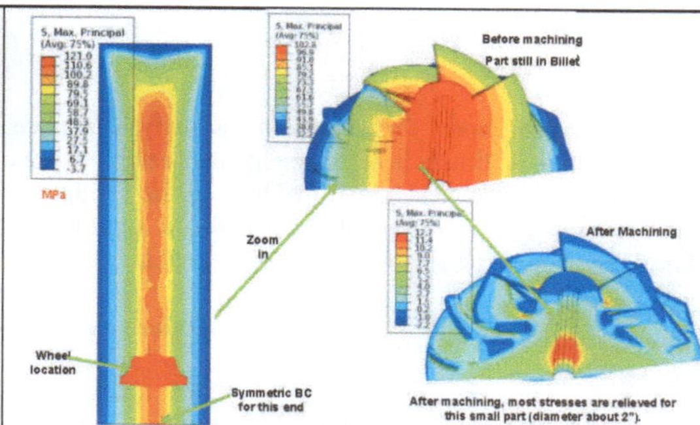

Fig. 4 Residual stress evolution in quenching and machining processes (no stretching)

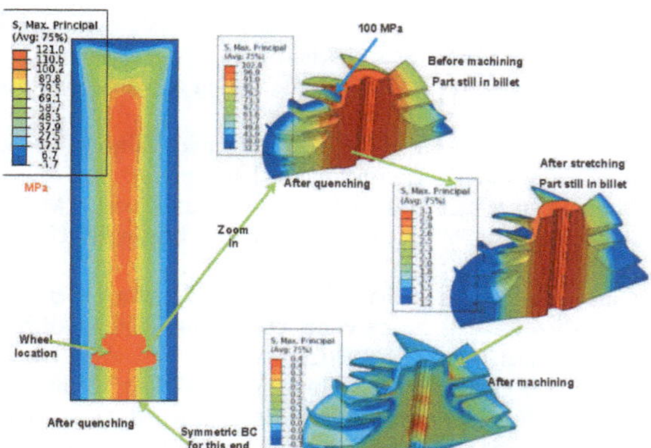

Fig. 5 Residual stress evolution in quenching, 1% stretching and machining processes

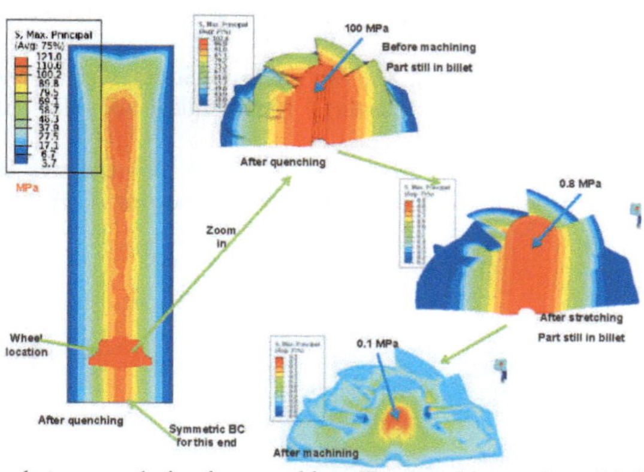

Fig. 6 Residual stress evolution in quenching, 2% stretching and machining processes

Summary

A computational model for residual stress analysis has been used in aluminum wrought alloy to understand and optimize heat treatment and stretching processes. It is found that stretching up to 2% after quenching can relieve majority of the residual stresses in the extrusion billet for this small component case. It is also noted that a significant amount of residual stress is relieved during machining, particularly for small components with simple geometry and thin walls.

References

[1] Li, K., Xiao, B., and Wang, Q., 2009, "Residual Stresses in as-Quenched Aluminum Castings," SAE International Journal of Materials & Manufacturing, **1**(1) pp. 725-731.
[2] Li, P., Maijer, D. M., Lindley, T. C., 2007, "Simulating the Residual Stress in an A356 Automotive Wheel and its Impact on Fatigue Life," Metallurgical and Materials Transactions B, **38**(4) pp. 505-515.
[3] Rose, A., Kessler, O., Hoffmann, F., 2006, "Quenching Distortion of Aluminum Castings-Improvement by Gas Cooling," Materialwissenschaft Und Werkstofftechnik, **37**(1) pp. 116-121.
[4] Lee, Y.L., Pan, J., Hathaway, R., 2005, "Fatigue testing and analysis: theory and practice," Elsevier Butterworth-Heinemann, pp. 402.
[5] Qigui Wang, Bowang Xiao, Cherng-chi Chang, David Paluch, "Modeling of Residual Stresses in Aluminum Castings", AFS Transactions 2013 (in press).
[6] Qigui Wang, Cherng-Chi Chang, David Paluch, Guihua Zhang, "Methods Of Predicting Residual Stresses And Distortion In Quenched Aluminum Castings," US Patent 8214182 B2 (July 3, 2012).
[7] Bowang Xiao, Qigui Wang, Parag Jadhav, Keyu Li, (2010), "An Experimental Study of Heat Transfer in Aluminum Alloy Casting during Water Quenching," Journal of Materials Processing Technology, volume:210, issue: 14, pp. 2023–2028.
[8] Bowang Xiao, Keyu Li, Qigui Wang, Yiming Rong, (2011) "Numerical Simulation and Experimental Validation of Residual Stresses in Water-Quenched Aluminum Alloy Castings, " Journal of Materials Engineering and Performance, DOI: 10.1007/s11665-011-9866-7, volume: 20, issue:9 , pp. 1648-1657
[9] Newman, M. L., Robinson, B. J., Sehitoglu, H., 2003, "Deformation, Residual Stress, and Constitutive Relations for Quenched W319 Aluminum," Metallurgical and Materials Transactions A, 34(7) pp. 1483-1491.
[10] Sehitoglu, H., Smith, T., Qing, X., 2000, "Stress-Strain Response of a Cast 319-T6 Aluminum Under Thermomechanical Loading," Metallurgical and Materials Transactions A, 31(1) pp. 139-151.
[11] Estey, C. M., Cockcroft, S. L., Maijer, D. M., 2004, "Constitutive Behaviour of A356 during the Quenching Operation," Materials Science & Engineering A, 383(2) pp. 245-251.

2nd World Congress on Integrated Computational Materials Engineering
Edited by: Mei Li, Carelyn Campbell, Katsuyo Thornton, Elizabeth Holm, and Peter Gumbsch
TMS (The Minerals, Metals & Materials Society), 2013

ICME APPROACH TO CORROSION PIT GROWTH PREDICTION

Kenneth D Smith, Lei Chen, Robert Darling, Thomas Garosshen, Mark Jaworowski, Susanne Opalka, Sonia Tulyani, and George Zafiris

United Technologies Research Center, 411 Silver Lane, East Hartford, CT 06118, USA

Keywords: Corrosion, Pitting, Multi-scale Modeling, Solution Equilibria

Abstract

Corrosion is a significant problem in a large number of aerospace and commercial applications. Prediction of expected pit size and distribution due to localized corrosion processes is essential in understanding product life. Until now, the approach to predict pit size has been based on statistical analysis of pits in exposure tests. In this work, a combined experimental and multi-scale modeling method is used to develop a physics-based approach to understanding the factors that affect pit growth. The integrated approach uses thermodynamic modeling to understand the pit solution chemistry, atomistic modeling to study the kinetics, and experimental electro-kinetic measurements to validate the model predictions. Each of these pieces feed into an analytical model of pit growth to determine the maximum theoretical size. Model assumptions, results and pit size predictions for model aluminum systems are discussed.

Introduction

The aluminum alloys used in many products are susceptible to localized pitting corrosion that can create very deep pits that either break through the material or create a point for mechanical stresses to induce fatigue failures. Pitting corrosion results from a breakdown in the passive oxide covering a metal. The anode inside the pit is fed by a large cathode, where the pit acidifies so that the cations are thermodynamically stable in the pit and anodic dissolution is favored. The various stages of pitting beginning with film breakdown and ending with pit death are beyond the scope of the present investigation, where the goal is to determine if there is a maximum, stable pit size for atmospheric corrosion.

The early description of the anode was developed by Galvele [1], who assumed transport limitations dominated pit growth and developed a model that accounts for the increase in pit acidification with depth through anodic metal dissolution reactions and hydrolysis reactions. A key finding was the pit stability product, *xi*. A critical pit stability value has also be determined and validated experimentally that delineates where significant changes in the pit chemistry occur.

In order to understand pit chemistry and growth, different approaches have been used to solve the coupled differential equations [2,3,4,5,6]. These models require accurate descriptions of the thermodynamic equilibrium, chemical kinetics, and transport phenomena. These models provide detailed descriptions of both the spatial temporal distribution of species and the potential throughout the pit. A simplified approach has also been developed [7] that attempts to determine the maximum possible pit depth. However, most of these approaches either have significant assumptions present or were developed for ideal cases, which significantly limits the ability to have a general model capable of predicting pit sizes for a variety of metallic systems and alloys that include many reactions and corrosion events.

Integrated Approach

Being able to predict the maximum possible pit depth is important for determining product service and maintenance intervals. The objective of this effort is to develop an integrated set of modeling tools that can be used to identify pit growth criteria as a function of time, chemistry, and environment, which can also be linked with other life-prediction tools to determine product life. A detailed, differential equation model that includes all the individual species and reactions present is being developed to characterize the influence of individual processes on corrosion or to understand the time evolution of the pitting process. Furthermore, this approach allows for the consideration of different alloys through the incorporation of thermodynamics and kinetic descriptions representative of each alloy system.

The overall scheme being developed involves inputs from experimental electrochemistry and computational chemistry. The first aspect is the solution thermodynamics, which identifies species present under different electrochemical and pH conditions. This information provides guidance for species stability or solubility at different conditions. The second aspect is the atomistic modeling of reaction kinetics. Besides thermodynamic limitations, kinetic limitations can limit the presence of some species in the pit. Through atomic modeling the objective is to predict the solution-based chemical kinetics to be incorporated into pit chemistry models. The third aspect includes experimental data through electrokinetic probe measurements. This information can be used to provide guidance on film thickness and other parameters that are required but not readily known from modeling. The fourth aspect is the pit model that couples the anode and cathode together with all relevant physics to predict pit growth. Together these four aspects provide the necessary framework and tools for pit growth predictions. Development of this approach uses a general framework for easy future expansion to any system of interest.

Solution Thermodynamics

Pitting-corrosion models require accurate thermodynamic descriptions of the soluble species and the precipitate phases that may occur. While parameterized thermodynamic databases containing Al species are available, it is necessary to account for water-coordination and non-complexed counter-ions in non-dilute, multi-component solutions with varying ionic strength. For example, the hydrolysis of the Al^{3+} ion involves coordination of inner and outer coordination shells of water with Cl^- counter-ions. Depending on the pH and the counter-ion concentration, the inner coordination shell water molecules can dissociate to release protons, leaving a hydroxyl ion to coordinate with Al forming $Al(OH)^{2+}$, $Al(OH)_2^+$, or $Al(OH)_3$ [5,8], which may also be present in the pit. A further simplifying assumption is the low solvated Al concentration present, making the formation of dimers and larger species negligible in the presence of NaCl [9]. In order to describe the thermodynamic parameters for pitting corrosion, atomic modeling was conducted to determine the structure and stability of these solute complexes and to obtain reliable quantum chemical calculations that can provide equilibrium constants for the corrosion pit model.

The clusters were modeled with Gaussian 03 [10] in three different cases: considering the Al as a bare cluster neutralized with Cl anion(s), as a cluster explicitly solvated with a single, inner hydration sphere, and as a cluster explicitly solvated with inner and outer hydration spheres. Gas-phase calculations were used to assess the thermodynamics of the Al clusters and extension to include solvent polarizable continuum model (IEFPCM) calculations is planned. The B3LYP/6–311G(p,pd) method is used for the analysis, following the work of Ashcraft, *et* al. [11]. Prediction of the standard state enthalpy of formation, $\Delta_f H°$, is used to determine the enthalpy of solvation, ΔH_{solv}, and the Gibbs free energy of solvation, ΔG_{solv}, which is related to the solubility products, K_{SP}, the dissociation constants, pK_a, and equilibrium constants, K_{eq}, for the clusters.

The effect of including explicit water molecule hydration on the Al complexes with Cl^- counter-ions is shown through the gas phase $\Delta_f H^\circ$ calculations in Table I, which shows the explicitly hydrated Al^{3+} ion to be most stable. With the complexes neutralized by Cl^- ions, the inner-most water coordination shell contained a combination of six water molecules or hydroxyl groups for stabilization. The Cl^- ions did not directly interact with the Al ion, instead interacting with the outer-most water-coordination shell present in the model. While the Cl^- ions did not substitute or displace any water molecules, there was an impact on the water configuration by decreasing the anharmonicity of water vibrations and decreasing the configurational degrees of freedom for the entire cluster. However, the exact proximity of the Cl^- ions to the solvated Al clusters needs further investigation.

Table I. Standard state enthalpy of formation, $\Delta_f H^\circ$, for the gas phase Al ion clusters neutralized with Cl^- counter-ions and varying hydration shells in kJ/mol.

Bare Cluster	Cl^- Counter-ions	Inner Water Shell	Outer Water Shell
Al^{3+} 5522.7	$AlCl_3$ -501.8	$Al(H_2O)_6Cl_3$ -2245.0	$Al(H_2O)_6(H_2O)_{12}Cl_3$ -5630.7
$[Al(OH)]^{2+}$ 2413.6	$Al(OH)Cl_2$ -631.9	$Al(OH)(H_2O)_5Cl_2$ -2065.8	$Al(OH)(H_2O)_5(H_2O)_{12}Cl_2$ -5443.1
$[Al(OH)_2]^+$ 275.3	$Al(OH)_2Cl$ -762.0	$Al(OH)_2(H_2O)_4Cl$ -1931.9	$Al(OH)_2(H_2O)_4(H_2O)_{12}Cl$ -5287.9
$Al(OH)_3$ -884.5	—	$Al(OH)_3(H_2O)_3$ -1770.2	$Al(OH)_3(H_2O)_3(H_2O)_{12}$ -5071.6

The fully hydrated Al^{3+} cluster neutralized by water is the most stable form. To evaluate the impact of water-coordination shells around this species, the enthalpy of stabilization, $\Delta H_{stabilization}$, was computed as the difference between the $\Delta_f H^\circ$ of the cluster and the $\Delta_f H^\circ$ of the cluster without the outer-most water-coordination layer plus the sum of the water molecules $\Delta_f H^\circ$ that were used to form the outer-most shell. The results show that stabilization by the inner water coordination shell is stronger per water molecule. This stabilization also increases with decreasing bare complex charge (or less hydroxyl groups coordinating with the Al ion). While the outer water shell does provide additional stabilization, its impact is less per water molecule. The next step is to understand how the hydrogen bonding effects are partitioned in the individual complexes and how these effects contribute to the cluster stability.

Corrosion Pit Modeling

A model for a single, isolated, corroding pit was developed based upon a macroscopic diffusion-reaction system that employs the quasipotential transform [12], which has been used to solve similar pitting models [13]. The quasipotential transform is applicable to steady-state systems where species transport can be described with the Nernst-Planck equation:

$$N_i = -D_i\left(\nabla c_i - z_i c_i \nabla \phi\right), \quad \phi = \frac{F\Phi}{RT} \tag{1}$$

where N_i is the molar flux of species i, D_i is the diffusion coefficient, c_i is the concentration, z_i is the charge number, ϕ is the potential, Φ is electric potential, F is Faraday's constant, R is the universal gas constant, and T is the temperature. The quasipotential reduces the coupled problem into a chemical part and a geometric part, where the chemical part is a differential-algebraic system of equations and the geometric part is Laplace's equation. A five-component system consisting of Na^+, Cl^-, Al^{3+}, $Al(OH)^{2+}$, and H^+ in water is considered. The homogeneous reaction of Al^{3+} and water forming $[Al(OH)]^{2+}$ and H^+ was also included in the model, as this appears to be the most dominate homogenous reaction present. No other reactions or equilibrium processes are considered. Charge neutrality is assumed to be valid locally throughout the pit. Although hydrogen evolution can be included in this formulation, it was assumed negligible and not included. For this model, the partial current density associated with the Al dissolution reaction is constant so that the pit will remain hemispherical. The numerical solution to the system of equations was completed using Matlab.

Figure 1 shows simulation results for the five species represented in a corroding Aluminum pit. The most significant concentration changes occur for Al^{3+} and Cl^-. There is relatively little Na^+ present within the pit. The change in pH occurs from the hydrolysis reaction, where the pH is 2.69 at the bottom and 2.86 at the top. The pH near the bottom of the pit is close to the pH of 3 expected to occur in this situation. However, the relatively low pH value at the top of the pit seems too low as the pH would be expected to be closer to neutral as the solution will eventually transition to the alkaline cathode on the surface. Inclusion of the full model including explicit cathode and any kinetic effects may be necessary to capture the actual chemistry transition.

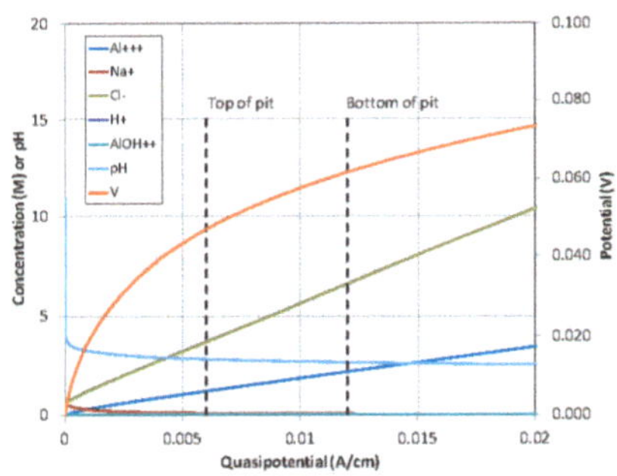

Figure 1. Concentrations and solution potential for the five component system.

The cathode can impose a limitation on atmospheric corrosion rates, which is not the case when Al corrodes in a large bath of electrolyte with a large counter-electrode. Assuming that the anodic and cathodic current densities are constant for a pit of specified radius, the limiting current density for oxygen reduction and outer radius on a circular cathode centered on the pit can be determined from the current balance. The current is assumed to move in the direction parallel to the surface of the cathode in the thin electrolyte film. A one-dimensional differential current balance in the film yields a current density that drops rapidly with increasing radius due to the cylindrical geometry.

The Ohmic drop associated with the electrolyte film can be determined from Ohm's law and the current density, which provides an estimate of the Ohmic potential drop on the cathode for any pit size. This analysis is also used to calculate the maximum pit depth, provided the maximum value of potential difference is known. This procedure relies on Ohm's law being applicable, which is true when concentration variations can be ignored. For a 100 μm pit with 50% RH, 0.32 S/cm electrolyte conductivity, and 5.7 μm electrolyte film thickness, the surface potential drop is much larger than the potential drop in the pit, which is typically less than 50 mV. The potential drop is found to increase monotonically with pit radius.

The voltage drops across the cell include Ohmic, kinetic, and transport losses. Because the anode and cathode are shorted, and there is no difference in solid phase potentials, the reversible potential is equal to the sum of the losses. The pitting potential on pure Aluminum is -0.591 V vs. SCE. Ignoring the Ohmic drop on the anode and the cathodic overpotential, the maximum allowable Ohmic drop is approximately 1.135 V; however, this value will vary with the compositions in the pit and on the external surface. Figure 2 shows how the pit radius and Ohmic drop on the cathode evolve with time. The pit is assumed to grow until the potential drop on the cathode reaches its maximum Ohmic drop. This occurs at a pit radius of approximately 350 μm for a pit at 50% RH with 200 μg/cm^2 salt deposition. While these predictions depend on humidity and NaCl surface deposition, a potential bounded estimate of maximum pit size for an isolated pit with and infinite cathode is determined, providing a starting point for use in life-prediction models.

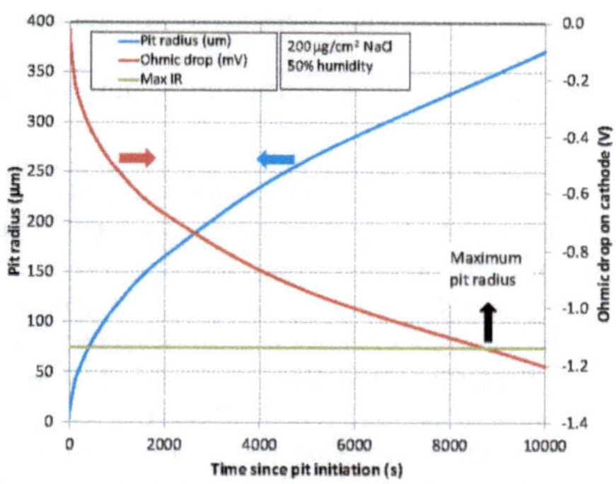

Figure 2. Transient pit growth occurring on Al at 50% RH with a 200 μg/cm^2 salt deposition.

Conclusions

From the solution thermochemistry, the Al^{3+} ion solvated by water is revealed to be the most stable form. The layers of water surrounding the Al^{3+} ion provide stabilization, which decreases with increasing number of layers. The understanding obtained from the ion salvation is necessary to develop equilibrium predictions of the ion species under various solution conditions. The next step is to convert the complex structural information between steps into equilibrium constants, dissociation constants, and solubility products. Combining the detailed thermochemistry of other ions found in alloys allows for the creation of complex phase diagrams and the influences of alloying species into the pitting analysis.

The basic framework of the generalized pitting model has been developed with results that are plausible. Further investigation in the pit growth mechanism is required based upon the extremely rapid pit growth rates predicted with the current model compared to experimental measurements. Additionally, the calculations done with the quasipotential may not be directly applicable when the anode and cathode are in close proximity, because there is no infinite source of Cl^- and no infinite sink of Na^+ available for a large pit corroding under atmospheric conditions. This model is being further developed to include additional hydrolysis reactions, hydrogen evolution, anode and cathode coupling, non-isolated pits, and microstructure effects. While the quasipotential approach is amenable to many of these effects, a more straightforward numerical solution using, for example, Comsol is preferred in the future as it makes inclusion of more complex system structure possible.

The integration of kinetics and experimental validation will also occur as the development continues. Specific kinetic events may be necessary for non-equilibrium processes that may occur, which will be evaluated through the use of atomic modeling of individual kinetic events. Experimental measurements are also required to help guide the modeling and provide validation for the model. With these two aspects included with the thermochemistry and the full pit growth model, it is possible to understand the complex chemistry occurring within the pit and provide guidance on ways to mitigate the effects of pitting on aerospace and commercial materials.

References

1. J.R. Galvele, "Transport Processes and the Mechanism of Pitting of Metals," *J Electrochem Soc*, 123 (1976) 464-474.
2. T.R. Beck, "Salt Film Formation During Corrosion of Aluminum," *Electrochimica Acta*, 29 (1984) 485-491.
3. M. Verhoff, R. Alkire, "Experimental and Modeling Studies of Single Pits on Pure Aluminum in pH 11 NaCl Solutions," *J Electrochem Soc*, 147 (2000) 1349-1358.
4. F. Cui, F.J. Persuel-Moreno, R.G. Kelly, "Computational modeling of cathodic limitations on localized corrosion of wetted SS 316L at room temperature," *Corr Sci*, 47 (2005) 2987-3005.
5. O. Guseva, P. Schmutz, T. Suter, O. von Trzebiatowski, "Modeling of anodic dissolution of pure aluminum in sodium chloride," *Electrochimica Acta*, 54 (2009) 4514-4524.
6. J. Xiao, S. Chaudhuri, "Predictive modeling of localized corrosion: An application to aluminum alloys," *Electrochimica Acta,* 56 (2011) 5630-5641.
7. Z.Y. Chen, R.G. Kelly, "Computational Modeling of Bounding Conditions for Pit Size on Stainless Steel in Atomspheric Environments," *J Electrochem Soc*, 157 (2010) C69-C78.
8. J. Duan and J. Gregory, "Coagulation by hydrolyzing metal salts," *Adv. Colloid Interfac.* 100–102 (2003) 475–502.
9. X. Jin, W. Yang, Z. Qian, Y. Wang, and S. Bi, "DFT study on the interaction between monomeric aluminum and chloride ion in aqueous solution," *Dalton Trans.* 40 (2011) 5052–58.
10. Gaussian 03, Revision E.01, M.J. Frisch, et al., Gaussian, Inc., Pittsburgh, PA (2003).
11. R.W. Ashcraft, S. Raman, and W.H. Green, "Ab initio aqueous thermochemistry: application to the oxidation of hydroxylamine in nitric acid solution," *J Phys Chem B* 111 (2007) 11968–83.
12. M.W. Verbrugge, D.R. Baker, J. Newman, "Dependent-Variable Transformation for the Treatment of Diffusion, Migration, and Homogenous Reactions. Application to a Corroding Pit," *J Electrochel Soc*, 140 (1993) 2530-2537.
13. M. Alkire, R. Verhoff, "Experimental and Modeling Studies of Single Corrosion Pits on Pure Aluminum in pH 11 NaCl Solutions. II. Pit Stability," *J Electrochem Soc* 147 (2000) 1349-1358.

2nd World Congress on Integrated Computational Materials Engineering
Edited by: Mei Li, Carelyn Campbell, Katsuyo Thornton, Elizabeth Holm, and Peter Gumbsch
TMS (The Minerals, Metals & Materials Society), 2013

STEEL – AB INITIO: QUANTUM MECHANICS GUIDED DESIGN OF NEW FE-BASED MATERIALS

Ulrich Prahl[1], Wolfgang Bleck[1], Alireza Saeed-Akbari[1,2]

[1] Institute of Ferrous Metallurgy, RWTH Aachen University, Aachen, Germany
[2] Department of Materials, ETH Zurich, Zurich, Switzerland

Keywords: ab initio modeling, deformation mechanisms, strain hardening, knowledge-based steel design

Abstract

This contribution reports the results of the collaborative research unit SFB 761 "Steel - ab initio", a cooperative project between RWTH Aachen University and the Max-Planck-Institute for Iron Research in Düsseldorf (MPIE) financed by the German Research Foundation (DFG). For the first time, it is exploited how ab initio approaches may lead to a detailed understanding and thus to a specific improvement of material development. The challenge lies in the combination of abstract natural science theories with rather engineering-like established concepts. Aiming at the technological target of the development of a new type of structural materials based on Fe-Mn-C alloys, the combination of ab initio and engineering methods is new, but could be followed quite successfully. Three major topics are treated in this research unit: a) development of a new method for material- and process-development based on ab initio calculations; b) design of a new class of structural materials with extraordinary property combinations; c) acceleration of development time and reduction of experimental efforts and complexity for material- and process-development. In the present work, an overview of the results of the first five years as well as an outlook for the upcoming three-year period is given.

Introduction

The design of new steels usually is based on three major principles: a) requests: the material has to meet specific requirements for the envisaged future application; b) tools: various experimental and numerical methods are applied for materials development; the best balance of effort and accuracy needs to be chosen for a specific materials design task; and c) constraints: steel, being an engineering material characteristically produced in large quantities, always needs to be processed according to the economic rules. Thus, the processes available today and in the future or the impacts of recycling and environment have to be considered already in the early stages of the material development.

Currently, a new class of steels with an austenitic matrix having Mn content of 15 to 35 wt.-% together with additional alloying elements like C, Si, Al is under investigation. The interest in these materials is due to the occurrence of different deformation mechanisms that can attribute to the extraordinary strain hardening and the resultant high formability and high strength. The deformation mechanisms are controlled by the stacking fault energy (SFE). This opens the stage for a new chapter of materials design as new modelling tools based on the numerical understanding of the electronic structure of metals can be used to identify the interaction of different alloy compositions. These ab initio or first principles methods are currently tested with respect to the accuracy of the property prediction. There is a good chance that this will lead to the first real physical based design of a new group of structural materials.

Materials

High Mn steels (HMnS) composed of single phase austenite or multiphase matrix with a large volume fraction of austenite can be divided according to the characteristic phenomena occurring during the plastic deformation into Transformation Induced Plasticity (TRIP) and Twinning Induced Plasticity (TWIP) steels. Furthermore, by adding Al, the group of Microband Induced Plasticity (MBIP) steels have also been developed. Recently, specific interest grows for easily processed medium Mn steels with a combined ferritic/austenitic matrix. Common to all these steels is the fact that they develop a much better combination of strength and ductility compared to the current industrially used group of Advanced High Strength Steels (AHSS). This can easily be demonstrated by comparing the ECO-index, the product of ultimate tensile strength and uniform elongation that is usually limited to max 20 GPa% in AHSS and typically exceeds this value significantly in HMnS. The research centre "Steel – ab initio" currently concentrates on the TWIP steels and the transition between the TWIP and the TRIP effect. Some of the investigated alloys are indicated in **Table 1** with respect to chemical composition and in **Figure 1** with respect to Mn content and ECO-Index.

Table 1: Alloying content of exemplary HMnS, mass contents in %.

	C	Mn	Si	P	S	Cr	Ni	Al
I	0.315	22.8	0.07	0.0073	0.0012	0.016	0.037	0.0054
II	0.573	23.2	0.17	0.0090	0.0002	0.310	0.019	0.0049
III	0.282	28.0	0.10	0.0084	0.0010	0.016	0.037	0.0054
IV	0.594	18.4	0.05	0.0071	0.0091	0.015	0.039	0.0045
VII	0.714	23.5	0.06	0.0071	0.0071	0.016	0.039	0.0059

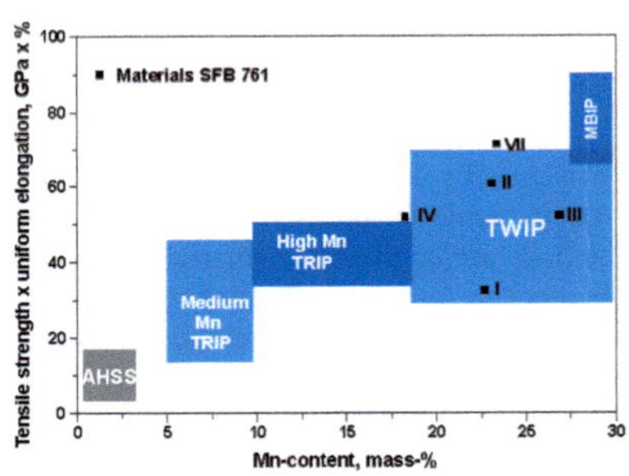

Figure 1: Characteristics of new steel concepts: schematic representation of Mn-content and the tensile properties [1].

Deformation Mechanisms in Low SFE Regime

Any deviation from the normal sequence of stacking of the atomic planes (as defined by the crystal structure) is called a stacking fault. A stacking fault is a two-dimensional crystal defect. The SFE is defined as the difference in energy per unit fault area between the faulted and the perfect structures. The SFE determines the plastic deformation behaviour of an fcc alloy, **Figure 2**. The dominant deformation mechanism below the SFE value of 20 mJ/m^2 is TRIP while above 20 mJ/m^2 the TWIP mechanism dominates [2,3]. A further increase of the SFE leads to reduced twinning activity by increasing the critical resolved shear stress (CRSS) for twinning. The fact that a CRSS is needed indicates that twinning is crystal orientation dependent and even at low SFE unfavourably oriented grains will remain twin free even at high deformation degrees. Dislocation slip occurs at any SFE, but dominates at low deformation degree and/or high SFE.

There is no clearly defined limit in SFE above which TWIP ceases. The interaction between the microstructural components resulting from the low-SFE and other constituents like solute atoms, precipitates, etc. (Figure 2) defines the mechanical response of HMnS.

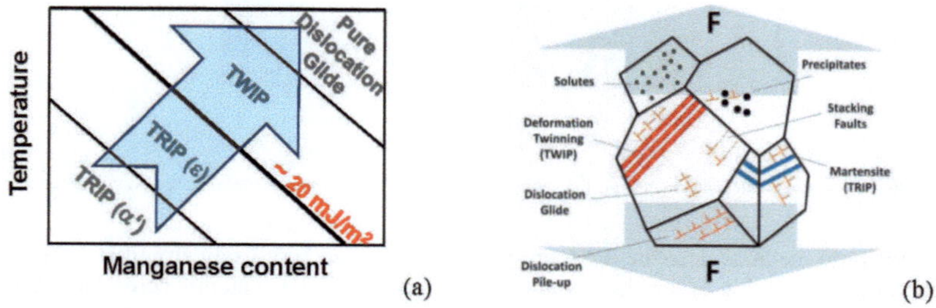

Figure 2: Deformation mechanisms in HMnS and the characteristic SFE for the TRIP/TWIP transition (a); strengthening mechanisms in this system (b) [1,4].

Material Modeling and Simulation

The so-called deformation mechanism maps can be developed by means of a range of thermodynamic calculations in which the SFE value is correlated with other parameters like the free energy change during the austenite to martensite transformation, and the interface energy between austenite and martensite, **Figure 3** [5]. Since a stacking fault can be viewed as a two atomic layers slice of hcp (ε martensite) within the fcc crystal, the SFE should be related to the difference in Gibbs energy between the hcp and fcc phases. This idea has been applied successfully by Allain et al. [3] and Cotes et al. [6], using the following equation:

$$\gamma_{SFE} = 2\rho\Delta G^{\gamma\rightarrow\varepsilon} + 2\sigma \tag{1}$$

where ρ is the number of moles of atoms per m^2 in one atomic layer, $\Delta G^{\gamma\rightarrow\varepsilon}$ is the difference in Gibbs energy between the hcp and fcc phases, and σ is the related interface energy. There is no independent way to determine the interface energy σ. This parameter has been estimated to be in the range of 5 to 15 mJ/m^2 [3,6,7] based on the measured SFE values, the observation of ε-martensite formation, and the activation of deformation mechanisms. The major contribution to the composition and temperature dependency of the SFE comes from $\Delta G^{\gamma\rightarrow\varepsilon}$. The Gibbs free energy of the fcc and hcp phases for pure elements can be taken from Calphad evaluations of the appropriate systems; $\Delta G^{\gamma\rightarrow\varepsilon}$ is then calculated by taking the difference. However, it should be noted that the hcp phase is not a stable phase in the Fe–Mn–C (or other steel systems) and its properties cannot be directly measured. Instead, the hcp phase can be transformed as ε-martensite from the fcc phase in the range 15 to 30 wt-% Mn. Data from this transformation are used to model the Gibbs energy of the hcp phase. The Gibbs energies of the hcp and fcc phases are equal at the so called T_0 line, which must be located between the M_s and A_s temperatures. Thus, the $\Delta G^{\gamma\rightarrow\varepsilon}$ from Calphad evaluations is closely related to the formation of ε-martensite, which is the kind of martensite formed via TRIP mechanism. Below 15% Mn, α-martensite forms instead of ε-martensite and above 30% Mn a magnetic transition in the fcc phase increases the stability of the fcc phase so that the ε-martensite formation is suppressed.

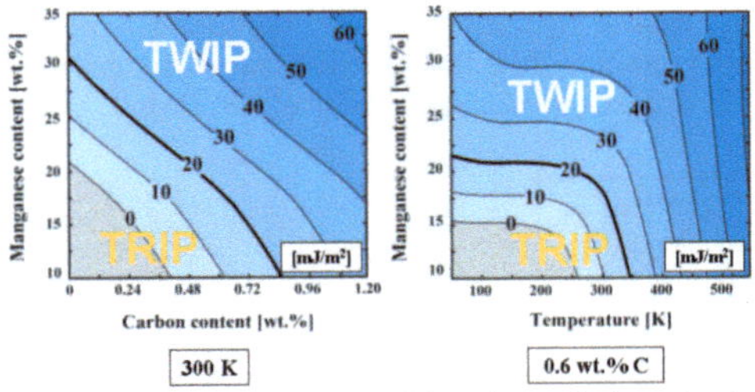

Figure 3: Mechanism maps indicating the relationship between deformation mechanisms, chemical composition, temperature and SFE [5,8,9].

Ab initio methods describe the properties of crystals without using any fitting parameter by the explicit description of the electrostatic interactions between the atomic cores and electrons based on the equations of Schrödinger and Dirac. The input data required for ab initio calculations is given by the atomic structure and the chemical type of the involved atoms. The density functional theory (DFT) is one of the most established ab initio methods providing an efficient way to calculate the electronic properties and the total energy of a material system. This scheme, however, describes the system in its ground state (at T = 0 K) and in practice relies on an approximated description of a small but crucial energy contribution which is due to quantum interaction effects between the constituent particles (the so-called exchange-correlation energy). In order to overcome the limitations of T = 0 K, the DFT calculations are combined with concepts of thermodynamics or with molecular dynamics (MD) simulations.

In steels, usually a random distribution of C atoms in the octahedral voids of the lattice is assumed. In HMnS, the question arises whether short-range atomic ordering might exist that contributes to a void selection according to the number of Fe and Mn atoms that forms the specific octahedral void. Ten different octahedral voids from pure Fe to pure Mn neighbour situation have been considered in one super cell of 16 Fe and 16 Mn atoms, thus representing an alloy with 50% Mn, **Figure 4** [10]. By ab initio modelling the whole structure of the super cell has been calculated including one C atom that is positioned within different voids. The reaction enthalpy has been calculated as a function of the coordination number of the Fe atoms forming the void. It becomes obvious that the smaller the coordination number the more stable is the compound, indicating the strong interaction of Mn and C. This coincides with the fact that stoichiometric manganese carbides are more stable than iron carbides indicating that the Mn-C bond is more attractive than the Fe-C bond.

The mechanical properties of HMnS are characterized by their pronounced strain hardening that is significantly different from the strain hardening behavior of bcc steels. Specifically, mechanically induced twinning leads to an adaptive and continuous refinement of the microstructure during the mechanical testing. Adaptive means, the strain hardening is dependent on the local amount of strain. The measured flow curves sometimes show characteristic turning points that lead to the local minima in the 1st derivative over strain which is the strain hardening curves. The microstructural observations show that during forming individual grains exist without twins together with grains with one or more active twin systems. Thus, the crystallographic nature of twinning has to be considered by means of Crystal Plasticity Finite Element Modeling (CPFEM) [11]. As the SFE and by this the selection of deformation mechanisms is strongly dependent on temperature, it is essential that a suitable model has to consider temperature and strain rate effects. Furthermore, the CPFEM model used distinguishes between the dislocation densities in the cell center and in the cell walls as well as the twin density [12,13].

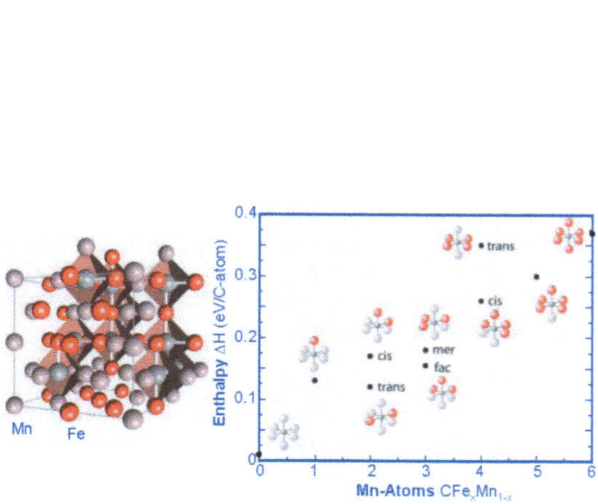

Figure 4: Supercell of 16 Fe (red) and 16 Mn (white) atoms with 10 different types of octahedral voids; enthalpy change by adding 1 C atom to the different voids [10].

From this, a modified free mean path λ is calculated that is a combination of the mean free path of a dislocation λ_{dis} and the distance between twins λ_{twin}:

$$1/\lambda = 1/\lambda_{dis} + 1/\lambda_{twin} \qquad (2)$$

λ_{twin} is calculated according to the equation

$$\lambda_{twin} = s\,(1-V_{twin})/V_{twin} \qquad (3)$$

with s medium thickness and V_{twin} volume fraction of twins. The results in **Figure 5a** indicate that by choosing a proper parameter set the temperature dependency can be properly derived. The main governing fit parameter is the twin density that needs to be determined accurately. This results in the demand for a proper quantitative identification method of twins in these steels.

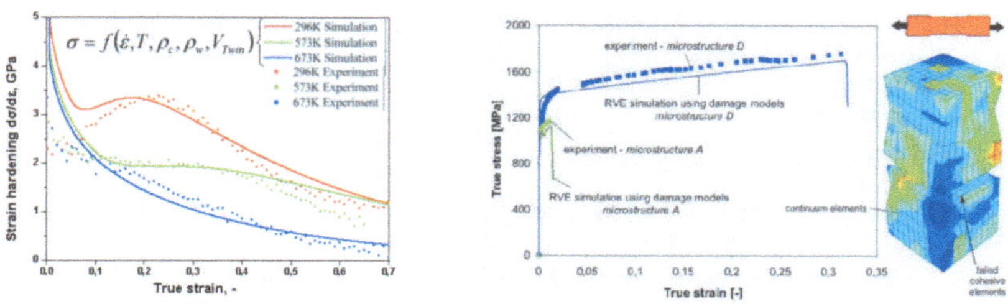

(a) (b)

Figure 5: a) Stress-strain-curves and strain-hardening curves as the function of temperature for alloy I; comparison of experimental results and CPFEM calculations.[11] b) failure modeling of dual phase steels based on CZM using model parameters as function of microstructure in terms of martensite fraction.[16]

Besides the description of strain hardening, an evaluation procedure for the prediction of microcracks as a result of applied mechanical loading is required. The concept of representative volume elements (RVE) is used to describe microstructures with their characteristic phases and defects within the framework of continuum mechanics. The RVE is generated as a submodel of macroscopic models and simplifies the complex microstructures by homogenization. The deformation tensor computed from the macromodel is transferred to the RVE as boundary conditions [14,15]. The RVE is arranged according to the volume fraction and morphologies derived from the real microstructure, **Figure 5b**. In TWIP steels, the interaction of different twin systems might result in an additional damage source especially when the overall strain hardening capacity is exhausted. In any case, the morphology of the phases, their spatial distribution, and the crystallographic constraints have to be considered. The experimental and numerical flow curves (derived from RVE calculation with periodic boundary conditions) have been precisely modeled for two microstructures: A = martensitic, D = dual phase ferrite + martensite. The failure mechanisms are accurately described as a competition between ductile failure due to a void mechanism according to the Gurson-Tveergard-Needleman (GTN) approach and a cohesive zone model (CZM) for the interface ferrite/martensite.

Conclusions

The high manganese steels show an excellent balance of strength and ductility due to continuous adaptive grain refinement during straining. The full exploitation of these properties request for a basic understanding of the physical phenomena involved. New numerical tools are applied and in some cases needed to be developed to predict the properties and parameters but also for improved understanding of involved mechanisms. Especially ab initio techniques help to elucidate the specific effects of these high alloy systems with the occurrence of short range ordering and deformation twinning. The crystallographic nature of the different deformation mechanisms requires crystal plasticity and new continuous damage mechanics modeling.

Acknowledgements

The authors gratefully acknowledge the financial support in the frame of the Collaborative Research Center SFB 761 – "Steel *ab initio*" by Deutsche Forschungsgemeinschaft (DFG).

References

[1] A. Saeed-Akbari, Mechanism Maps, Mechanical Properties, and Flow Behavior in High-Manganese TRIP/TWIP and TWIP Steels, Doctoral Thesis, Shaker Verlag, Aachen, ISBN: 978-3-8440-0016-0, 2011.

[2] Sato, K.; Ichinose, M.; Hirotsu, Y.; Inoue, Y.: ISIJ Int., 29 (1989), 868-877.

[3] Allain, S.; Chateau, J.-P.; Bouaziz, O.; Migot, S.; Guelton, N.: Mater. Sci.Eng. A 387-389 (2004), 158-162.

[4] Saeed-Akbari, A.; Mishra, A.K.; Mayer, J.; Bleck, W.: Metall. Mater. Trans. A, vol. 43A (2012), 1705-1723.

[5] Saeed-Akbari, A.; Imlau, J.; Prahl, U.; Bleck, W.: Metall. Mater. Trans. A, 40A (2009), 3076-3090.

[6] Cotes, S.M.; Guillermet, F.; Sade, M.: Metall. Mater. Trans. A, 35A (2004), 83-91.

[7] Kim, J.-C.; Han, D.-W.; Baik, S.-H. Lee, Y.-K.: Mater. Sci. Eng., A378 (2004), 323-327.

[8] Saeed-Akbari, A.; Mosecker, L.; Schwedt, A.; Bleck, W.: Metall. Materi. Trans. A, 43A (2012), 1688-1704.

[9] Saeed-Akbari, A; Schwedt, A.; Bleck, W.: Scripta Mater. 66 (2012) 1024–1029.

[10] V. Appen, J.; Dronskowski, R.: Steel Res. Int. 82 (2011), 101-107.

[11] Steinmetz, D.R.; Jäpel, T.; Wietbrock, B.; Eisenlohr, P.; Gutierrez-Urrutia, I.; Saeed–Akbari, A.; Hickel, T.; Roters, F.; Raabe, D.: Acta Mater., vol. 61, No. 2 (2013), 494–510.

[12] Roters, F.; Eisenlohr, P.; Hantcherli, L.; Tjahjanto, D.D.; Bieler, T.R.; Raabe, D.: Acta mater. 58 (2010), 1152-1211.

[13] Roters, F.; Eisenlohr, P.; Bieler, T.R.; Raabe, D.: Wiley-VCH Weinheim, 2010, ISBN-13: 978-3-527-32447-7.

[14] Uthaisangsuk, V.; Prahl, U.; Bleck, W.: Proc. Eng., 1 (2009), 171-176.

[15] Uthaisangsuk, V.; Münstermann, S.; Prahl, U.; Bleck, W.; Schmitz, H.-P.; Pretorius, T.: Comp. Mater. Sci. 50 (2011), 1225-1232.

[16] Twardowski, R.; Steinmetz, D.; Prahl, U.: Steel Res. Int., In Press.

2nd World Congress on Integrated Computational Materials Engineering
Edited by: Mei Li, Carelyn Campbell, Katsuyo Thornton, Elizabeth Holm, and Peter Gumbsch
TMS (The Minerals, Metals & Materials Society), 2013

MICROSTRUCTURE MEDIATED DESIGN OF MATERIAL AND PRODUCT

Ayan Sinha[1], Janet K. Allen[2], Jitesh Panchal[1], and Farrokh Mistree[2]
[1]Department of Mechanical Engineering
Purdue University, West Lafayette, Indiana, USA
[2]The Systems Realization Laboratory@ OU
The University of Oklahoma, Norman, OK, USA

KEYWORDS: Microstructure mediated design; Robust design; Inductive design exploration method

ABSTRACT

In this paper, the construct of microstructure-mediated design is explored by framing a multiscale system with the appropriate aspects of the material microstructure, followed by multiscale material modeling, and then engineering the microstructure using the Inductive Design Exploration Method, to achieve the product specifications. As the microstructure represents the limiting interface between structure-property relations including system performance and process-structure relations, we have adopted the phrase *microstructure mediated design*. We illustrate the efficacy of this construct via the integrated design of a submersible and an Al-based matrix composite.

1. FRAME OF REFERENCE - MICROSTRUCTURE MEDIATED DESIGN

Because of the cost of modeling and simulation, it is essential that top-down (inductive) design requirements guide the bottom-up (deductive) process of computer aided materials engineering. Such a systems-based integrated approach enables tailoring materials to meet targeted performance requirements. We augment the notions of Olson's materials design hierarchy and explain the challenges in this system-based inductive approach; see Figure 1 (left). The materials design hierarchy is decomposed as a set of mappings, i.e., processing-structure (PS) relations; structure-property (SP) relations and property-performance (PP) relations [1]. The arrows are unidirectional for the PS and SP relations indicating limited computational inversion due to non-linear, nondeterministic and non-equilibrium characteristics of the behavior of a material. PP relations may be invertible and this is the basis for selecting materials for a design. The key to materials engineering lies in effectively modeling the microstructure acting as an interface between the PS and SP relations as it strongly influences physical, mechanical and chemical properties. Also, the material microstructure needs to be engineered so that constraints on the material properties, imposed by the geometric structure, are satisfied. Thus, the microstructure acts as a decisive interface in the materials design hierarchy, therefore, we adopt a microstructure-mediated design (MMD) centered approach for integrated computational materials engineering (ICME).

The second important feature in Figure 1 (left) is the lateral transformations at each level of the hierarchy to favor design exploration while introducing different aspects of uncertainty. We classify sources of uncertainty as natural uncertainty (NU), model structure uncertainty (MSU), model parameter uncertainty (MPU) and propagated uncertainty (PU) (See [2] for detailed definitions). The dotted arrows linking the MSU models indicate that although it is possible to use the simulation models for design space exploration, the requirement of a complex computational infrastructure linking different simulation and analysis software makes it infeasible. This drives us to use metamodeling techniques to permit efficient inversion, though with an increased MPU component. The metamodels enable reversibility through the hierarchical material domains but a fourth type of uncertainty, i.e., PU needs to be considered for exploring the hierarchical degrees of freedom associated with material composition and process modifications. Thus for successful ICME we frame a multiscale system, develop hierarchical models, i.e., multiscale models, use an inductive approach to find feasible processing routes in order to achieve performance targets and, decide the most suitable processing route among feasible solutions. We use an Inductive Design Exploration Method (IDEM) [3], which combines inductive (top-down) engineering with deductive (bottom-up) science for efficient design space exploration while minimizing uncertainty. For finding the best solution among feasible processing routes we employ the compromise Decision Support Problem (cDSP) [4]. In our approach, we move from the top-down in Olson's hierarchy using IDEM after completing the bottom-up (deductive) analysis of simulation models. We achieve robustness by trading off the degree of system performance with the degree of reliability based on MSU associated with system models in the cDSP while quantifying NU and MPU in IDEM. PU is mitigated by developing exact constraint boundaries at the various scales. IDEM and cDSP together comprise the multiscale design strategy for MMD. We believe considering uncertainty is vital due to the infancy of materials design techniques and design space exploration should be performed in a way that guarantees robustness. Systematic uncertainty management is a key separating factor of MMD from other materials design techniques employing microstructure information. We illustrate the efficacy of the MMD construct via the integrated design of autonomous underwater vehicle (AUV) and Al-based metal matrix composites (MMCs). Due to space considerations, we refer to [5, 6, and 7] for further details of the models considered. We look at four steps associated with MMD of AUV using in-situ Al MMCs in Section 2. In Section 3 we discuss the MMD solution and present closing thoughts.

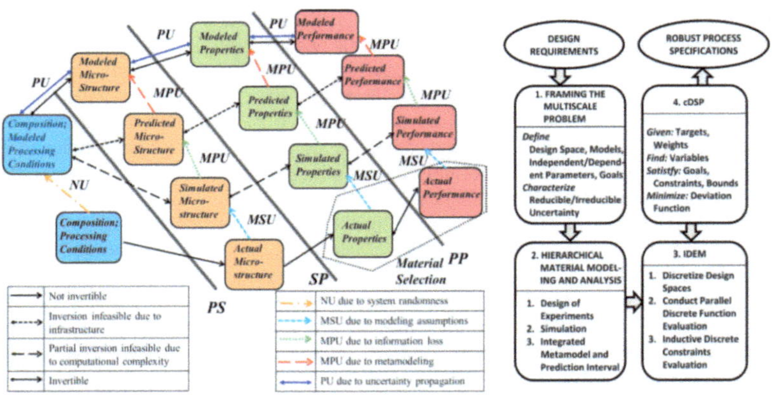

Figure 1: Hierarchical mapping in materials design (Left) and strategy for MMD (Right)

2. MULTISCALE DESIGN

In Figure 1 (right), we instantiate the 4 steps relevant to MMD for ICME. The multiscale system for AUV design is framed briefly. Please refer to [6, 7] for details of constitutive models for developing the PS, SP and PP linkages. The steps of IDEM can be found in [3] and the formulation of cDSP can be found in [4]. IDEM is implemented in MATLAB™ and the cDSP is implemented in JAVA™ with a MATLAB™ builder. Based on the materials processing steps involved and mechanical design requirements, the interconnected modules that constitute the MMD process chain can be captured in 3 Modules (Figure 2, left). MODULE 1 provides the simulated microstructure after processing (PS). The resulting mechanical properties are estimated in MODULE 2 (SP), whereas MODULE 3 maps the required mechanical properties based on the system design considerations (PP). In this application, the strength is principally determined by the sizes, shapes and distribution of TiB_2 precipitates – in other words the microstructure. The AUV design can be modified in two ways, namely, 1) by changing the processing conditions to modify the MMC microstructure and 2) by changing the geometry of the shell which puts constraints on required mechanical properties. Hence, the microstructure needs to be designed concurrently with the shell so as to satisfy the imposed constraints on material properties.

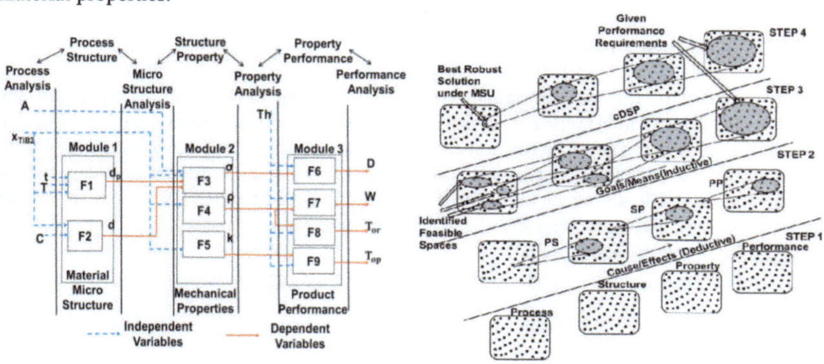

Figure 2: The multiscale system for MMD (left) and steps for MMD (right)

Given the complexity inherent in the design process chain, we represent the independent and dependent variables, the models and the information flow between the different modules, in Figure 2 (left). *F1, F2, F3* and *F6* represent simulation models used to design the autonomous underwater vehicle (AUV) while *F4, F5, F7, F8* and *F9* represent theoretical or empirical models considered for design. The inputs to *F1* are the volume fraction of TiB_2 (x_{TiB2}), time of reaction (t) and temperature of processing in degree K (T). The output of *F1* is the average TiB_2 particle size (d_p) which is an input to MODULE 2. The independent inputs to *F2* are volume fraction of TiB_2 (x_{TiB2}) and cooling rate (C) and the output of MODULE 1 (*F2*) is the average grain size (d) of microstructure which is another input to MODULE 2. *F3* gives the yield stress (σ), *F4* gives density (ρ) and *F5* gives the heat transfer coefficient (k) as output. The model for yield stress (σ) receives inputs from the outputs of MODULE 1 (d_p and d) along with the independent inputs of volume fraction of TiB_2 (x_{TiB2}) and percentage of area fraction (A). The only input to models for density (ρ) and heat transfer coefficient (k) is the independent variable of volume fraction of TiB_2 (x_{TiB2}). The performance variable of depth of operation (D) is evaluated in *F6*; weight of the outer shell (W) is evaluated in *F7*, time of operation (T_{or}) for the submersible is evaluated in *F8* and temperature of operation (T_{op}) is evaluated in *F9*. The independent parameter, thickness of the shell (T_h) is an input to all the models, i.e., *F6, F7, F8* and *F9*. The dependent parameters are density (ρ) to the simulation which determines weight (*F7*) and the module which determines time of operation (*F8*) while yield stress (σ) is an input to depth *(F6)*. The feasible design spaces are inductively passed (using IDEM) from MODULE 3 to MODULE 2 and subsequently to MODULE 1 of design after completing the deductive analysis. In Figure 2 (right), we also depict the 4 steps necessary in order to find the solution. (i) We first set the bounds and variability of the processing variables, the constraints and targets for the system goals along with ranges and resolution for the process-structure-property-performance spaces. (ii) Second order response surface models are developed along with interval estimates. (iii) IDEM is then employed to control the microstructure within feasible bounds. Ranged feasible property, structure and processing spaces are evaluated inductively in order to achieve performance targets. (iv) The ranges and constraints evaluated in the IDEM process are input to the cDSP along with target HD-EMI (a metric indicating the degree of reliability that it will satisfy design space constraints or performance bounds if it undergoes a shift in the output range due to uncertainty). These are used to decide the best processing route among multiple feasible solutions by trading off the HD-EMI achievements.

3. RESULTS AND DISCUSSION

For a full summary of the models, ranges, resolution, targets and constraint values for the dependent and independent parameters in the multiscale system, see Tables 1 and 2 in Sinha, Bera, and colleagues [6]. The discretization resolution for IDEM is set to twice the variability. Model variability due to MPU is determined using response surface confidence intervals for the simulation models and is set equal to ±2.5% for the theoretical models. The response surface models [8] for the simulation models *(F1, F2, F3* and *F6)* are evaluated using data from [5].

Response Surface Methodology for *F1*

$$d_p(\mu m) = 1.42 + 0.16x_{TiB_2} + 0.61T + 1.37t - 0.20x_{TiB_2}{}^2 + 0.09x_{TiB_2}T + 0.54Tt + 0.27x_{TiB_2}t$$

Response Surface Methodology for F2

$$d(\mu m) = 3.50 - 0.54x_{TiB_2} - 0.34C$$

Response Surface Methodology for F3

$$\sigma(MPa) = 243.33 - 16.44x_{TiB2} - 2.21d_p - 5.88d + 10.30A - 1.12\, x_{TiB2}d_p + 4.50d_pA$$

Response Surface Methodology for F6

$$D(m) = 1000(2.75 + 3.31T_h - 1.04\sigma - 1.36\, T_h^2 + 1.17\sigma^2$$

Based on ranged performance targets, mapping models and HD-EMI constraints, ranged sets of property, structure and process variables are inductively determined along with the constraint boundaries. Exact constraint boundaries are identified in a top-down manner using the bisection method to avoid PU. Independent variables (*t* and *A* in this example) are selected by using the value which maximizes the number of discrete feasible sets in the corresponding domain. A dummy model is introduced for variables input to multiple domains (e.g., the volume fraction x_{TiB_2} is an input to both PS and SP domains). All HD-EMI values are constrained to be greater than 1, ensuring that the ranges of discrete property points are robust against modeled NU and MPU. Higher HD-EMIs indicate that the output range is further away from constraint boundaries and has a larger margin for error in the mapping model due to MSU of the estimate of the output range. Feasible and robust design points are identified in the property, structure, and processing design space based on the interlinked mapping models (see Figure 3).

The robust space of the volume fraction of TiB_2 (processing variable) lies within the ranges [0.0225, 0.035]; [0.045, 0.0525]; [0.06, 0.0675] and [0.075, 0.0775]. This indicates that the achieved space of the volume fraction of TiB_2, [0.0225, 0.035]; [0.045, 0.0525]; [0.06, 0.0675]; [0.075, 0.0775], thickness of shell, 10.75 mm, and area fraction, 30%, guarantees satisfactory submersible performance while maintaining all quantifiable uncertainty (MPU and NU) and its propagation within bounds. We see from Figure 3 (top left) that higher yield strength σ values and lower density ρ values are favorable for the design and are associated with higher HD-EMI values. It can also be concluded from Figure 3 (top right) that a lower grain size (d) and lower TiB_2 particle sizes (d_p) yields higher HD-EMI values and hence are favorable. From Figure 3 (left and right bottom) it can also be seen that a higher volume fraction of TiB_2 yields higher HD-EMI values and hence more favorable structures of the composite. In the figures, the black diamond points are constraint boundary points. The absence of constraint boundary points along an axis (e.g., thermal conductivity, k) indicates that the entire range meets the desired requirements subject to modeled uncertainty. IDEM makes it possible for a designer to determine ranged sets of processing variables robust against modeled MPU, NU and PU. The most suitable processing route is determined by using the cDSP by compromising the HD-EMI attainment and hence robustness against MSU. The deviation function for the cDSP is formulated using HD-EMI value attainment and the robust solution for the multiscale system, i.e., values of the design variables, is calculated by minimizing the deviation of the HD-EMIs from the target values (10 for all models and equal weights). While only one scenario for robust design specification is analyzed, the cDSP provides for setting different target values based on provision for potential MSU and hierarchical optimization of deviation functions and other scenarios can be set as deemed appropriate by the system level designer. The results of the design exploration are shown in Table 1. The achieved HD-EMI values are infinite when no constraint boundaries exist in the output direction. The HD-EMI value is -1 when the mean output lies outside the ranged specifications.

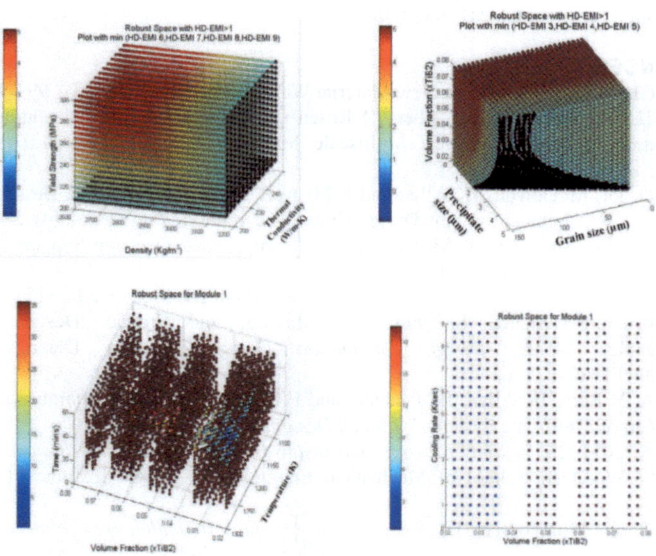

Figure 3: Robust design space for MODULE 3 (Top Left), MODULE 2(Top Right), MODULE 1- Precipitate size (Bottom Left), MODULE 1-Grain size (Bottom Right). The colorbars indicate HD-EMI values.

Table 1: cDSP Solutions

F1 (d_p)	*F2* (d)	*F3* (σ)	*F4* (ρ)	*F5* (k)	*F6* (D)	*F7* (W)	*F8* (T_{OR})	*F9* (T_{OP})
HD-EMI$_1$	HD-EMI$_2$	HD-EMI$_3$	HD-EMI$_4$	HD-EMI$_5$	HD-EMI$_6$	HD-EMI$_7$	HD-EMI$_8$	HD-EMI$_9$
∞	∞	12.90	4.21	∞	12.79	18.05	4.21	5.45

DESIGN VARIABLES				PERFORMANCE			
x_{TiB_2}	T (K)	C (K/sec)	t (min)	D (m)	W (kg)	T_{OR} (hr)	T_{OP} (C)
0.0775	1073	0.20	21	4551.4	15.64	13.45	17.50

In closing, a robust design approach for MMD is investigated by coupling IDEM and cDSP by making system level performance insensitive to MPU, MSU, PU and NU without eliminating its sources. Ranged sets of process variables are identified for MMD using IDEM while maximizing design freedom and considering quantifiable uncertainty (MPU, NU and PU). A singular robust solution is identified by compromising product performance for achieving robustness against MSU using cDSP and HD-EMI metric. The work presented in this paper constitutes one of the most complete applications of uncertainty management, by successfully managing it over three modules and nine analysis models in the PS, SP and PP domains for MMD. The MMD approach holds promise in early design space exploration for integrated computational materials engineering.

ACKNOWLEDGMENTS

F. Mistree and J.K. Allen gratefully acknowledge support from the NSF Grant CMMI 1258439 and the L.A. Comp and John and Mary Moore Chairs, respectively. J.H. Panchal acknowledges support from NSF Grant CMMI 1042350.

REFERENCES

1. Olson, GB. 2000. "Designing a New Material World." *Science,* 288(5468). 993-998.
2. Choi, HJ, DL McDowell, JK Allen, D Rosen and F Mistree. 2008. "An Inductive Design Exploration Method for Robust Multiscale Materials Design." *J Mech Design.* 130(3). 031402.
3. Choi, HJ, DL McDowell, JK Allen and F Mistree. 2008. "An Inductive Design Exploration Method for Hierarchical Systems Design Under Uncertainty," *Eng Opt,* 40(4). 287-307.
4. Vadde, S, JK Allen and F Mistree. 1994. "Compromise Decision-Support Problems for Hierarchical Design Involving Uncertainty." *Computers & Structures.* 52(4). 645-658.
5. Sinha, A, A Srivastava, S Ghosh, JH Panchal, JK Allen, DL McDowell and F Mistree. 2009. "Microstructure-Mediated Integration of Material and Product Design – Undersea Submersible," *ASME Design Automation Conference*, San Diego, CA. Paper DETC2009/DAC-87276.
6. Sinha, A, N Bera, JK Allen, JH Panchal, and F Mistree. 2013. "Uncertainty Management in the Design of Multiscale Systems." *J Mech Design,* 135(1). 011008.
7. Sinha, A. 2011. *Uncertainty Management in the Design of Multiscale Systems.* MS Thesis. The GW Woodurff School of Mechanical Engineering, Georgia Institute of Technology, Atlanta.
8. Myers, R, and Montgomery, D. 1995. *Response Surface Methodology: Product and Process Optimization Using Designed Experiments.* Wiley. New York.
9. Choi, HJ, R Austin, JK Allen, DL McDowell, F Mistree and DJ Benson. 2005. "An Approach for Robust Design of Reactive Powder Metal Mixtures Based on Non-deterministic Micro-scale Shock Simulation." *J Computer-Aided Materials Design,* 12(1). 57-85.

2nd World Congress on Integrated Computational Materials Engineering
Edited by: Mei Li, Carelyn Campbell, Katsuyo Thornton, Elizabeth Holm, and Peter Gumbsch
TMS (The Minerals, Metals & Materials Society), 2013

Virtual Prototyping of Lightweight Designs Made with Cold and Hot Formed Tailored Solutions

Eren Billur[1], Harald Porzner[2], Taylan Altan[1]

[1]Center for Precision Forming, The Ohio State University, 339 Baker System, 1971 Neil Ave., Columbus, OH 43210 USA.
[2]ESI North America, 32605 W. 12 Mile Road, Suite 350, Farmington Hills, MI 48334 USA.

Keywords: Hot Forming, Tailored Parts, Finite Element Simulation.

Abstract

Tailored cold and hot formed solutions are the key to the lightweight construction of the future vehicles. Only tailored solutions allow the combination of soft and hard zones for the best possible crash performance and minimum weight. The safety cage of future vehicles may completely be designed with hot formed components. In order to achieve the desired properties of the component, a heat treatment operation is necessary, and it comprises a significant part of the hot forming process. This poses additional challenges to the automotive industry as the heat treatment process is a function of many parameters and depends strongly on the forming process. In addition, only perfectly heat-treated parts will fulfill specifications.

Virtual Prototyping makes sure that component manufacturing and assembly processes yield to the designed tolerances and crash performance. ESI GROUP aims at providing all necessary tools for virtual prototyping of lightweight constructions designed and assembled with tailored solutions, in particular hot forming. This paper outlines what is needed for realistic virtual prototyping, and what is the status of the simulation solution. Validated realistic engineering examples are used to illustrate the capabilities in the field of virtual die design, forming, quenching, cooling channel engineering, assembly and product performance.

Introduction to Hot Forming of High Strength Steels

Higher strength steels can help reduce weight by reducing the blank thickness. However, as strength increases, formability decreases (which leads to splits), and the force required for forming increases. Maintaining dimensional tolerances is also an issue due to severe springback, Figure 1 [1].

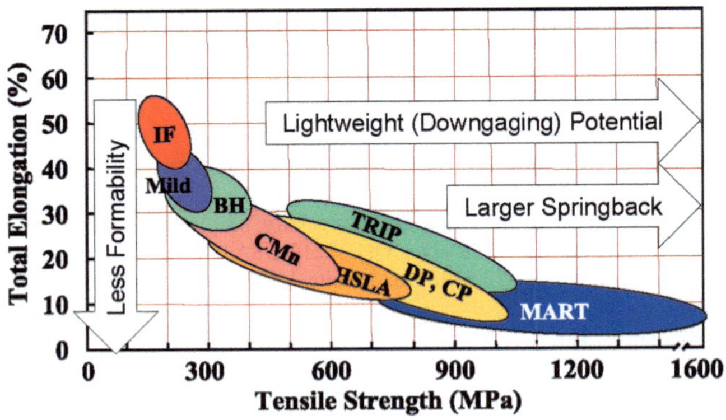

Figure 1: "Banana Curve": higher strength steels have lightweight potential, but limited formability and hard to control springback [1].

Hot forming (also known as press hardening or hot stamping) is a relatively new technology which allows ultra-high strength steels (typically 22MnB5) to be formed into complex shapes, which is not possible with regular cold stamping operations [2]. This is achieved by two methods [3]: (1) Indirect Process: the blank is formed, trimmed, and pierced in cold condition (i.e., state 1 in Figure 2). It is later heated and quenched in a die to get high strength properties; (2) Direct Process: the unformed blank is heated in a furnace, formed in hot condition (state 2 in Figure 2), and quenched in the die to achieve the required properties.

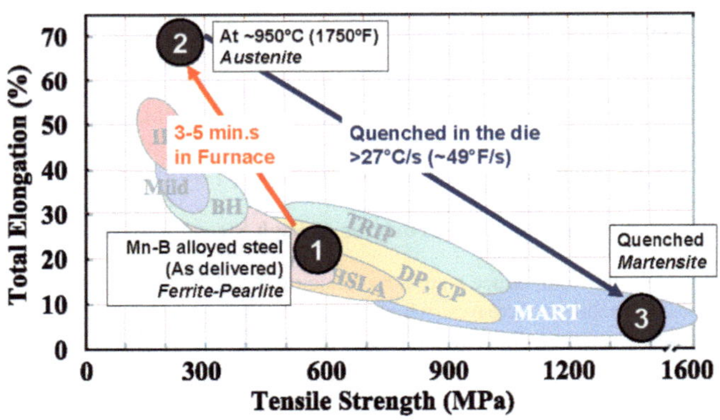

Figure 2: Summary of hot forming process.

Need for Tailored Properties

Tailored parts have been commonly used in the automotive industry to reduce the weight of components, simply by either: 1) eliminating the need for reinforcement and/or 2) to reduce the thickness in low-load areas. Tailored blanks can be made by, welding (Tailor Welded Blanks: TWB) or rolling (Tailor Rolled Blanks: TRB). TWBs can be composed of (1) two or more different thicknesses, (2) two or more different alloys (i.e., typically high strength steel with mild steel), or (3) a combination of both. TRBs are made with the same material with variable thickness profiles [4-6].

With hot stamping, generally ultra-high strength (in the order of *UTS > 1400 MPa (>200 ksi)*) is achievable. However, high strength causes several problems, especially in automotive applications: (1) strength reduces the elongation and thus the energy that can be absorbed and (2) welding of very high strength steel to mild steel creates a heat affected zone which initiates cracks. Use of TWB's and TRB's in hot stamping is recent technology, applied since 2008 and 2006, respectively [4, 7]. Hot stamping makes a new type of "Tailored Blank" possible, one where the final microstructure (and therefore the strength and elongation properties) can be tailored in a single blank, Figure 3c [5].

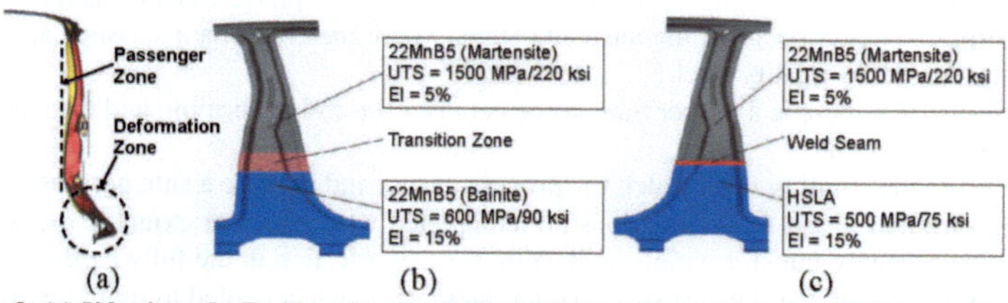

Figure 3: (a) Side view of a B-pillar after crash [8]. (b) Tailored properties or (c) TWBs may improve elongation in the deformation zone [5].

Challenges in Modeling and Analyses

Typical cold forming operations require one flow stress curve, anisotropy coefficients (r_0, r_{45} and r_{90}), friction parameters and yield criterion (Hill 48, Barlat 90, etc.). However, hot forming processes require more input parameters since there are: 1) mechanic, 2) thermic, 3) metallurgic and 4) fluid mechanic fields involved in the process and are all interrelated as illustrated in Figure 4 [9].

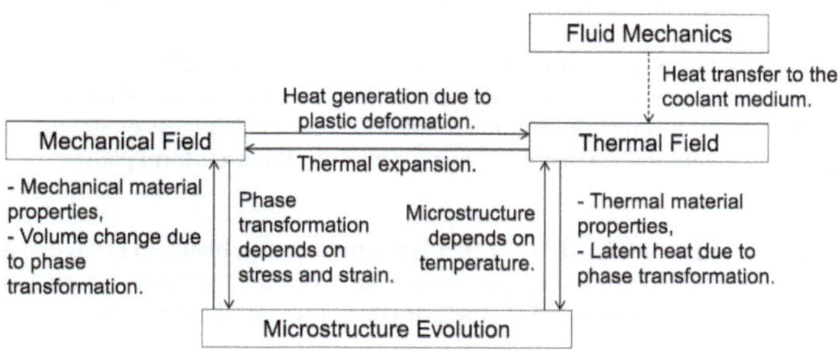

Figure 4: Multiphysics problem of hot forming [9].

Simulation of hot forming process can be divided into 4 stages. In each stage, different physics are involved. The gravity stage is modeled as implicit and only uses a mechanical field. Holding and forming stages have both mechanical and thermal fields - forming has to be completed while the blank is still at the austenite phase and no phase transformation is desired. The quenching stage has thermal and microstructure evolution fields. Fluid mechanics is not included in quick simulations so that costly couplings are avoided. Once the die face is designed by FEA using shell elements, cooling channels can be designed with volume elements.

Engineering Goals with Virtual Prototyping

Engineers can accomplish the following goals by means of virtual hot forming engineering:

- Ensure that no cracks or wrinkles occur during the forming process.
- Ensure that the forming process occurs in a good process window. This is in particular important for Zn coated blanks.
- Ensure that the forming process is completed before phase transformations start. The forming should occur in the austenite state, where material characteristics are similar to standard draw quality steel.
- Ensure that there is a proper time range between the end of forming and the start of phase transformations.
- Ensure that cooling rates match the process design and provide a safe process window.
- Ensure that phase transformations are completed when the part exits the die. Otherwise, ongoing phase transformations will cause severe distortion in the fully hardened part.
- Ensure that the part has the specified hardness value when cooled to room temperature.
- Minimize the part cycle time.
- Optimize material usage.
- Eliminate or minimize additional trimming operations.
- Optimize size and location of cooling channels.

Virtual Press Shop: Hot Stamping and Tailored Parts

Hot forming dies typically cost more than those for cold stamping due to (1) complex cooling channels drilled in, (2) expensive hot forming tool materials with high thermal conductivity (such as H11, H13 and their variations), and (3) more intense die spotting. Particularly with tailored quenching dies where dies in contact with "hard zones" are water cooled and areas in contact with "soft zones" are cartridge heated, late changes in tooling are expensive.

Virtual press shop can help discover if the dies and the process will yield useful parts. A useful part can be defined as one without (1) cracks/splits, (2) necking (excessive thinning), and (3) wrinkles. In tailored hot forming, in addition to these defects, (1) hardness or strength has to meet design requirements and (2) distortion may occur due to non-uniform phase transformations.

In the latest version of PAM-STAMP (2012), it is possible, with a blank, to (1) assign different thicknesses,(2) assign different material properties, and (3) model weld lines , as shown in Figure 5. Dies are automatically compensated for the extra thickness. All these features are compatible with hot forming simulations.

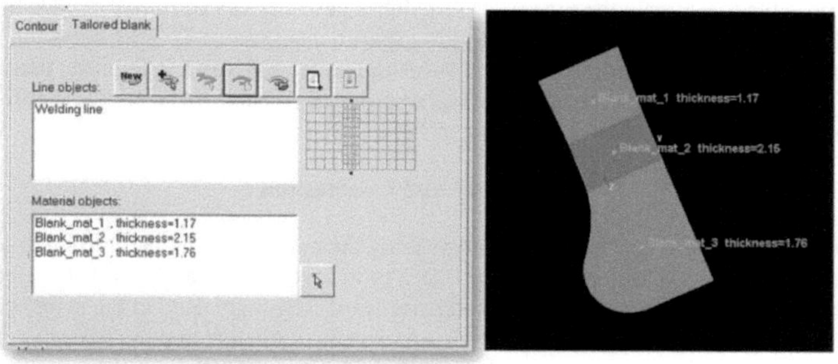

Figure 5: Modelling of Tailor Welded and Tailor Rolled Blanks in PAM-STAMP 2012.

As discussed earlier, hot forming adds a new, alternative method, in addition to TWB and TRB, to tailor the final part, by tailoring the microstructure. Process was designed based on Numisheet problem, and all the tools were divided into two assumed portions: heated zone (450°C) and cooled zone (20°C). By virtual try-out, it can be concluded that this part cannot be hot formed with a one-piece blankholder as the final part would have wrinkles in the soft-zone and cracks in the hard-zone. As seen in Figure 6, a two-piece blankholder was found to solve the problems and make a part without wrinkles or cracks.

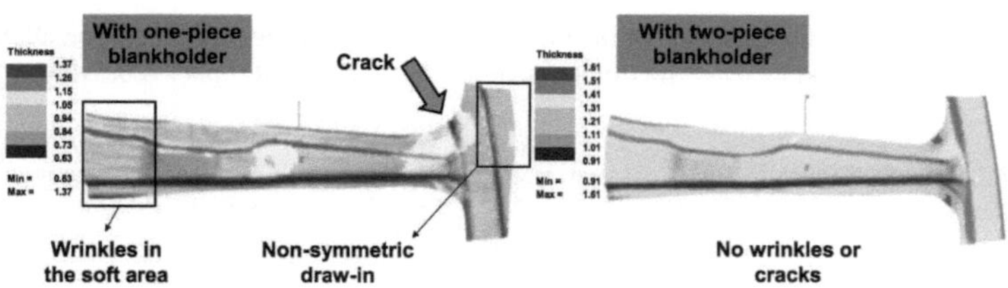

(a) (b)

Figure 6: A tailored part simulation: (a) with one-piece blankholder, (b) with two-piece blankholder.

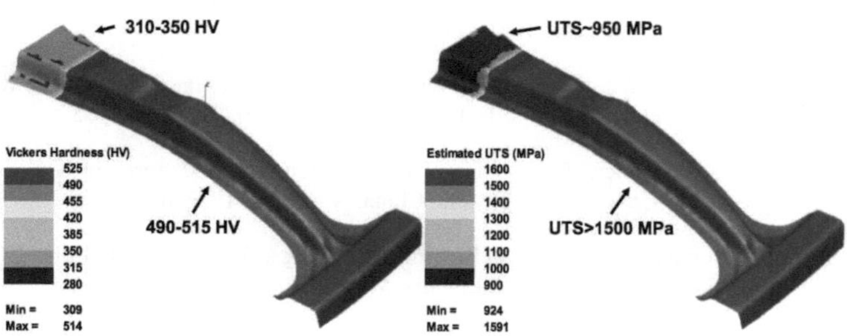

(a) (b)

Figure 7: Properties of the final part: (a) hardness and (b) estimated UTS distribution.

It is also possible to estimate the properties in tailored part, such as Hardness and UTS, without physically making the parts and performing the experiments (Figure 7). This helps the manufacturing engineer to optimize the process parameters (e.g., die temperature, press force at the bottom, quenching cycle time, etc.) for higher productivity and ensuring that the final properties are as required.

Summary and Conclusions

Regulations in safety and fuel efficiency are pushing the automotive industry to lightweight solutions more than ever. Tailored parts are one way to save weight, and TWBs have been used extensively in cold formed parts. Hot formed parts are increasingly used in the safety cages but face several challenges, including the lack of experienced engineers in process design and the need for tailored solutions to improve energy absorption characteristics.

Possible problems in mass production may be identified and solved by use of "Virtual Prototyping", which can save costly reworking of the dies and shorten the lead-time of the final product. With the advances in software engineering, it is now possible to control, adapt, and transport quantities between engineering disciplines. This is a key to concurrent engineering which helps to further reduce the overall cost.

References

1. J.M., Hall, "Evolution of Advanced High Strength Steels in Automotive Applications", (presented at Great Designs in Steel, Livonia, MI, USA, May 19th, 2011).

2. H. So, et al., "An investigation of the blanking process of the quenchable boron alloyed steel 22MnB5 before and after hot stamping process", *Journal of Materials Processing Technology*, 212, (2012), 437-449.

3. H. Engels, O. Schalmin and C. Müller-Bollenhagen, "Controlling and Monitoring of the Hot-Stamping Process of Boron-Alloyed Heat-Treated Steels", Proceedings of New Developments in Sheet Metal Forming Technology, pp. 135-150.

4. M. Rehse, "Flexible Rolling of Tailor Rolled Blanks", (presented at Great Designs in Steel, Livonia, MI, USA. March 8th 2006).

5. H. Karbasian and A. E. Tekkaya, "A review on hot stamping", Journal of Materials Processing Technology, 210, (2010), 2103-2118.

6. D. Guo, "Body Light Weight and Cost Control", Proceedings of the FISITA 2012 World Automotive Congress, Lecture Notes in Electrical Engineering 196 (2013) 977-985.

7. E. Hilfrich and D. Seidner, "Crash safety with high strength steels", (presented at International Automotive Congress, Shengyang, China, November 1st, 2008).

8. B. Macek, "Optimization Side Crash Performance Using a Hot-Stamped B-Pillar", (presented at Great Designs in Steel, Livonia, MI, USA, March 8th, 2006).

9. H. Porzner, et al., "Virtual Prototyping: Hot Forming Engineering with Virtual Manufacturing", (presented at AP&T Press Hardening Next Step Seminar, Dearborn, MI, USA, September 19, 2012)

2nd World Congress
on Integrated Computational Materials Engineering (ICME)

Process Optimization

2nd World Congress on Integrated Computational Materials Engineering
Edited by: Mei Li, Carelyn Campbell, Katsuyo Thornton, Elizabeth Holm, and Peter Gumbsch
TMS (The Minerals, Metals & Materials Society), 2013

MULTISCALE MODEL FOR NON-METALLIC INCLUSIONS/STEEL COMPOSITE SYSTEM USING DATA SCIENCE ENABLED STRUCTURE-PROPERTY LINKAGES

Akash Gupta[1], Ahmet Cecen[2], Sharad Goyal[1], A. K. Singh[1], S. R. Kalidindi[3]

[1]TRDDC-TCS Innovation Labs, Tata Consultancy Services Ltd., 54 B, Hadapsar Industrial Estate, Pune, Maharashtra, 411013, India
[2]Department of Mechanical Engineering and Mechanics, Drexel University, Philadelphia, PA 19104, USA
[3]Woodruff School of Mechanical Engineering, Georgia Institute of Technology, Atlanta, GA 30332, USA

Keywords: Steel, Inclusions, Multiscale model, Data science, 2 point statistics, ICME

Abstract

Cleanliness is a major concern for steel manufacturers. Therefore, they constantly strive to modify and reduce non-metallic inclusions in the final product. Performance and quality of final steel sheet is strongly influenced by composition, morphology, type, size and distribution of inclusions in steel sheet. The aim of current work is to critically evaluate the versatility of a new data science enabled approach for establishing objective, high fidelity, structure-property correlations that are needed to facilitate optimal design of the processing path to realize enhanced performance of the final product.
A 2-D finite element based micro-mechanical model was developed to simulate, the effect of various spatial configurations and geometries of hard and soft inclusions in a steel matrix system, on the final properties of processed sheet. From each microscale simulation macroscale parameters such as yield strength, effective hardening rate, localization propensity, and plasticity index, were extracted. A large number of microstructures were evaluated using the micro-mechanical model. A reduced-order representation was extracted for the selected ensemble of microstructures using principal components of their 2-point statistics. These objective measures of the microstructure were then linked with the macroscale parameters listed above using regression methods. The extracted structure-property correlations are presented on this paper.

Introduction

Steelmakers constantly strive to modify and reduce the non-metallic inclusions to meet the ever-increasing demands on the quality of steel. Non-metallic inclusions are foreign substances, (defects) mainly oxides, sulphides and oxy-sulphides, which are embedded in steel matrix. The inclusions are generated during upstream steelmaking operations and get embedded in the steel matrix during solidification in a caster. They disrupt the homogeneity of structure, so their influence on the mechanical and other properties can be considerable. Performance and quality of final steel sheet is strongly influenced by composition, morphology, type, size and distribution of inclusions. For example, during rolling, hard inclusions lead to interface delamination, voiding and crack formation while soft inclusions get elongated. Many authors have studied the inclusions present in a matrix. In the classical model of Eshelby [1], elastic field in and around elastic ellipsoidal inclusion in an elastic matrix was determined. Using the finite element method (FEM), a few authors have attempted to study void formation and cracks originating from inclusion/matrix interfaces during the process of hot rolling [2,3] and cold rolling [4].

The critical need for a multiscale model for optimizing the performance of non-metallic inclusions/steel composite system in the ICME framework has been demonstrated by the author in recent work [5]. This multiscale model along with providing guidelines for design of rolling operation also provides constraints to design of ladle and tundish process due to its integration with CFD and thermodynamic models of upstream processes in framework [5].

In this paper, a new data science enabled approach is explored for establishing objective and high fidelity multiscale structure-property linkages. The data science approach uses PCA (Principal Component Analysis) representation of 2-point statistics of microstructures. 2-point statistics describe the first-order spatial correlations between the constituent distinct local states in microstructure. These are usually recovered by throwing vectors of all sizes and orientations into the material microstructure. It has been demonstrated in [6] that the complete set of 2-point correlations carry all of the information needed to uniquely reconstruct most eigen microstructures to within a translation and/or an inversion.

Methodology

Micro-mechanical Model

A 2-D finite element based micro-mechanical model was developed in commercial finite element package ANSYS. Multiple hard or soft inclusions with different geometries, sizes and spatial configurations were distributed in steel matrix and uniaxial compression tests were simulated for the composite system. Steel matrix was considered as square of size 250 μm and radius of spherical inclusions were in size range of 5-30 μm. For inclusions of other shapes (square, triangle, rhombus, platelet) individual sizes were calculated assuming that they have area equal to that of spherical inclusion. The constitutive equation for inclusion and steel matrix were taken from [7]. Inclusion/matrix interface was considered as weak interface. Periodic boundary conditions were applied to the system and a uniform mesh with 6400 elements was generated. From each microscale simulation, macroscale parameters were extracted and used in establishing structure-property correlations. These parameters consisted of effective bulk averaged properties such as yield strength (YS), effective hardening rate (SH) and defect-sensitive properties such as localization propensity (LP), plasticity index (PI). LP is defined as volume fraction of matrix elements where strain is greater than 1.25 times the applied strain, and PI is defined as the ratio of average inclusion strain to the average matrix strain.

To generate data for application of data science approach 385 synthetic microstructures were constructed in MATLAB by assuming random spatial distribution of inclusions of different shapes and sizes in matrix. Maximum inclusion volume fraction in system was fixed at 10%. Further, micro-mechanical model was run for each microstructure and data generated was analyzed using the data science approach described next.

Data Science Approach

A novel data driven approach was applied on the data generated from simulations of microstructures and macroscale parameters. For each two phase microstructure consisting of inclusions and steel matrix, 2-point statistics using FFT (Fast Fourier Transform) was generated. The concept of a microstructure function is central to the approach employed in this work. Mathematically, a microstructure function $m(x, h)$ can be defined to represent the probability density associated with finding local state h at spatial location x [6]. In this work, 2-point autocorrelation of the inclusion phase was used to quantify the microstructure in a statistical framework. After obtaining 2-point statistics, PCA analysis was done on the dataset. PCA is one of the widely used techniques for dimensionality reduction. The motivation behind dimensionality reduction is that most of the microstructure descriptors are highly correlated. PCA creates a new series of uncorrelated (orthogonal) variables called principal components (PCs), which are sorted by decreasing variance. For each microstructure, a reduced-order representation was extracted using principal components of its 2-point statistics. Microstructure measures obtained after PCA are linked with the macroscale parameters extracted from each microscale simulation using linear regression methods to establish structure-property correlations.

Linear regression analyses were performed in this work using polynomial functions; the error associated with each data point was expressed as the absolute residual:

$$E = |F - f^p(PC_1, PC_2 \dots PC_n)| \tag{1}$$

where, F is the observation and $f^p(PC_1, PC_2 \dots PC_n)$ denotes a pth-order polynomial function of n PCs. In ordinary least squares (OLS) regression, the polynomial coefficients are established by minimizing the residual sum of squares (RSS) in the entire dataset. The measure of error and the extracted polynomial fit depend critically on the selection of both p and n. Although higher values of p and n will always produce a lower value of the error, they do not necessarily increase the fidelity of fit. This is because the higher values of p and n may lead to over-fitting and can produce erroneous estimates in any subsequent applications. For quantifying the robustness of the polynomial-fits, the following specific measures were adopted: (i) Mean absolute error of fit, $\bar{E}$ and Standard deviation, σ. (ii) Mean absolute error, E_{CV}, and Standard deviation, σ_{CV}, of leave-one-out cross validation (CV). CV is one of the many ways to provide a better selection of the parameters p and n. This technique involves the training of a polynomial fit K times, while leaving one data point out of the test set each time. Given a large K, for an over-fitted polynomial, the exclusion of a single data point will cause significant change in the coefficients, whereas for a good fit this change will be negligible. (iii) The law of succinctness (Ockham's razor) suggests that out of all viable fits, the least complex one (the one with the lowest number of fit parameters) should be used. Above three measures are used to arrive at the best linkages without over-fitting the data points.

Results and Discussions

Case studies were performed for the case of all hard and all soft inclusions, since these two are the extreme cases. Figure 1 and 2 are representative simulation results obtained from the micromechanical model. Figure 1 shows the contour of equivalent plastic strain for case study 1 conducted for multiple hard inclusions present in steel matrix at T= 900 °C and $\dot{\varepsilon} = 5\ s^{-1}$. Strain localization and void formation at inclusion/matrix interface can be clearly observed. Figure 2 shows the contour of equivalent plastic strain for an example simulation in case study 2 performed for multiple soft inclusions present in steel matrix at T= 1000 °C and $\dot{\varepsilon} = 5\ s^{-1}$. No void formation is seen for the case of multiple soft inclusions; instead they just get elongated. These results are consistent with experimental results in published literature [2-4].

PCA analysis was done on the dataset generated after calculating 2-point statistics of all microstructures. Figure 3 plots the cumulative explained variance obtained from PCA. The score plot PC1 vs PC2 is shown by Figure 4.

In case study 1, two macroscale parameters YS and LP were obtained and regression analysis was done to obtain structure-property correlation. Procedure for quantifying the robustness of polynomial fit was followed to avoid overfit. Complete set of regression results will be shown only for LP parameter for this case study.

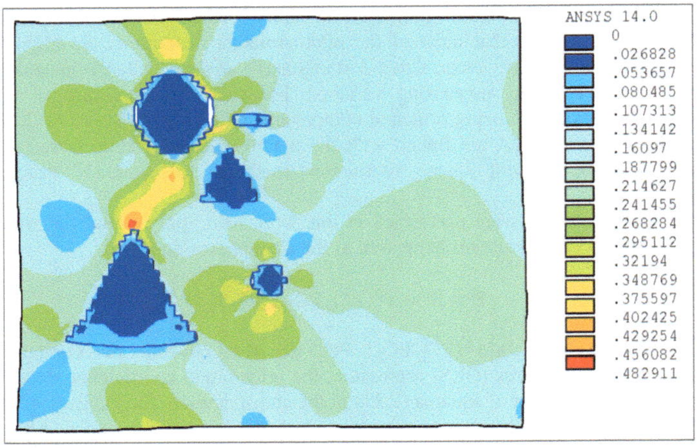

Figure 1. Contour plot of equivalent plastic strain for the case of multiple hard inclusions (case study 1) present in steel matrix at T= 900 °C and $\dot{\varepsilon} = 5\ s^{-1}$.

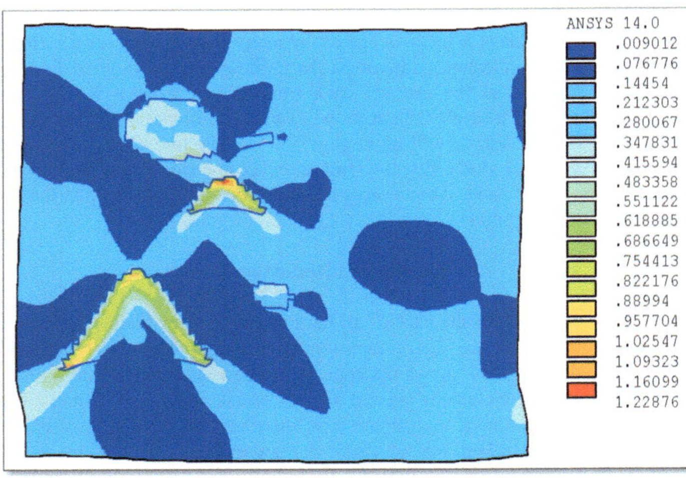

Figure 2. Contour plot of equivalent plastic strain for the case of multiple soft inclusions (case study 2) present in steel matrix at T= 1000 °C and $\dot{\varepsilon} = 5\ s^{-1}$.

Figure 5(a) shows error of fit for OLS (Ordinary Least Square) regression for LP parameter with bars representing, $\bar{E}$, and lines representing, σ. Increasing degree of polynomial ($1 \leq p \leq 5$) and increasing number of PCA weights ($1 \leq n \leq 5$) were explored in regression analysis. Figure 5(b) shows error of cross validation for OLS regression for LP parameter with bars representing, E_{CV}, and lines representing, σ_{CV}. Based on Figures 5(a) and 5(b) and Ockham's razor, best linkage of

5th order polynomial with 2 PC's is identified. R-squared value obtained for LP parameter for this combination is 0.94. Similar analysis was done for YS parameters and best fit was obtained for 5th order polynomial with 3 PC's and R-squared value of 0.97. Regression plots for LP and YS parameter are shown by Figure 6(a) and 6(b) respectively.

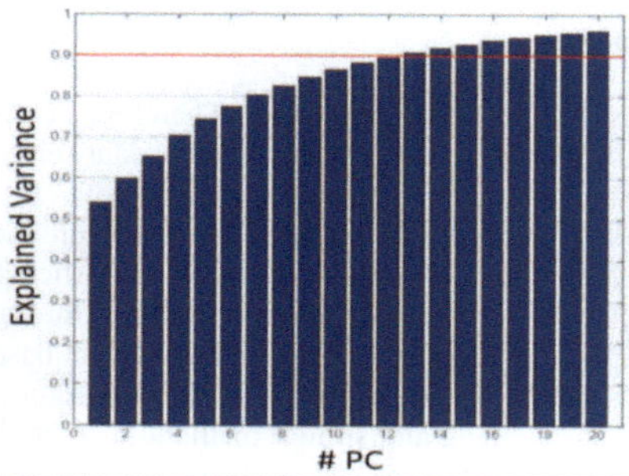

Figure 3. Cumulative explained variance of principal components

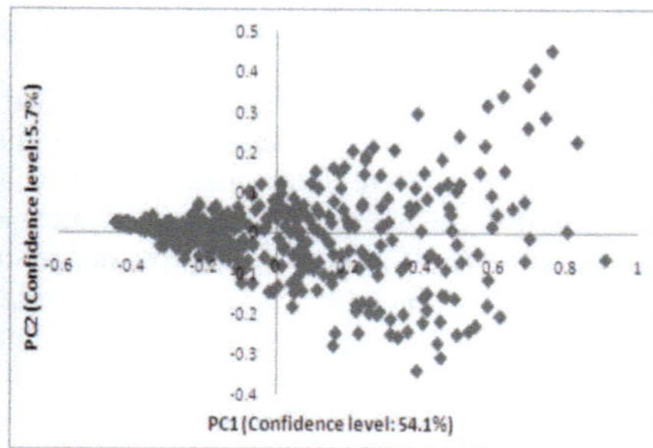

Figure 4. Score plot of PC1 vs PC2

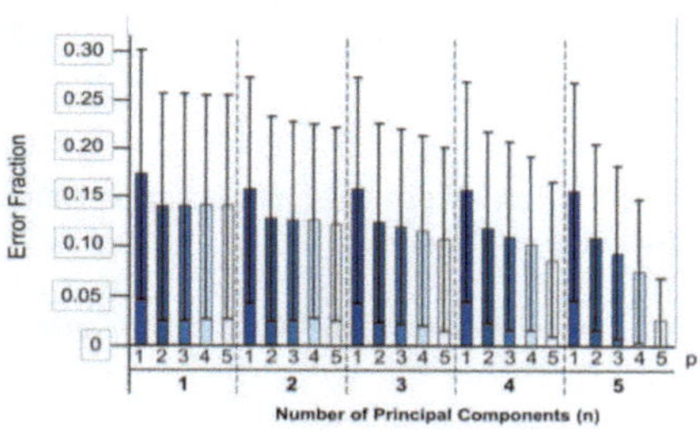

(a)

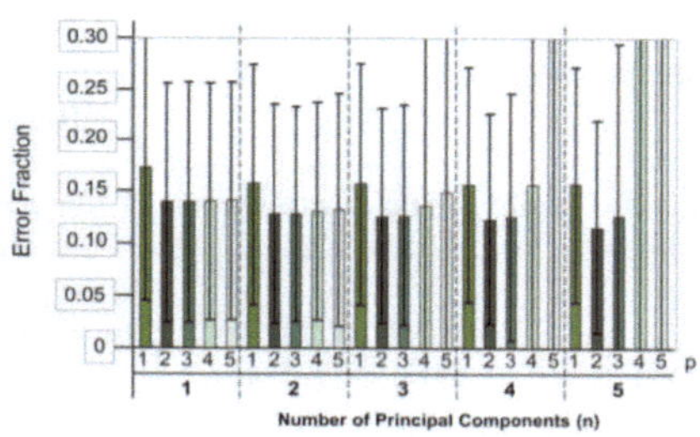

(b)

Figure 5. Plots for LP parameter with p(polynomial degrees) and n(number of PCs) showing (a) Error of fit (b) Error of cross validation

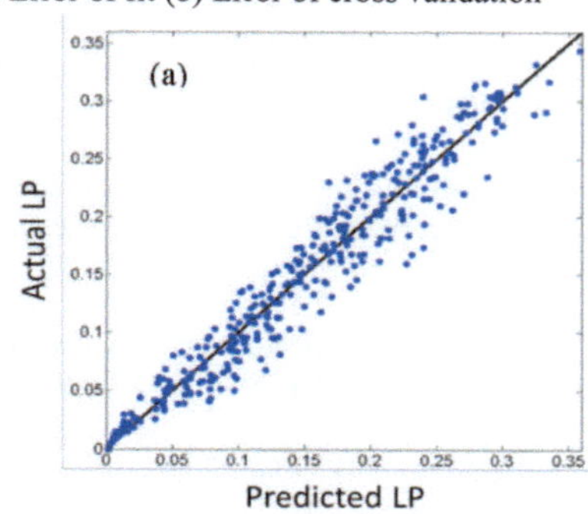

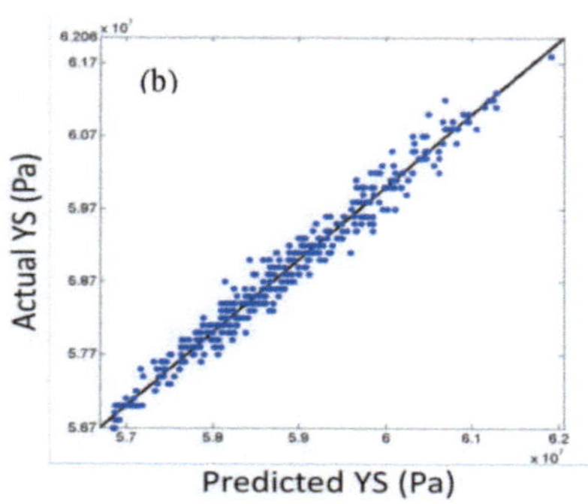

Figure 6. Regression plots for (a) LP parameter (R^2 =0.94) and (b) YS parameter(R^2 =0.97)

For any new microstructure LP or YS parameter can be directly estimated from the equation obtained after regression analysis and can be used as structure-property linkages for multiscale model of rolling. Similarly, for case study 2 macroscale parameters YS and SH were obtained with good regression fits.

Summary

Inclusions have different shapes, sizes, properties and spatial distributions. Therefore, a data science based approach is applicable to this problem. The central ideas of new data-science based approach were demonstrated and validated through a specific case study. It is demonstrated that the data-driven, fully automated protocols developed in this study produce computationally efficient high fidelity linkages needed to facilitate process design. In particular, these linkages would prove valuable in designing hot and cold rolling operations through consideration of the acceptable level of inclusions that can be tolerated in the final product based on quality and processing requirements. Though this work is done on synthetic microstructure it proves the usefulness of data based approach for this problem, and similar methodology can be applied for real two phase microstructures of non-metallic inclusions and steels.

Acknowledgements

Encouragement and support from TCS CTO, Mr. K Ananth Krishnan, and TRDDC Process Engineering Lab Head, Dr. Pradip, are gratefully acknowledged. AC and SK acknowledge support from N00014-11-1-0759.

References

1. J.D. Eshelby, "The determination of the elastic field of an ellipsoidal inclusion and related problems," Proc. R. Soc. Lon. A: Math. Phys. Sci. 241, (1957), 376-396.

2. C. Luo, "Evolution of voids close to an inclusion in hot deformation of metals," Comput. Mater. Sci. 21. (2001), 360–374.

3. Esa Ervasti, Ulf Stahlberg, "Void initiation close to a macro-inclusion during single pass reductions in the hot rolling of steel slabs: A numerical study," Journal of Materials Processing Technology 170 (2005), 142–150.

4. H. Yu, et al., "Behaviour of inclusions with weak adhesion to strip matrix during rolling using FEM," Journal of Materials Processing Technology 209, (2009), 4274–4280.

5. Akash Gupta et al., "Integrated modeling of steel refining, casting and rolling operations to obtain design set points for quality steel sheet production," Proceedings of Numiform, Shenyang, China 2013.

6. David T. Fullwood, Stephen R. Niezgoda, Surya R. Kalidindi, "Microstructure reconstructions from 2-point statistics using phase-recovery algorithms," Acta Materialia 56 (2008) 942–948.

7. G.Bernard, P.V.Ribound, G.Urbain, "Investigation of the plasticity of oxide inclusions," Rev. Metall. Cahiers d'Informations Techniques 78, (1981), 421-433.

2nd World Congress on Integrated Computational Materials Engineering
Edited by: Mei Li, Carelyn Campbell, Katsuyo Thornton, Elizabeth Holm, and Peter Gumbsch
TMS (The Minerals, Metals & Materials Society), 2013

A MULTI-SCALE, MULTI-PHYSICS OPTIMIZATION FRAMEWORK FOR ADDITIVELY MANUFACTURED STRUCTURAL COMPONENTS

Tahany El-Wardany, Mathew Lynch, Wenjiong Gu, Arthur Hsu, Michael Klecka, Aaron Nardi, Daniel Viens

United Technologies Research Center, 411 Silver Lane, East Hartford, CT, 06108, USA

Keywords: Topology Optimization, Additive Manufacturing, Optimization framework

Abstract

This paper proposes an optimization framework enabling the integration of multi-scale / multi-physics simulation codes to perform structural optimization design for additively manufactured components. Cold spray was selected as the additive manufacturing (AM) process and its constraints were identified and included in the optimization scheme. The developed framework first utilizes topology optimization to maximize stiffness for conceptual design. The subsequent step applies shape optimization to refine the design for stress-life fatigue. The component weight was reduced by 20% while stresses were reduced by 75% and the rigidity was improved by 37%. The framework and analysis codes were implemented using Altair software as well as an in-house loading code. The optimized design was subsequently produced by the cold spray process.

Introduction

AM also referred to as 3D printing, is a layer-by-layer technique of producing objects directly from a digital model. AM technology enables low-cost product assembly and the building of any number of products with complex shapes /geometries, complex material compositions and designed property gradients [1,2]. The objective of current research is to integrate AM processes and design exploration methods to synthesis of shapes, geometric meso-structures, and microstructures to achieve desired performance [3]. Manufacturing constraints are defined based on the capabilities and limitations associated with the AM processes such as: speed of build, accuracy, surface geometry, tolerances, wall thickness and feature size, material properties, and range of materials. Any AM optimization framework systematically arranges and merges design and analysis tools for a preliminary design stage of the fabricated component.

In general, a multi-objective optimization framework such as that published in [4,5,6] is required to design AM components which are governed by mutually interacting physical phenomena to achieve the required performance. The developed framework is based on the functional decomposition of AM processes. It also identifies generic sub-functions and various physical principles that support the conceptual design process and thus aid in decision-making in the early stages of design.

This paper presents a developed optimization framework that uses optimization techniques to produce novel designs well-suited to the cold spray AM process. A topology optimization (TO) methodology suited for AM processes are used to redesign a highly loaded bracket to replace the traditional sheet metal forming. A multi-physics programming scheme for the conceptual design will be discussed. During the conceptual design, multiple loadings, multilevel AM constraints, weight and fatigue stress constraints are coupled to settle the associated difficulties in considering the whole structure as a pre-defined design domain. The commercial Altair® software package of Hyperworks including OptiStruct and HyperStudy has been used. The primary objective is to integrate the design disciplines and AM process requirements.

Challenges of design optimization with additive manufacturing

While AM offers an array of benefits, it also requires consideration of a unique set of manufacturability decisions. AM entails new constraints that are not present in conventional manufacturing as follows:

- *Workpiece support design* - AM requires a support structure onto which material can be deposited. The support design and its removal after deposition is an important issue in AM.
- *Material deposition restrictions*- Figure 1 shows the cross-section of the material deposited using cold spray process. The different shapes are examples of the possible material cross-section after deposition. The figure also indicates that the angle of material deposit has to be included when designing a component made by AM.

Figure 1 Powder depositions by cold spray and the resulting material formation

- *Deposition nozzle clearance from the component* - The geometry of the component and the incumbent support design must account for the nozzle physical space and movement.
- *Finishing* - A rough surface condition is obtained after the deposition of the powder. Post machining is required and must be included in the design constraints.
- *Manufacturing cost* - Manufacturing cost can be included in the objective function of the optimization framework.

Optimization Framework for additive manufacturing applications

A framework for additive topology optimized manufacturing processes has been developed, where design and analysis tools are collectively deployed for a preliminary design stage of a component fabricated by AM. The framework, shown in Figure 2, consists of two main stages; the first stage is associated with the concept stage, while the second defines the steps for detailed design. The concept stage considers design constraints, AM manufacturability constraints and objectives for TO. The result from TO is interpreted by the designer into a CAD model and FEA is performed to determine the initial part performance and stress state. While the concept stage relies heavily on the inputs of a design engineer with aid from TO, the design stage relies more heavily on automated software tools.

The design from TO must be parameterized into a vector of design variables $x = (x_1, x_2, \ldots, x_n)$. The goal of the Design Optimizer is to find the optimal x^* according to an objective function $f(x)$ such that all constraints are satisfied. A detailed evaluation (e.g., FEA for stress analysis) of the concept part is performed to identify the critical areas where a wider design space would be beneficial.

Functional grading may also be folded into this approach. The parameters that govern the grading such as discrete decisions about the materials or their thickness would be specified as decision variables which would then be optimized.

Once the set of design variables is determined, the objective function $f(x)$ needs to be expressed entirely in terms of x. Similarly, constraints need to be expressed completely in terms of the design variables. The set of constraints may be partitioned into

- Explicit constraints that will be evaluated during the iterative design optimization process.
- Implicit constraints that will be set aside from the Design Optimizer.

Only after the design optimization has outputted an optimized design x*, a 'verification step' will verify that the implicit constraints are satisfied. If one or more implicit constraints are not satisfied, a surrogate constraint will be added to the set of explicit constraints that aims to avoid violating the implicit constraint. The Design Optimizer is re-run with both the explicit and surrogate constraints. The Design Optimizer considers a different solution in every iteration and makes calls to one or more functional modelsto evaluate the performance of a given solution (e.g., stress, displacement). The Design Optimizer explores the design space and normally terminates with an optimal solution that optimizes a given objective function while satisfying the (explicit) constraints. The Design Optimizer may have an "inner loop" where a certain set of design variables are considered and an "outer loop" where another set of design variables are considered. The inner loop optimization might be performed by OptiStruct and a custom outer loop may be used.

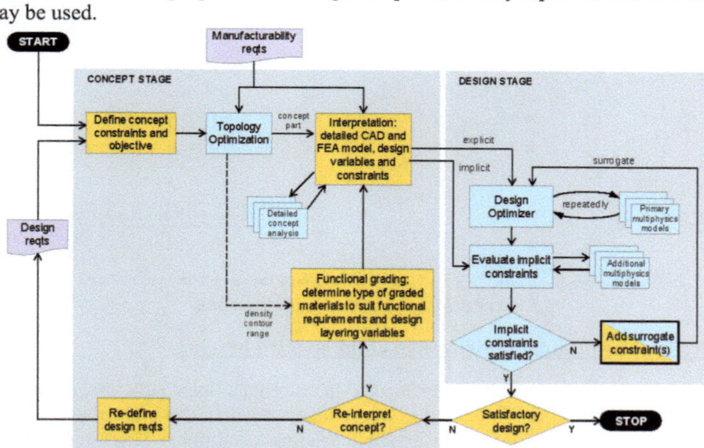

Figure 2 Flowchart of multi-scale / multi-physics optimization framework - Yellow boxes are human based steps, blue boxes are automated steps & pink boxes are requirements.

Multi-physics Optimization

Numerous tools are available to evaluate various forms of physical behavior and can be integrated within an optimization framework. Given that there is no universal multi-physics analysis tool, HyperStudy software will be used to call several different multi-physics / multi-models to be run in tandem at every iteration. With this setup, HyperStudy allows user-level, solver-neutral, multi-disciplinary analysis including exploration, study, and optimization. Figure 3-a shows its architecture scheme that can be used to integrate multiple models for design optimization, while Figure 3-b presents an approach that will be used in the case study presented in the next section.

The design process begins with the formulation of functional requirements and performance constraints and then continues with conceptual design, optimization and finally detailing of the component. Different components are sized for applied load and are optimized for weight or fatigue with consideration of other factors. The factors that influence the design directly arise from performance requirements, component layout, selected material and methods of additive manufacturing. Requirements in design optimization for additive manufacturing process can be divided into the following performance constraints:

- Structure performance constraints (allowable stresses, weight, stiffness, loading, fatigue performance, thermal load, deformation and distortion, dynamic behavior, mesh selection)
- Properties of materials produced by AM processes (corrosion resistance, bonding strength, mechanical and thermal properties)
- AM constraints (design of support structures, build accuracy, surface finish and z-direction mechanical properties, minimum feature size constraint, overhang constraint)

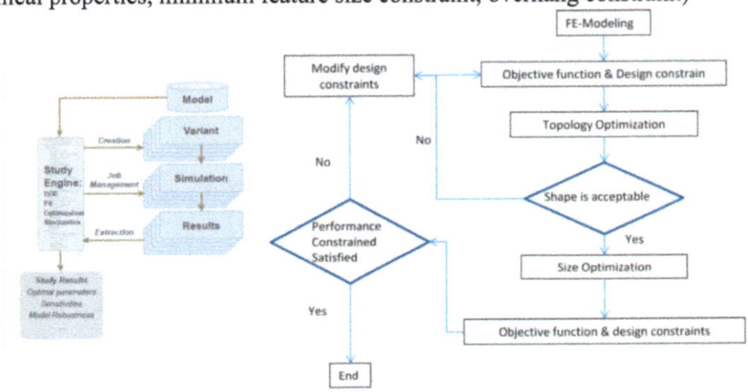

a) HyperStudy architecture scheme *b) Flow chart of multi-models integration approach*

Figure 3 Multi-physics Optimization

Case Study: Topology / shape optimization of the selected bracket

The proposed framework was implemented to redesign a failed structural mount (bracket) using Hyperworks, ABAQUS FE, and an in-house load software. The bracket is used to fixture components to a structural shell under dynamic loading conditions. Sheet metal forming is currently used to fabricate the bracket. Figure 4 shows the bracket design and applied dynamic load, which causes the bracket face to deform leading to stress concentration along bolted areas.

A finite element model using shell elements was developed to verify the current design and identify areas of high stress and failure. Nonlinear geometrical elastic-plastic analysis with contact interactions between mount surface and bracket was used to predict stress state and deformation. The dynamic loading used in the analysis is shown in Figure 4. The mounting surface was modeled as a rigid body and the face connection is assumed to be structurally critical. Figure 5 shows the prediction of stress gradient or transition at the major failure location. A picture of the failed bracket is also presented illustrating that the failure location correlates well with the FE predictions.

The new optimization framework was based on the FE results, the requirements for the design, material selection, and cold spray constraints. Figure 6 illustrates the approach used to combine AM and topology optimization to design and manufacture the selected bracket.

The first step is to take the given design requirements and to extract the subset of constraints that applies at the concept level, which are 1) geometry constraints (design space); 2) interface constraints; 3) minimum feature size constraints; 4) load conditions for which the performance constraints must be satisfied; and 5) performance constraints. The general process flow starts by defining a region whose entire volume is eligible to participate in the load path. Then the region is meshed for FEA, loads and specifications are given, and the TO is allowed to run.

The design space (Figure 6-1) is an envelope of material any subset of which is subject to possible removal. The setup of the design space is the first opportunity to incorporate AM constraints. The result is the topological optimum (Figure 6-2), the material which most efficiently carries the load. The OptiStruct software package can enforce certain manufacturing constraints specified by the user in a TO run, which will be modified to incorporate AM constraints. The result from the TO step provides a starting point for design interpretation.

The design interpretation (Figure 6-3) is subject to the judgment and experience of the designer. The rules for interpreting TO results for AM vary depending on the process, as shown in Figure 6 the final design of the part is controlled by the AM constrains. The development and implementation of the cold spray design rules control the design interpretation of the TO result. The FE analysis of this design indicates a reduction in the maximum stresses by 40% and the design was used as the basis for subsequent shape optimization conducted with the OptiStruct software. The use of the interpreted design for fine tuning is often called "shape optimization (SO)". SO begins with a solid model meshed for FEA (Figure 6-4). The designer identifies parameters of the meshed model to optimize. Then, the designer develops morphs to the mesh, literally stretching or compressing the FEA mesh to increase or decrease dimensions of features. Multiple morphs can be applied simultaneously using the HyperMorph tool inside the HyperMesh software. Design responses are also included by setting constraints such as maximum allowable stress, and setting an optimization objective such as minimum part weight. OptiStruct conducts SO by treating each individual morph as a design variable and finding the optimum application of each morph to achieve the objective without violating the constraints. A full finite element simulation is performed at each iteration of SO and a further reduction in maximum stresses led to a total reduction of up to 75% with a mass reduction of up to 21% from the original configuration.

Furthermore an LS-DYNA multiple particle deposition FE model is also developed to identify optimum cold spray process parameters (speed, feed, angle of deposition). The optimized parameters are then used to produce the new designed bracket.

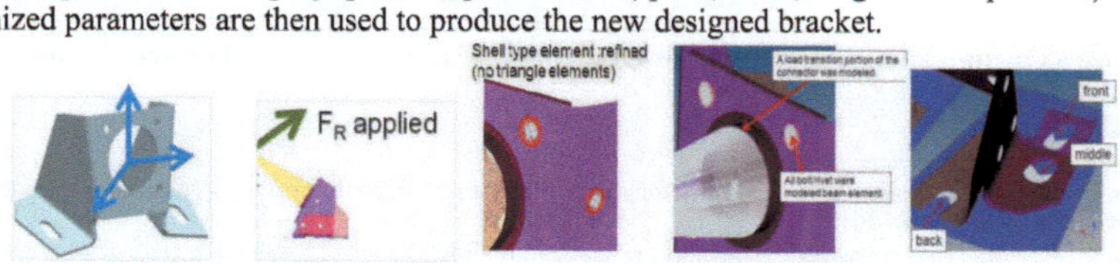

Figure 4 Failed bracket load distribution, mesh configuration and assumptions

Conclusions

This paper presents a multi-physics programming scheme for the conceptual design of additive topology optimized manufacturing. An optimization framework was developed that uses TO techniques to produce novel designs well-suited for AM cold spray process. During the conceptual design of the component with multiple loadings, both the multilevel AM constraints, part weight and fatigue stress constraints are coupled to settle the associated difficulties when the whole structure of the component is considered as a pre-defined design domain. The part weight was reduced by up to 21%, peak stresses were reduced by up to 75%.

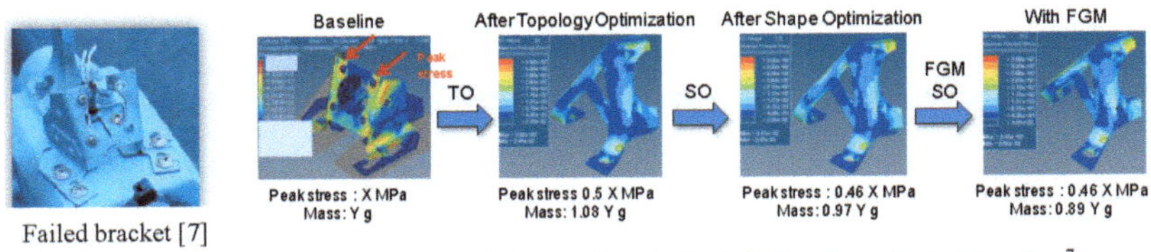

Figure 5 Predicted failure positions and optimized design for selected bracket[7]

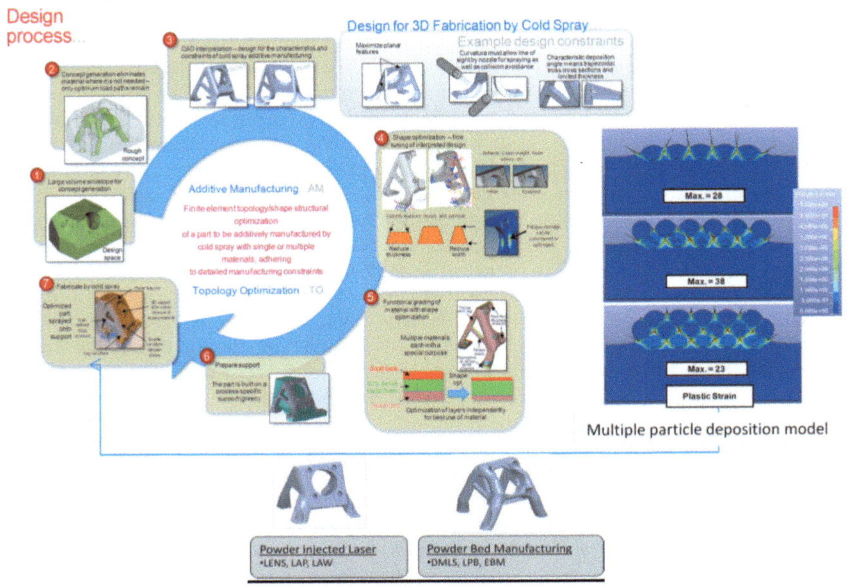

Figure 6 Optimization framework for cold sprayed additively manufactured component

Reference

[1] Rosen, D. W., 2007, 'Computer-Aided Design for Additive Manufacturing of Cellular Structures', Computer-Aided Design & Applications, vol. 4(5), pp. 585-594.
[2] Rosen, D. W.; Atwood, C.; Beaman, J.; Bourell, D.; Bergman, T.; Hollister, S., 2004, 'Results of WTEC additive / Subtractive Manufacturing Study of European Research', Proc. SME Rapid Prototyping & Manufacturing Conference, paper #TP04PUB211, Dearborn, MI. http://wtec.org/additive/welcome.htm for the complete report
[3] Chen Chu, Greg Graf and David W. Rosen, 2008, 'Design for Additive Manufacturing of Cellular Structures', Computer-Aided Design and Applications, CAD Solutions, LLC, http://www.cadanda.com
[4] Ho-Jun Lee, Jae-Woo Lee, Jeong-Oog Lee, 2009, 'Development of Web services-based Multidisciplinary Design Optimization framework' Advances in Engineering Software, vol. 40, pp: 176–183
[5] Fernando P. BernardO and Pedro M. Saraiva, 1998, 'Robust Optimization Framework for Process Parameter and Tolerance Design', AIChE Journal, vol. 44/ 9, pp: 2007-2017
[6] Jaroslaw Sobieszczanski-Sobieski, Raphael T. Haftka, 1996, 'Multidisciplinary aerospace design optimization - Survey of recent developments', AIAA 34th Aerospace Sciences Meeting and Exhibit, Reno, NV
[7] http://www.aviation.dla.mil/externalweb/userweb/AviationEngineering/Engineering/valueengineering/AFCAT/castandforged.asp - Photo source

2nd World Congress on Integrated Computational Materials Engineering
Edited by: Mei Li, Carelyn Campbell, Katsuyo Thornton, Elizabeth Holm, and Peter Gumbsch
TMS (The Minerals, Metals & Materials Society), 2013

OPTIMAL PROCESS CONTROL THROUGH FEATURE-BASED STATE TRACKING ALONG PROCESS CHAINS

Melanie Senn[1], Norbert Link[1], Peter Gumbsch[2]

[1]Karlsruhe University of Applied Sciences, Moltkestr. 30; Karlsruhe, 76133, Germany
[2]Fraunhofer IWM, Wöhlerstr. 11; Freiburg, 79108, Germany

Keywords: state tracking, optimal control, manufacturing process, process chain

Abstract

The optimal control of a manufacturing process aims at control parameters that achieve the optimal result with least effort while accepting and handling uncertainty in the state space. This requires a description of the process which includes a representation of the state of the processed material. Only few observable quantities can usually be measured from which the state has to be reconstructed by real-time capable and robust state tracker models. This state tracking is performed by a mapping of the measured quantities on the state variables which is found by nonlinear regression. The mapping also includes a dimension reduction to lower the complexity of the multi-stage optimization problem which is approximately solved online. The proposed generic process model provides a universal description that can be adapted to specific data from simulations or experiments. We show the feasibility of the generic approach by the application to two deep drawing simulation models.

Introduction

In manufacturing engineering, the modeling of processes allows to gain insight and hence to determine the optimal control parameters virtually instead of executing experiments that are associated with high material and personnel efforts. The coupling of models of different scales facilitates the consideration of the detailed material behavior in the description of the entire workpiece. On the contrary, the linking of individual processes into a process chain enables a comprehensive view including the interactions of the single processes on the workpiece which can then be assessed and controlled purposefully. Multi-scale modeling is combined with the holistic view of a process chain in order to improve the prediction of the workpiece properties for the optimization of the single processes and the entire process chain [1, 2]. However, the applied analytical material models are not suitable for a real-time capable online process control. The results of the analytical material models can on the other hand serve as a database for statistical learning methods which we use to instantiate a process state tracker.

The optimization of individual processes or entire process chains can be divided into two classes. In offline optimization (class 1), the optimal control parameters are determined offline (before runtime) and are then applied online (at runtime). In optimization class 2, online feedback control takes uncertainties into account and adapts the control parameters accordingly. A virtual optimization of the local forces of a segment-elastic blank holder over segments and time has been realized for a deep drawing process from simulations [3]. The objective function is formulated by minimizing the sheet thinning. The optimization is thereby constrained by the forming limit diagram. An online control that determines the optimal

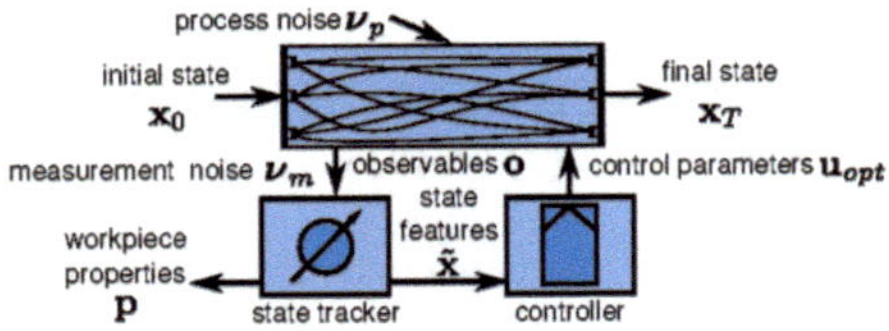

Figure 1: Generic process modeling with state tracker and controller

blank holder force values in real time has been implemented for a deep drawing process by an analytical description taking the limits for wrinkling and tearing into account [4]. However, future uncertainties are neglected. The optimization of single processes and entire process chains is addressed in [5]. Past uncertainties are incorporated by measurements, but future uncertainties are not taken into account. We have introduced an approach for the optimization of process chains considering future uncertainties in the optimization model [6]. In this work, we concentrate on the online feedback control of a single process under uncertainty.

Generic Process Modeling

The generic process model as depicted in Figure 1 provides a universal description in terms of a state tracker and a controller. The process quantities and their dependencies can be adapted to specific data from simulations or experiments by statistical learning methods. The detailed procedure of generic process modeling and its application is described in [7].

The process is modeled as a dynamic system which is characterized by the state of the processed material that cannot be measured at runtime with reasonable effort. Instead, observable quantities which are related to the state can be monitored online. The process state can then be reconstructed from the observables by a real-time capable state tracker model. The state tracker can also deduce the workpiece properties. It must be robust with respect to the measurement noise in the observables. In order to realize an efficient process control, only the process-characteristic state features are extracted from the possibly high-dimensional state. From these state features, the controller can determine the optimal control parameters by a given objective function and a model of the state transition.

For deep drawing, the process quantities of the generic process model are given as the local flow stresses (von Mises stresses) in the workpiece as the state, as well as forces, displacements and strains in workpiece and tools as the observables. The blank holder force corresponds to the control parameter and varying lubrication conditions (friction coefficient) can be seen as the process noise. The workpiece properties are characterized by the earing profile.

In the following sections, we describe how the state tracker and optimal controller models are instantiated from two deep drawing finite element models. Unfortunately, we cannot apply the generic approach for state tracking and control consistently to the detailed micromechanical model due to the high effort for generating a large number of samples (simulation runs) for statistical learning. The state tracker is applied to the complex micromechanical model for which a small set of samples is created by varying the process parameter from simulation to simulation. The optimal control is realized based on the phenomenological model of low model complexity. A large set of samples is thereby generated by varying the control parameter at multiple decision points in time (for manipulation) for each simulation.

State Tracking

The state tracker model is instantiated by deterministic simulations from the micromechanical model without inducing any stochastic influences.

Micromechanical Model

The micromechanical model [8] is of high complexity due to its 3D geometry and the consideration of crystal plasticity. Anisotropic elasticity and plasticity and isotropic hardening characterize the material behavior that represents a DC04 steel. A state tracker model is realized from 100 samples that are generated by variation of the blank holder force in the range of [60, 140] kN and the friction coefficient in the range of [0.08, 0.12].

State Tracker Model

The objective of state tracking is to model the dependency $\mathbf{f}_t$ between the temporal evolution of observables $\mathbf{o}_{t_0}, \ldots, \mathbf{o}_t$ and the current state variables

$$\mathbf{x}_t = \mathbf{f}_t(\mathbf{o}_{t_0}, \ldots, \mathbf{o}_t). \tag{1}$$

Statistical learning methods for regression and dimension reduction [9] are used for the implementation of real-time capable and robust state tracker models in our work. Regression allows to model functional dependencies from given sets of input and output data. Dimension reduction assists in reducing the complexity of the regression relation between the possibly high-dimensional input and output quantities on the one hand and extracts low-dimensional features that characterize the regression relation on the other hand. In this work, we present an integrated approach that combines nonlinear regression with nonlinear dimension reduction to extract low-dimensional features which can be used for an efficient process control or to deduce the workpiece properties. The integrated approach in terms of Principal Function Approximators (PFAs, for details see [7]) is based on Artificial Neural Networks (ANNs) [9]. PFAs enable a feature-based state tracking by dividing the relation of observables and state variables into two sub relations $\mathbf{f}_t^o$ (map from observables to features) and $\mathbf{f}_t^s$ (map from features to state). This representation can be used to predict the process state, and at the same time, to derive the low-dimensional features that can be further processed at runtime.

Realizations and Results

The feature-based state tracking is realized with a PFA to predict the high-dimensional stresses in the workpiece at 234 locations at process end from the temporal evolution of 14 observables in the workpiece (forces, displacements, strains) of 51 past time steps. The PFA is based on an ANN with three hidden layers to represent the high complexity of the regression relation. The low-dimensional features are extracted from the middle hidden layer to deduce the workpiece properties with a separate ANN with one hidden layer. In order to create the state tracker model by statistical learning methods, 80 % of the data are used for training, while the remaining 20 % are taken as test data in a five-fold cross validation to ensure the independence of specific data. The robustness of the state tracker model has been investigated in [7] by inducing measurement noise into the input.

Prediction results are visualized for a blank holder force of 64212 kN and a friction coefficient of 0.08 at process end. Figure 2a shows the state tracker model results (von Mises stresses). The magnitudes of the relative errors (between state tracker and simulation) in

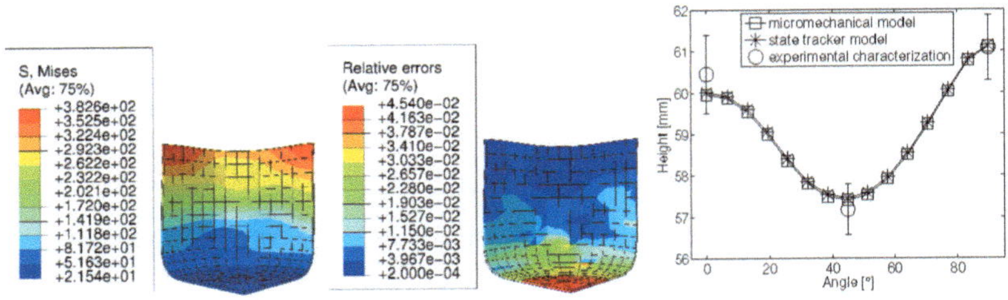

(a) von Mises stresses (state tracker)
(b) Relative error (state tracker, simulation)
(c) Earing profile (state tracker, simulation, experiment)

Figure 2: Results for state tracking at process end

the range of [0, 0.0454] in Figure 2b confirm the good model quality. The relative errors averaged over the 20 test samples result in a mean and maximum value of 0.0082 and 0.1292.

The predicted earing profile of the state tracker model is depicted in Figure 2c together with the simulation results and the experimental characterization. The relative errors (between state tracker and simulation) averaged over the 20 test samples yield a mean value of 0.0040 and a maximum value of 0.0278.

Optimal Control

The optimal controller models are instantiated by simulations from the phenomenological model. Since the simulations contain only deterministic data, the uncertainty (process noise) is modeled by samples drawn from a normal distribution with mean $\mu = 0$ and standard deviation $\sigma = 1$ for each considered state dimension. The superposition of the deterministic data from simulations and the induced uncertainty yields stochastic paths in the state space.

Phenomenological Model

The phenomenological model of low complexity is characterized by a simple 2D axisymmetric geometry. The material behavior is described by isotropic elasticity and plasticity and isotropic hardening of a Fe-28Mn-9Al-0,8C steel [10]. To create an optimal controller by statistical learning, 768 samples are generated by variation of the blank holder force (control parameter) at five decision points in time. At $t_1 = 0s$ admissible values are [80, 90, 100] kN, where at $t_2 = 0.2s, t_3 = 0.4s, t_4 = 0.6s, t_5 = 0.8s$ possible choices are [70, 80, 90, 100] kN.

Optimal Controller Models

The objective of optimal control is to find an optimal process path in the state space (and associated optimal control parameters) under stochastic influences. Optimality is defined by the minimum total costs, which are composed of control-relevant efforts and the quality-related end-product value. In contrast to classical control which is implemented by tracking a predefined optimal path in state space, the optimal path is here defined on the current state considering future uncertainties. This can be formulated as a multi-stage dynamic optimization problem with interconnected time steps in terms of the Bellman Equation

$$J_t(\mathbf{x}_t) = \min_{u_t \in U_t} \{ J_{Pt}(\mathbf{x}_t, u_t) + \langle J_{t+1}(\mathbf{x}_{t+1}) \rangle \}. \tag{2}$$

J_{Pt} describes the local transformation costs (production effort) to transfer the process from the current state $\mathbf{x}_t$ to the successor state $\mathbf{x}_{t+1}$ with the control parameter u_t. J_{t+1} corresponds to the successor costs that depend on the successor state $\mathbf{x}_{t+1}$. The successor costs comprise the local transformation costs of the remaining time steps $J_{Pt+1}, \dots, J_{PT-1}$ and the final costs J_F to evaluate the workpiece quality at process end T based on the final state $\mathbf{x}_T$. The future uncertainty is taken into account by the expectation operator $< \cdots >$.

The Bellman Equation can then be solved for each time step by Dynamic Programming (DP) [11] or Approximate Dynamic Programming (ADP) [11] depending on the complexity of the problem. While classical DP uses discrete state spaces, continuous state spaces are applied in ADP. The multi-stage optimization problem is solved by stepping backward in time performing an exhaustive search on all possible combinations of states and control parameters in classical DP and Backward ADP [7]. This results in an exponential growth of the complexity with an increase in the dimensions of the state or the control parameters. This explains the importance of compact state and control parameter spaces which can be achieved by dimension reduction methods or expert knowledge. However, in Forward ADP [7], one steps forward in time on various created stochastic paths (e.g. by a simulator) and thereby concentrates on a limited state space taking averages on the stochastic quantities. With each additional iteration (stochastic path), the complexity grows only linear.

Our concept of an optimal controller comprises models for

1. the state transition $\mathbf{g}_t$ to the successor state $\mathbf{x}_{t+1} = \mathbf{g}_t(\mathbf{x}_t, u_t, \boldsymbol{\nu}_{Pt})$ that depends on the current state $\mathbf{x}_t$, the control parameter u_t and the process noise $\boldsymbol{\nu}_{Pt}$
2. the costs $J_t(\mathbf{x}_t)$ that depend on the current state $\mathbf{x}_t$ (Bellman Equation).

Realizations and Results

The optimal controllers are implemented by DP and ADP based on a low-dimensional state space (3 dimensions) that has been extracted from the original high-dimensional state space (400 dimensions) by expert knowledge. The realized optimal controllers comprise a discrete backward optimization by classical DP (1), a continuous backward optimization by Backward ADP (2) and a continuous forward optimization by Forward ADP (3). The models for the state transition, the costs and the optimal control parameters are implemented by lookup tables in combination with a k-Nearest Neighbor predictor [9] to interpolate between the discrete table values in the first approach. In the second approach, batch-learning ANNs (trained by a batch of samples) are used to approximate the state transition and the costs, whereas incremental-learning ANNs (trained by single samples) describe the costs in the third approach. In the two ADP approaches, the Bellman Equation has to be solved online.

The different optimal controllers are compared in Table I by their relative cost difference averaged over 100 stochastic paths. The relative cost difference is characterized as the difference of the total costs (production efforts and final costs) of the evaluated approach and the absolute minimum costs (deterministic case neglecting noise) divided by the absolute minimum costs. For the selected configurations of the models, the first approach yields the highest efficiency (least relative cost difference). In general, however, the suitable approach

Table I: Comparison of optimal controllers

Control approach	1. Classical DP	2. Backward ADP	3. Forward ADP
Relative cost difference (in %)	4.96	8.43	8.85

has to be selected based on the complexity of the problem. For problems with continuous state space under uncertainty, the compact ADP representation should be preferred.

Conclusion

We have presented a generic concept for feature-based state tracking and optimal control of a manufacturing process. Using dimension reduction and compact approximations, we obtain real-time capable and robust state tracker and control models. The feasibility of the generic approach is shown by the application to two deep drawing simulation models. The realized state tracker model yields good prediction results and is robust against measurement noise. The optimal controllers compensate the process noise such that the total costs are optimal on average (evaluated on multiple stochastic paths under uncertainty). The presented results sound promising, whereas the application to real production machines has to be investigated.

References

[1] D. Helm et al. "Microstructure-based description of the deformation of metals: Theory and application". *JOM*, 63 (4), (2011), 26–33.

[2] G. Gottstein, ed. *Integral Materials Modeling* (Wiley VCH, 2007).

[3] K. M. Wurster et al. "Procedure for Automated Virtual Optimization of Variable Blank Holder Force Distributions for Deep-Drawing Processes with LS-Dyna and optiSLang". *Weimarer Optimierungs- und Stochastiktage 8.0* (2011).

[4] Y. Song and X. Li. "Intelligent Control Technology for the Deep Drawing of Sheet Metal". *Proceedings of the International Conference on Intelligent Computation Technology and Automation* (2009), 797–801.

[5] M. Dijkman and G. Goch. "Distortion Compensation Strategies in the Production Process of Bearing Rings". *Materialwissenschaft und Werkstofftechnik*, 40 (5-6), (2009), 443–447.

[6] M. Senn et al. "A System-Oriented Approach for the Optimal Control of Process Chains under Stochastic Influences". *AIP Conference Proceedings 1389* (2011), 419–422.

[7] M. Senn. "Optimale Prozessführung mit merkmalsbasierter Zustandsverfolgung". Ph.D. thesis, Karlsruhe Institute of Technology (2013).

[8] T. Böhlke, G. Risy, and A. Bertram. "Finite element simulation of metal forming operations with texture based material models". *Modelling and Simulation in Materials Science and Engineering*, 14 (3), (2006), 365–387.

[9] T. Hastie, R. Tibshirani, and J. Friedman. *The Elements of Statistical Learning* (Springer, 2009).

[10] J. D. Yoo, S. W. Hwang, and K.-T. Park. "Factors influencing the tensile behaviour of a Fe-28Mn-9Al-0.8C steel". *Material Science and Engineering*, 508 (1-2), (2009), 234–240.

[11] D. P. Bertsekas. *Dynamic Programming and Optimal Control* (Athena Scientific, 2005).

2nd World Congress on Integrated Computational Materials Engineering
Edited by: Mei Li, Carelyn Campbell, Katsuyo Thornton, Elizabeth Holm, and Peter Gumbsch
TMS (The Minerals, Metals & Materials Society), 2013

APPLICATION OF ICME METHODS FOR THE DEVELOPMENT OF RAPID MANUFACTURING TECHNOLOGIES

T.Maiwald-Immer[1], T.Göhler[1], A.Fischersworring-Bunk[1], C.Körner[2], F.Osmanlic[2], A.Bauereiß[2]
[1]MTU Aero Engines GmbH, Dachauer Str. 665, München, 80995, Germany
[2] University Erlangen-Nürnberg, Institut WTM, Martensstr. 5, Erlangen, 91058, Germany

Keywords: additive manufacturing, lattice-Boltzmann, Process chain simulation, absorption coefficient in powder bed

Abstract

Rapid manufacturing technologies are lately gaining interest as alternative manufacturing method. Due to the large parameter sets applicable in these manufacturing methods and their impact on achievable material properties and quality, support of the manufacturing process development by the use of simulation is highly attractive. This is especially true for aerospace applications with their high quality demands and controlled scatter in the resulting material properties. The applicable simulation techniques to these manufacturing methods are manifold. The paper will focus on the melt pool simulation for a SLM (selective laser melting) process which was originally developed for EBM (electron beam melting). It will be discussed in the overall context of a multi-scale simulation within a virtual process chain.

Introduction

Bailing out the optimisation potential of the SLM process and guaranteeing aviation specific requirements, it is fundamental to understand allowable process parameter variations without losing part integrity and stable mechanical properties. The development of this production process leads to more complex parts whilst maintaining quality requirements for: surface, porosity, size fidelity, strength, and reproducibility of the production process. Furthermore, the part specific development time is limited. In this context, it becomes increasingly essential to understand which process parameter (e.g. beam power and velocity) is responsible for the observed effects. The objective within this development is the identification of the optimal parameter set that realises the best results. To manage this task, MTU Aero Engines GmbH and their partners develop a multi-scale simulation chain for the description of microstructure evolution and resultant mechanical properties and is based on a two-level simulation approach. The continuum or macroscopic level is used to describe properties and to assess the design requirements of the final part, e.g. distortion, life capability based on global material properties. This reflects the state of the art using standard FEM solver. For this reason, the simulation chain development focuses on the second simulation level. The objective of this level is the introduction and the prediction of local material properties. Therefore, the prediction of the process related microstructure is fundamental. Consequently, the simulation chain should be able to describe the interdependence between local microstructure and processing (Fig. 1). To realize this, three tools were chosen describing these complex interactions by using a bottom-up approach. The first tool simulates the interaction between the laser beam and the powder bed. The second tool is used to characterise the solidification process in detail, considering e.g. the dendritic structure and the crystal orientation formation. The last tool describes the local strength based on the microstructure evolution. To adopt these tools to future machinery generations and further materials, all modelling approaches based on describing the underlying physical effects. In addition it is also necessary to define simple, well-functioning interfaces between these tools.

This paper will focuses on the melt pool simulation of the SLM process as all occurring effects and material properties formation depend on the local temperature field and gradient.

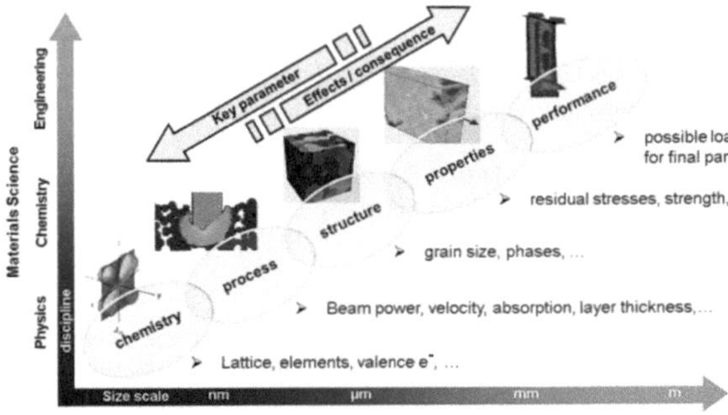

Fig.1: Multi-scale simulation chain for the SLM production process [1]

Description of the used Lattice-Boltzmann-Model (LBM)

The numerical method that is chosen to describe the melt pool physics should be able to simulate the thermodynamic and hydrodynamic incompressible transport behaviour including melting and re-solidification. Additionally, the single-phase continuum conservation for mass, momentum and energy equations have to be fulfilled, which are given by [2]:

$$\nabla \boldsymbol{u} = 0\,, \qquad \frac{\partial \boldsymbol{u}}{\partial t} + (\boldsymbol{u}\cdot\nabla)\boldsymbol{u} = -\frac{1}{\rho}\nabla p + \nu\nabla^2\boldsymbol{u} + \boldsymbol{g}\,, \qquad \frac{\partial E}{\partial t} + \nabla\cdot(\boldsymbol{u}E) = \nabla\cdot(k\nabla \mathrm{E}).$$

The used LBM is a bottom-up approach. To determine macroscopic fluid dynamics, it describes the collective behaviour of particles by microscopic kinetic equations [3]. Compared to molecular dynamics and conventional CFD, the advantages of the used method are an easy handling of complex geometries, parallel computing, low numerical dissipation, adequate simulation domain and innovative description of the free surface. The continuous Boltzmann equation is developed for a low Mach number and a finite set of velocities ξ_i in a space-filling lattice structure (D2Q9). Space and time are discretized coherently [2, 4, 5]. To solve the macroscopic transport equation two distribution functions are defined at each lattice site *i*. f_i represents the density and h_i the energy density distribution function in i-direction. The macroscopic quantities are given by [2, 3, 4]:

$$\rho = \textstyle\sum_i f_i\,, \qquad \rho\boldsymbol{u} = \sum_i \boldsymbol{e}_i f_i\,, \qquad E = \sum_i h_i.$$

Using the Bhatnagar-Gross-Krook operator as a simplification of the collision operator the two lattice Boltzmann equations for fluid dynamics and heat conduction can be written as [2, 4, 6]:

$$f_i(\boldsymbol{x} + \xi_i\Delta t, t + \Delta t) - f_i(\boldsymbol{x}, t) = \frac{\Delta t}{\tau_f}\left(f_i^{eq}(\boldsymbol{x}, t) - f_i(\boldsymbol{x}, t)\right) + \boldsymbol{F}_i\,,$$
$$h_i(\boldsymbol{x} + \xi_i\Delta t, t + \Delta t) - h_i(\boldsymbol{x}, t) = \frac{\Delta t}{\tau_h}\left(h_i^{eq}(\boldsymbol{x}, t) - h_i(\boldsymbol{x}, t)\right) + \boldsymbol{\Phi}_i.$$

$\boldsymbol{F}_i$ is the external force (e.g. the gravity $\boldsymbol{g}$), $\boldsymbol{\Phi}_i$ is the energy source that corresponds to the laser beam energy. The equation set is solved numerically in two steps: collision and streaming [2, 4, 5]. The powder bed is generated by a rain flow algorithm which adapts the density to experimental results (Fig. 2a, see section 4) [4]. One feature of the used tool is the free surface lattice Boltzmann model which enables a relatively simple treatment of the free surface boundary conditions. This offers a high computational efficiency without sacrificing the underlying physics. This model is able to capture the description and movement of the fluid-gas interface using five cell types: gas, interface, fluid, solid and wall cells (Fig. 2b).

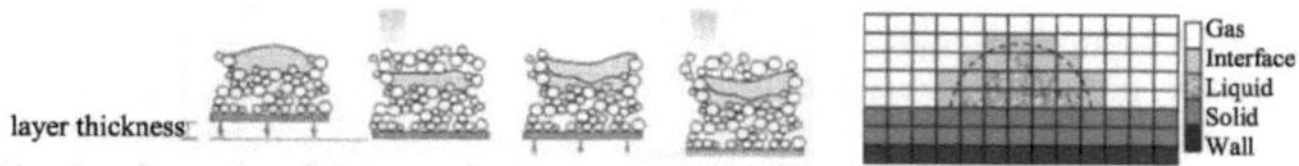

Fig. 2: schematic of: (a) powder bed generation algorithm, (b) free liquid-gas interface

The calculation of the dynamic wetting angle and the curvature in the interface cells are included [2, 4, 5]. The model does not solve the movement of the gas phase. Thus, one limitation of the model is that it can solely be used in systems where the flow of the gas phase is negligible. Consequently the distribution functions in the gas phase are unknown and have to be reconstructed so that the boundary conditions are fulfilled. This reconstruction takes the momentum and the heat fluxes into account [4]. All necessary temperature depending material parameters experimentally determined by Keene [7], Pottlacher et al. [8] and Overfelt et al. [9] are implemented. Important parameters are for example thermal conductivity, heat capacity, density, enthalpy, viscosity and surface tension. The above outlined numerical scheme and the implemented material parameters enable the physically based description of most of the relevant effects in the process. This comprises the generation of the powder layer, melting and solidification, melt pool dynamic, capillary forces (Plateau-Rayleigh-effect), gravity, wetting, heat conduction and convection, including Marangoni convection. A comprehensive validation and verification of the model, the algorithm and the boundary conditions were done for SEBM (selective electron beam melting) [2, 4, 5, 7]. Some necessary aspects for the adaption to the SLM process and the validation experiments are described in the following sections.

Absorption of the Laser Beam Radiation

For the simulation of SLM the correct description of the interaction between laser beam radiation and metal powder bed is particularly important. The absorption coefficient defines how much energy is fed into the system. This influences the size (deep, length, width), lifetime and dynamic of the melt pool as well as the temperature gradient in the material. The absorption coefficient of a planar surface depends on the angle of incidence, the temperature, the material and the wavelength of the beam radiation. It can be calculated based on the density of conduction electrons N_e, the plasma frequency ω_P and the collision frequency between electrons f_c [11]. The real ε_1 and the imaginary ε_2 part of the permittivity can be calculated using ω_p, f_c, and the frequency of the laser beam radiation ω_L (approx. 1,07µm for Nd:YAG and Yb:YAG laser):

$$\varepsilon_1 = 1 + \frac{\omega_p^2}{\omega_L^2 \cdot f_e^2}, \quad \varepsilon_2 = \frac{f_e}{\omega_L} \cdot \frac{\omega_p^2}{\omega_L^2 \cdot f_e^2}.$$

The index of refraction n and the extinction coefficient k can be determined by using these variables:

$$n = \left(0{,}5 \cdot \left(\sqrt{\varepsilon_1^2 + \varepsilon_2^2} + \varepsilon_1\right)\right)^{0{,}5}, \quad k = \left(0{,}5 \cdot \left(\sqrt{\varepsilon_1^2 + \varepsilon_2^2} - \varepsilon_1\right)\right)^{0{,}5}.$$

To obtain the absorption coefficient depending on incidence angle, temperature, and material this variables have to be inserted into the Fresnel term [12]. It is not necessary to consider different polarizations of the radiation (e.g. oscillation parallel or perpendicular to the surface) because the laser radiation is circular polarized. In this case the absorption coefficient can be determined by using the Pythagoras theorem. The calculation results, displayed in the following diagrams (Fig. 3), point out a minimal influence of temperature and incidence angle. Relevant values are located between 35 and 45 percent. According to the Mie-theory, the laws of geometrical optics are valid for a particle diameter in the range of 10 to 100μm and a wavelength of 1,07μm [13]. From this follows, that the description of the angle dependence via Fresnel term for the surface of powder particles is permitted. This theoretical results are also confirmed by Anderson et al. [15] and Sainte-Catherine et al. [16].

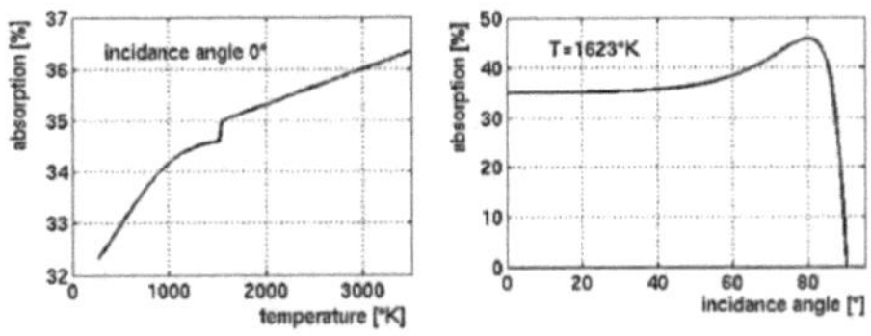

Fig. 3: influence of temperature and incidence angle on absorption coefficient

The effect of multi reflection in the powder bed is often discussed in literature. In this context, the literature indicates a value for the absorption coefficient that is up to twice as high [13, 14]. Phenomena like multi reflection, beam trapping and oxidation on the surface are responsible for such high values. There are several reasons why it is not necessary to take multi reflection effects in this simulation into account. For example, it can be shown by heat conduction models or by the LBM that the main section of the beam diameter interacts directly with the melt pool. Only the boundary area of the beam diameter could interact with individual powder particles and heat them up. Furthermore, the high absorption coefficients are measured at powder layers where the beam power is not sufficient to melt the particle where effects such as multi reflection and beam trapping could occur. Consequently, multi reflection is only important for key-hole welding, sintering and for a very short time when the beam is switched on. In the LBM the absorption coefficient is integrated as a function of temperature and incidence angle. Additionally, it is calculated for each discrete surface segment. Multi reflection effects are not taken into account.

Validation Experiments

Validation comprises the evaluation of density of the used powder bed, the calculated absorption coefficient and the simulated melt pool geometry and dynamics. Additionally, it must be clarified whether the simulation is able to capture all effects detected by the experiments. Therefore, crucial experiments e.g. powder packing analysis, single track scanning and high-speed-recording are done. Incident-light microscopy is used for the powder packing study which delivers a density between 52 and 58 percent (Fig. 4a). A maximum density of 66 percent is achieved under high compression. By means of single track analysis for a planar surface and powder bed (varied layer thickness), the melt pool and its dependence on track energy are investigated. Identified effects are Rayleigh capillary instability (Fig. 4b), melt pool pinning on particles driven by wetting (Fig. 4c), convection caused by, inter alia, surface tension (Fig. 4d) and evaporation. The investigated parameter field is shown in Fig. 5. Within this parameter field key-hole-welding as well as heat conduction welding occurs. Furthermore, areas can be identified where beam power or beam velocity are too high or too low. Based on the following requirements: connection to underlying layers, minimal key-hole-welding effects and a homogeneous melt pool, a region of possible parameter combinations is identified in the center of the investigated parameter field (Fig. 5).

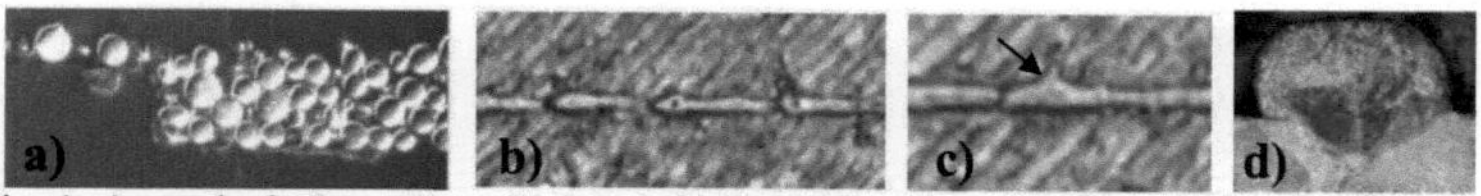

Fig. 4: a) powder bed at a 90µm high stair, b) Rayleigh effect, c) melt pool pinned on particle, d) melt pool cross-section (shape caused by surface tension)

The analysis of the high-speed-recordings points out a very high dynamic and lifetime of the melt pool. This dynamic and lifetime depends strongly on energy density input, track energy, the wetting angle, the surface tension and the material feed in the powder bed.

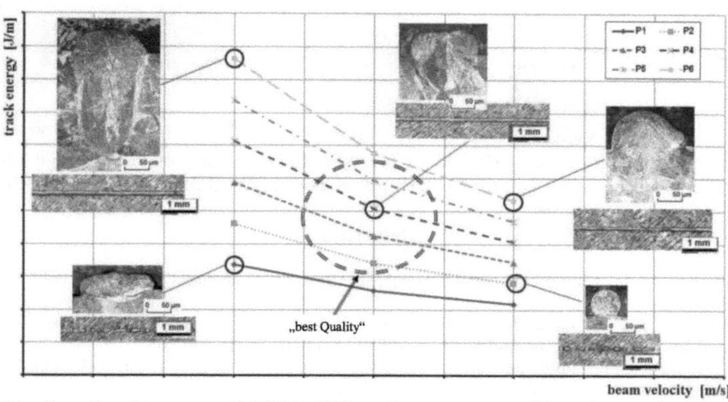

Fig. 5: analyzed parameter field (six different beam powers and three beam velocities)

Results

Within the range of heat conduction welding and the numerical resolution limit of 5 µm, a good correlation of the simulation results with the experimental investigations can be found. Furthermore, the simulation is able to represent the identified effects: Rayleigh capillary instability, melt pool pinning, convection, high dynamic and observed lifetime of the melt pool. Some examples are shown in Fig. 6. The simulation results are displayed in grey scale pictures and the dark grey areas indicate the liquid melt pool. The best results are achieved for a planar surface (Fig. 6a). This confirms also the calculated absorption coefficient. Currently, the simulation is not able to represent evaporation, which is becoming increasingly important with higher power densities. This explains the low melt pool depth at higher track energies shown in Fig. 6d. All investigations are done with a constant value of the surface tension. The implementation of their temperature dependency is still in progress.

Conclusion

The LBM provides a good description of the melt pool physics and of the interaction between laser beam and powder bed. The transfer of the energy coupling by electron beam to the laser beam is successfully completed. The influence and implementation of all parameters is not investigated conclusively but still in progress. Based on the ICME-level evaluation done by Cowles et. al. [17], it can be assumed that Level 4 is achievable. The accomplished intermediate results are a base of operation for the following simulation tools within the simulation chain. We would like to thank the DFG and the CRC814, SFB814, project B4 for funding.

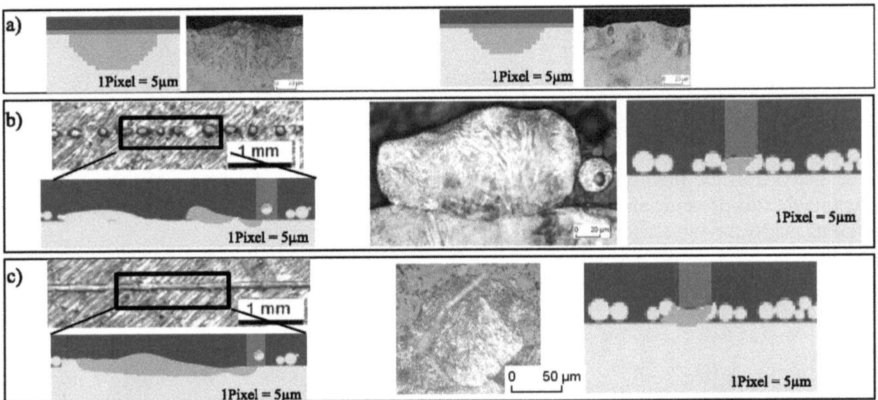

Fig. 6: (a) cross section for planar surface left: P2v1, right: P2v2; b) and c) on the left longitudinal cut, on the right the cross-section: (b) P2v1; (c) P3v2; (P…beam power (P3 = 1,35*P2), v … beam velocity (v2 = 1,5*v1))

References

1. GB Olsen, "Computational Design of Hierarchically Structured Materials." *Science 29*, Vol.277 (1997), 1237-1242
2. E. Attar, C. Körner, "Lattice Boltzmann model for thermal free surface flows with liquid-solid phase transition." *International journal of Heat and Fluid Flow*, 32 (2011), 156-163
3. S. Succi, *The Lattice Boltzmann Equation for Fluid Dynamics and Beyond.* (New York, published in the United States by Oxford University Press Inc., 2001)
4. E. Attar, "Simulation of Selective Electron Beam Melting Processes." (Ph.D. thesis, Erlangen-Nürnberg University, 2011)
5. C. Körner, M.Thies, T.Hofmann, N.Thürey and U.Rüde "Lattice Boltzmann model for Free Surface Flow for Modeling Foaming." *Journal of statistical Physics*, 121 (10/2005), 179-196
6. R.Zhang, H.Fan and H.Chen, "A lattice Boltzmann approach for solving scalar transport equations." *Phil. Trans. R. Soc.*, 369 (2011), 2264-2273
7. M. Thies, "Lattice Boltzmann Modeling with Free Surfaces Applied to Formation of Metal Foams." (Ph.D. thesis, Erlangen-Nürnberg University, 2005)
8. B.J. Keene, "Review of data for the surface tension of pure metals." *International Materials Reviews*, 38 (1993), 157-191
9. G. Pottlacher, H. Hosaeus, B. Wilthan, E. Kaschnitz, A. Seifter, „Thermophysikalische Eigenschaften von festem und flüssigem Inconel 718." *Thermochimica Acta*, 382 (2002), 255-267
10. R.A.Overfelt, C.A. Matlock and M.E. Wells, „Viscosity of Superalloy 718 by the Oscillating Vessel Technique." *Metallurgical and Materials Transactions*, 27B (1996), 698-701
11. S. Anders, "Numerische Simulation des Energieeintrages zur Modellierung einer Laserstrahlschweißung." (Ph.D. thesis, Bauhaus-Universität Weimar, 2005)
12. I. Lazanyi, L.Szirmay-Kalos, "Fresnel Term Approximations for Metals.", *UNION Agency-Science Press*, (2005)
13. W. Meiner, "direktes Selektives Laser Sintern einkomponentiger metallischer Werkstoffe.", (Ph.D. thesis, RWTH Aachen University, 1999)
14. C. Wagener, *Untersuchung zum Selektiven Lasersintern von Metallen.* (Berichte aus der Produktionstechnik, Frauenhofer Institut, Shaker Verlag, Band 11,2003)
15. M. Anderson, R.Patwa, Y. C. Shin, " Laser-assisted machining of Inconel 718 with an economic analysis." *International Journal of Machine Tools & Manufacture,* 46 (2006), 1879–1891
16. C. Sainte-Catherine(I), M. Jeandin, D. Kechemair, J.P. Ricaud, L. Sabatier, "Study of Dynamic Absorptivity at 10.6 µm (CO_2) and 1.06 µm (Nd:YAG) Wavelengths as a Function of Temperature." *Journal de Physique*, Vol. 1 (1991), 151-157
17. B. Cowles, D. Backman and R. Dutton, "Verification and Validation of ICME methods and models for aerospace applications." *Integrating Materials and Manufacturing Innovation*, 05/2012

2nd World Congress on Integrated Computational Materials Engineering
Edited by: Mei Li, Carelyn Campbell, Katsuyo Thornton, Elizabeth Holm, and Peter Gumbsch
TMS (The Minerals, Metals & Materials Society), 2013

ANALYTICAL MODELING AND PERFORMANCE PREDICTION OF REMANUFACTURED GEARBOX COMPONENTS

[1]Raja V. Pulikollu, [1]Nathan Bolander, [2]Sandeep Vijayakar, [3]Matthew D. Spies
[1]Sentient Science Corporation, 672 Delaware Ave, Buffalo, NY 14209
[2]Advanced Numerical Solutions LLC (ANSOL), 3962 Brown Park Drive, Suite C Hilliard OH 43026 USA
[3]US Army Aviation Development Directorate – AATD, Ft. Eustis, Virginia

Keywords: Rotorcraft, Wind Turbines, Gearbox, Remanufacturing, Fatigue models, Mixed-elastohydrodynamic lubrication model, Multi-body Dynamics

Abstract

Gearbox components operate in extreme environments, often leading to premature removal or overhaul. Though worn or damaged, these components still have the ability to function given the appropriate remanufacturing processes are deployed. Doing so reduces a significant amount of resources (time, materials, energy, manpower) otherwise required to produce a replacement part. Unfortunately, current design and analysis approaches require extensive testing and evaluation to validate the effectiveness and safety of a component that has been used in the field then processed outside of original OEM specification. To test all possible combination of component coupled with various levels of potential damage repaired through various options of processing would be an expensive and time consuming feat, thus prohibiting a broad deployment of remanufacturing processes across industry. However, such evaluation and validation can occur through Integrated Computational Materials Engineering (ICME) modeling and simulation. Sentient developed a microstructure-based component life prediction (CLP) tool to quantify and assist gearbox components remanufacturing process. This was achieved by modeling the design-manufacturing-microstructure-property relationship. The CLP tool assists in remanufacturing of high value, high demand rotorcraft, automotive and wind turbine gears and bearings. This paper summarizes the CLP models development, and validation efforts by comparing the simulation results with rotorcraft spiral bevel gear physical test data. CLP analyzes gear components and systems for safety, longevity, reliability and cost by predicting (1) New gearbox component performance, and optimal time-to-remanufacture (2) Qualification of used gearbox components for remanufacturing process (3) Predicting the remanufactured component performance.

1. Introduction

The manufacture of gears and bearings is a very energy and resource intensive process. This is true especially for high value added components used in precision engineered gearboxes used in energy, automotive, and aerospace applications. The process begins with refining the metal ore through smelting, forming billets, machining and forging of components, heat treating, and ends with surface finishing. As shown in Figure 1, most of the energy is spent in the early stages, not in the later stages. Finishing accounts for only 5% of the energy consumed in the manufacturing process. Repairs and refurbishing of components which rely on re-applying such later stage processes will significantly reduce energy consumption. The result is that refurbishing a component through applying only a finishing process realizes a 95% reduction in manufacturing energy consumption.

For instance, high value and high demand spiral bevel gears are widely used in the tail rotor drive trains of most rotorcraft. The loads associated with the tail rotor drive train are generally much more variable than those in the main rotor drive train primarily resulting from maneuvers. These variable load operating conditions may cause severe or slight surface damage to the gears. The slight surface damaged gears are rejected during inspection and order is placed to obtain new parts, which is expensive ($40,000 - $80,000) and time consuming (6-9 months lead time).

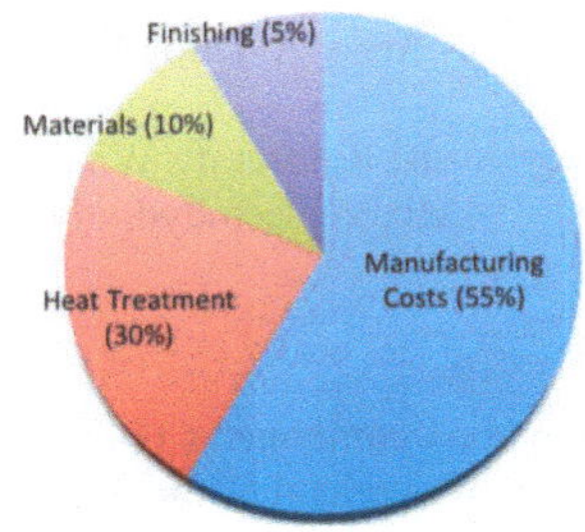

Figure 1 – Breakdown of energy costs associated with the manufacture of gears and bearings

An effort was conducted by the US Army ADD-AATD to investigate an emerging advanced superfinishing process to repair and put back the rejected parts in service [1][2]. Army performed series of overload tests on the remanufactured rotorcraft drive train AMS 6265 (AISI 9310) spiral bevel gears to evaluate the endurance limit. These tests were performed at the NAWCAD Propulsion System Evaluation Facility (PSEF), Patuxent River, MD. The testing was conducted in the Helicopter Drive Systems (HeDS) Test Facility, a primary element of PSEF, under the direction of the Army ADD-AATD. However, physical testing is time consuming, expensive, and the results are somewhat limited to a particular configuration of the gearbox. The ability to conduct virtual life testing using Integrated Computational Materials Science Engineering (ICME) modeling and simulation techniques employing probabilistic methods will have great potential to reduce the time and cost associated with the testing of complex gear systems.

Sentient's microstructure-based virtual modeling and performance simulation method, DigitalClone can accelerate research and deployment of the remanufacturing techniques by reducing physical test costs. Further advantage of the DigitalClone analyzed in this paper include ensuring that the components are removed prior to catastrophic damage, enabling such a cost savings due to remanufacturability and avoiding service/operational downtime issues.

2. Microstructure-Based DigitalClone Life Prediction Technology

Sentient developed DigitalClone component life prediction (CLP) models to predict the fatigue life of the remanufactured (repaired/refurbished) component, thus quantifying the benefits of such processing and demonstrating the feasibility of such a tool. This was achieved by modeling the gear design-manufacturing-microstructure-property relationship, as shown in Figure 2.

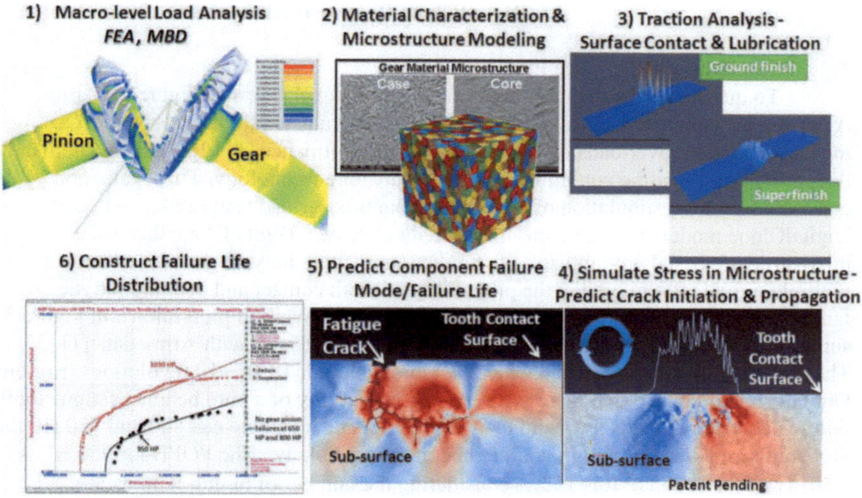

Figure 2: Microstructure-based DigitalClone life prediction approach for remanufacturing

DigitalClone integrates a 3-D finite element (FE) model, mixed-elastohydrodynamic (EHL) model, microstructure model, probabilistic methods, and fatigue damage model in the simulation process. Figure 2 shows ANSOL's Calyx multi-body dynamic analysis (global FEA) model, representative of a full scale gear mesh (Step 1) subjected to service operating conditions. It is computationally intensive to perform fatigue damage analysis on global models, as a complete microstructure needs to be modeled. Hence, a representative volume element (RVE) for the gear was defined (see the microstructure model shown in Step 2 of Figure 2) to simulate fatigue damage. This sub-model represents the high stress regions (critical sites for crack initiation) in the global FE model. Surface pressure and traction profiles (step 3) generated by the mixed-EHL solver are used to load the model segment in a dynamic solution. Mixed-EHL model can predict the influence of surface finish (ground finish vs. superfinish) on asperity interactions and pressure profiles. This approach can be used to perform micro-stress analysis at the grain level and update micro-stress fields as fatigue damage accumulates and propagates (Step 4). This modeling approach is capable of estimating fatigue damage on a cycle-by-cycle basis to provide accurate lifing analysis. Total fatigue life is determined at the end of simulation, when a component fails (Step 5). A new RVE is generated for each fatigue life simulation to capture the variability in material microstructure (grain size, orientation, and inclusions), material micro properties (hardness), surface finish (roughness statistics), and residual stresses (carburization process) using probabilistic methods. This will provide the fatigue life distribution (Step 6). The DigitalClone model predicts crack initiation and propagation life by modeling the gear component material microstructure and variations in the fatigue lives that are vital for the design and analysis of a gearbox.

3. DigitalClone Demonstration

3. 1 Remanufactured Gear Life Prediction

To qualify remanufactured AMS 6265 spiral bevel gear set (Figure 3), Army ADD-AATD conducted series of physical tests and evaluated the effects of short to moderate duration overloads on the bevel pinion performance [1] [2]. Sentient simulated the physical test loading conditions using DigitalClone technology. This evaluation consisted of virtual simulation of the actual dynamic overload testing of bevel gears. DigitalClone modeled the gear mesh and predict stresses. DigitalClone then used microstructure based probabilistic life prediction methodology (Figure 2) to conduct multiple virtual tests to predict the probability of tooth contact and bending fatigue failures (POFC and POFB) over a wide range of loads. Weibull plots shown in Figure 3 summarize the results of the virtual testing that correlated well with Army data [1] [2]. The simulations were suspended when the pinion exceeded 1.0E+10 revolutions (run-out). DigitalClone predicted that there is a very low probability of a pure bending fatigue tooth failure occurring at loads levels up to 800 HP. At some load between 800 and 950 HP the probability of a bending fatigue failure becomes more likely. The POFB for 950 HP is 2.50% and for 1030 HP it is 15.2% considering the number of pinion failures at $\leq$ 1.0E+10 revolutions. In the case of contact fatigue, the simulation results indicated that the high contact pressure at the edge-of-contact in pinion (Figure 3) lead to micro-pits formation and eventually resulted in tooth fracture, in agreement with Army ADD-AATD physical test observations. The damage is likely the result of the lack of adequate tip relief at the very high overload conditions.

3.2 Improve Gearbox Performance and Trade Analysis

DigitalClone technology can also be used to simulate the life of gear and bearings under a wide variety of surface finish, residual stresses and other scenarios that would otherwise be too expensive to physically test. This would assist in improving gearbox performance and performing trade analysis.

Figure 4a shows the influence of surface finish on surface pressure/traction profile and gear fatigue life. Sentient's mixed-EHL model accurately predicted the influence of surface finishing methods (Superfinish, Root Mean Square Height Rq: 3.0 μin and Ground finish, Rq: 21.63 μin) on traction profiles. It is evident from Figure 4(a) that improving the surface profile through a superfinishing process reduced asperity interaction, thereby improved the fatigue resistance over ground finish gears. The contact fatigue life of superfinish gear was about 1.81 times greater than the ground finish gears. Figure 4(b) compares surface fatigue life predictions of carburized gear (surface compressive stress of 145 Ksi), and carburized and shot peened gear (surface compressive stress of 174 Ksi). The mean fatigue life of shot peened gear is 2.0 times higher than carburized gear due to higher surface and subsurface compressive stresses. This analysis demonstrates that DigitalClone can be used as the alternative to physical testing.

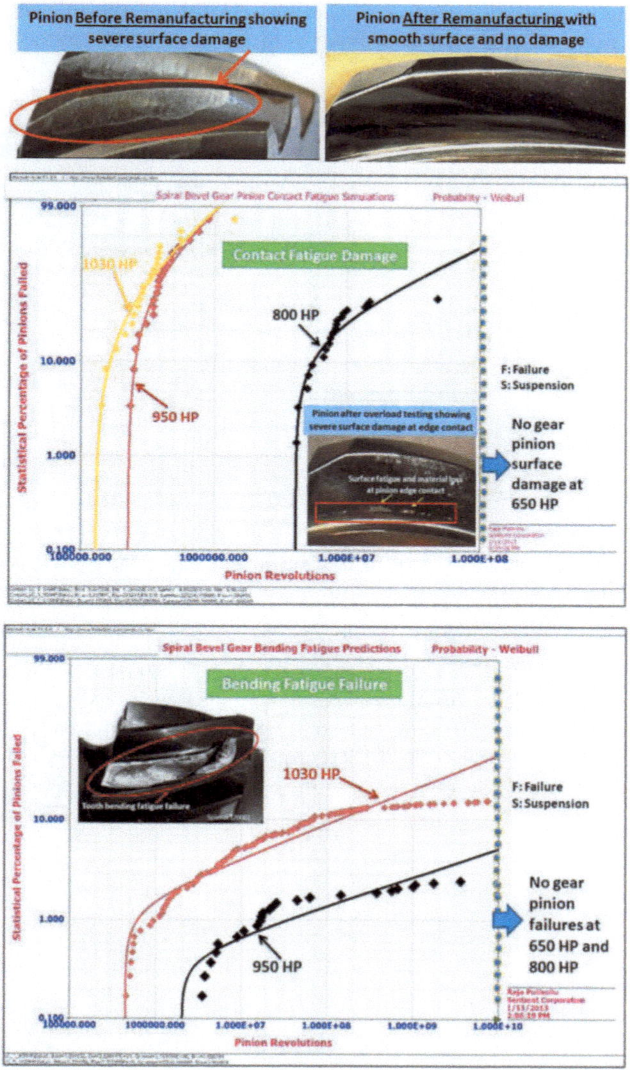

Figure 3: DigitalClone predictions of percentage of remanufactured AMS 6265 pinions failed vs. pinion revolutions for four different Horse Powers. DigitalClone predicted run-outs (no fatigue damage) up to 800 HP, in agreement with Army endurance test results [1] [2].

4. Summary

Sentient's DigitalClone technology assists in remanufacturing of high value, high demand rotorcraft and wind turbine gearbox components. The benefits of this technology includes (1) Reduce physical testing using virtual testing and provide cost savings, (2)

Decrease the energy, material resources, and costs associated with manufacturing, (3) Ensure product performance is maintained/improved as a result of the process modification or enhancement.

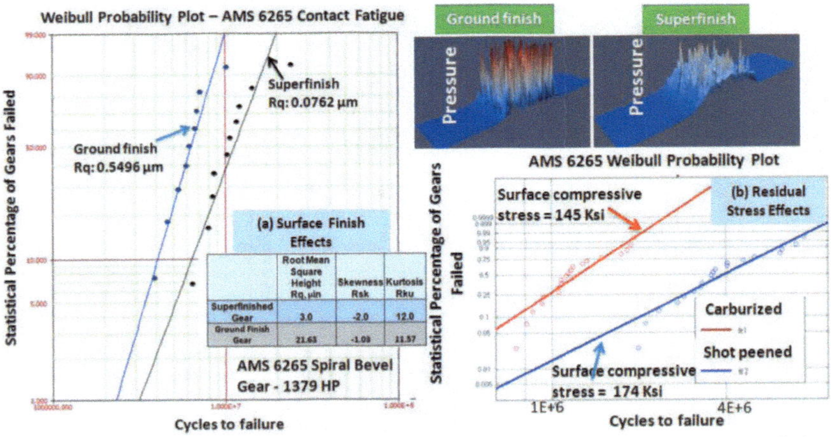

Figure 4: DigitalClone simulates the effects of surface finish, residual stresses on fatigue life that would otherwise be too expensive to physically test

5. References

1.E. C. Ames, Repair of High Value/High Demand Spiral Bevel Gears by Superfinishing, American Helicopter Society 67th Annual Forum, Virginia Beach, VA, May 3-5, 2011
2. L. Haney and E. C. Ames, MH-60K Tail Rotor Drive System High Load Endurance Test, Report No: NAWCADPAX/TR-2012/75, April 2012

2nd World Congress on Integrated Computational Materials Engineering
Edited by: Mei Li, Carelyn Campbell, Katsuyo Thornton, Elizabeth Holm, and Peter Gumbsch
TMS (The Minerals, Metals & Materials Society), 2013

DESIGN OPTIMIZATION OF TRANSMISSION OF SI/SIO_2 AND GE/SIO_2 MULTILAYER COATINGS

K. Iqbal [1]*, J. Sha[1] and A. Maqsood[2]

[1]State Key Laboratory of Structural Analyses for Industrial Equipment, School of Aeronautics and Astronautics, Dalian University of Technology, 116024 Dalian, China
[2]Centre for Emerging Sciences, Engineering & Technology, Islamabad, Pakistan

Keywords: ICME, Optical design, Si/SiO_2, Ge/SiO_2

Abstract

The development of computational tools for materials engineering requires physical phenomena that must be captured and integrated with various materials-related disciplines, and organization types. This integration discriminates between the materials development cycle and the product development cycle, and reduces the length of the materials development cycle from 10 years to 2 years. TFCalc software was used for optical designing .This article includes some mathematical models such as the spectral distribution of Si/SiO_2 and Ge/SiO_2 coatings transmittance.

Introduction

The exploration of the benefits and promise of ICME is an accomplishment in the field of computational materials. ICME was described [1] as "the integration of materials information, captured in computational tools, with engineering product performance analysis and manufacturing-process simulation".The development of computational tools for materials engineering has lagged behind the development of such tools in other engineering fields because of the complexity and sheer variety of the materials, and physical phenomena that must be captured. In spite of these scientific challenges, the computational tools for materials engineering are now reaching the level of maturity where they will have a substantial impact if they can be integrated into the product development process [1]. This integration is the basis for the emerging paradigm of ICME. ICME promises to eliminate the growing mismatch between the materials development cycle, and the product development cycle by integrating materials computational tools and information with the sophisticated computational analytical tools already in use in engineering fields other than materials. ICME will be transformative for the materials discipline, promising a reduction in the length of the materials development cycle from its current 10 years to 2 years in the best scenarios. As described later in this article, includes some mathematical models such as the spectral distribution of Si/SiO_2 and Ge/SiO_2 coatings transmittance.

Modeling

The optimization of multilayer coatings was done using thin film design TFCalc software [2] to minimize the reflectance of a Si and Ge substrates in the wavelength range 3-5μm. The wavelength region of 3-5 μm is often called the mid-infrared region [3]. Typical examples of infrared transmission media include a vacuum, the atmosphere and optical fibers. In the atmosphere, absorption by H_2O, CO_2 and other elements take place at specific wavelengths [3, 4]. The bandwidth of 3-5 μm, which is based on heat radiation, is sometime called the atmospheric window and is often used for remote sensing applications [3]. Criteria for a potential

candidate for multilayer structures included deposition conditions, smooth surface, low stress and correlation with modeling works [5]. Also minimization of reflection was very important for these studies, since it reduced the total energy reaching the sensor [6].
The optimization fitting process of the optical constants and thickness determination of thin films requires [2]:

- The definition of a merit function (MF).
- Having approximate initial values for the parameters (such as n and thickness) and the range for each variable based on some physical constrains.
- An optimization algorithm that minimizes the merit function.

The TFCalc software offers for designing and manufacturing optical thin film coatings, can maintain the symmetry of a design while it's being optimized, and helps the designer locate the designs that minimize the merit function, In the TFCalc software, the substrate, as well as the incident and exit mediums, was selected from the substrate database, and the thickness of the substrate was considered a massive layer. The optical properties of the substrates and materials were stored as dispersion formulas of complex refractive index (n-ik) versus wavelength. The physical system modeled by TFCalc software is shown in Figure 1, and the merit function (MF) was used is given in Equation 1.

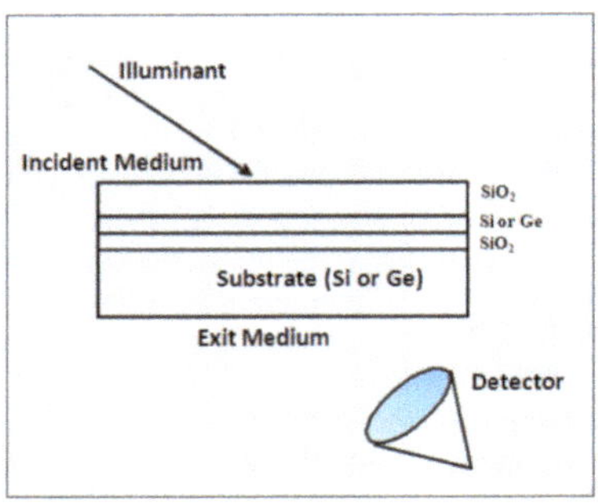

Figure 1. The physical system modeling by TFCalc software

$$F = \left[\frac{1}{m} \sum_{j=1}^{m} \left| \frac{I_j D_j C_j - T_j}{N_j Tol_j} \right|^k \right]^{1/k} \qquad (1)$$

Where m is the number of targets, k is the power of the method, I is the intensity of the illuminant, D is the efficiency of the detector, T is the desired target value, C is the computed value (of reflectance, transmittance, at the target wavelength, angle and polarization), Tol is the tolerance for a target, and N is the normalization factor for the target. For simplicity, we use I = D= Tol = 1.0.

SiO_2 was used as the low index material, while the high index materials were Si and Ge, and their refractive indices were calculated by the Sellmeier dispersion equation and the modified Sellmeier dispersion equation [7, 8] as given in Equation 2 and Equation 3, respectively, and the values of the constants are shown in Table 1 and Table 2.

$$n^2 - 1 = A_1\lambda^2 / (\lambda^2 - C_1^2) + A_2\lambda^2 / (\lambda^2 - C_2^2) + A_3\lambda^2 / (\lambda^2 - C_3^2) \quad (2)$$

$$n^2 = A + \frac{B\lambda^2}{(\lambda^2 - C)} + \frac{D\lambda^2}{(\lambda^2 - E)} \quad (3)$$

Table I. Parametric values of Sellmeier equation for SiO_2 and Si

S. No.	Material	A_1	A_2	A_3	C_1 (μm)	C_2 (μm)	C_3 (μm)
1	SiO_2	0.696	0.408	0.897	0.068	0.116	9.89616
2	Si	10.668	0.003	1.541	0.301	1.135	1104.0

Table II. Parametric values of modified Sellmeier equation for Ge

S. No.	Material	A	B	D	C (μm^2)	E (μm^2)
1	Ge	9.2695	6.7474	0.2139	0.44	3873.2

The model was designed for a 4.2μm wavelength due to the fact that carbon dioxide takes part in reducing the infrared radiations at the 4.2μm wavelength [3-6]. The total thickness of the Si/SiO_2 film layers and Ge/SiO_2 film layers were determined [Table 3] using two equations, $n^2_f = n_{sub} \times n_a$ and $d = \lambda_0 / (4\ n_f)$ [3-5], where n_f, n_{sub}, n_a and λ_0 are refractive indices of the film layer, substrate, and incident medium (usually air) and designed wavelength, respectively.

Table III. Film layer thickness of Si/SiO_2 and Ge/SiO_2

	Film layer thickness (nm)		
	Si	Ge	SiO_2
Si/SiO_2	571		429
Ge/SiO_2		200	800

With the accurate refractive index and extinction coefficient data of SiO_2, Si and Ge, we have then associated SiO_2 with Si layer , and SiO_2 with Ge layer so as to obtained antireflection coatings with high performance on Si and Ge substrates in 3-5μm, consisting of three layers. From Figure 2 (a) and 2 (b), we can see that the optical performance of each infrared single sides AR coatings is satisfying [6-8]. The average transmittance was achieved between 69% and 70% for the Si/SiO_2 film layers, and between 60% and 90% for the Ge/SiO_2 film layers.

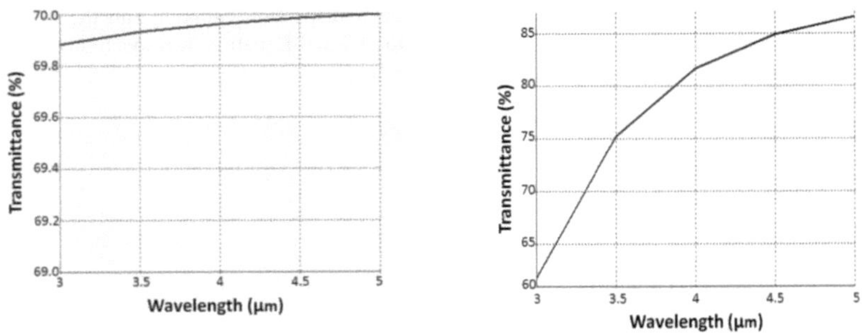

Figure 2. Transmittance % vs. wavelength (µm) profile of (a) Si/SiO_2 layers and (b) Ge/SiO_2 lay

Conclusions

This article provides a vision for ICME, an analysis of technological barriers, and a review of case studies. Alternate layers of Si and SiO_2, and Ge and SiO_2 were used to achieve the required spectral performance. By this arrangement, a refractive index of between 1(air) and 1.52 (glass) was achieved. In the future, design of Si or Ge optical window for the 4.2 µm wavelength open new possibilities in astronomical filter and bio-optical devices.

References

[1] G. Spanos, D.J. Rowenhorst, A.C. Lewis, and A. Geltmacher, MRS-Bulletin, 33 (2008), 597-602.

[2] TFCalc Manual, Thin Film Design Software for Windows Version 3.0 (Software Spectra, Inc., Portland, Ore., 1995).

[3] G.H. Rieke, "History of Infrared Telescope and Astronomy," Experimental Astronomy, 25 (2009), 125-141.

[4] G.C. Macfarlane, T.P. Mclean, J.E. Quarrington and V. Roberts, "Fine Structure in the Absorption-Edge Spectrum of Ge," Physical Review, 108 (1957), 1377-1383.

[5] R.M.A. Azzam, and N.M. Bennett, Ellipsometry and Polarized Light, North-Holland, Amsterdam, 1977.

[6] H.B. Briggs, "Optical Effects in Bulk Silicon and Germanium," Physical Review, 77 (1950), 287.

[7] I.H. Malitson, "Interspecimen Comparison of the Refractive Index of Fused Silica," Journal of the Optical Society of America, 55 (10) (1965), 1205.

[8] W. Primak, "Refractive index of silicon," Applied Optics, 10 (4) (1971), 759-763.

2nd World Congress on Integrated Computational Materials Engineering
Edited by: Mei Li, Carelyn Campbell, Katsuyo Thornton, Elizabeth Holm, and Peter Gumbsch
TMS (The Minerals, Metals & Materials Society), 2013

THE STUDY ON THE INDUCTION HEATING SYSTEM: THE ESTABLISHMENT OF ANALYTICAL MODEL WITH EXPERIMENTAL VERIFICATION AND THE PHENOMENOLOGICAL STUDY ON THE PROCESS FROM SIMULATION PERSPECTIVE

Tianxing Zhu[1], Feng Li[1], Xuekun Li[1, 2], Yiming (Kevin) Rong[1, 2, 3]

[1]Department of Mechanical Engineering,
Tsinghua University, Beijing, 100084, China
[2]Beijing Key Lab of Precision/Ultra-precision Manufacturing Equipment and Control
Tsinghua University, Beijing, 100084, China
[3]Department of Mechanical Engineering,
Worcester Polytechnic Institute, Worcester, MA, 01609, US

Keywords: Induction heating, Analytical model, Phenomenological study

Abstract

Induction heating is frequently used in the metalworking industry to heat metals for hardening, soldering, brazing, tempering and annealing. Compare to traditional heating methods like flaming or oven, the induction heating is a unique process that generated heat within the workpiece by its electromagnetic and thermal material properties. But due to the complex nature of induction heating process, the use of simulation to analyze the induction heating process could become very advantageous both in design and economic aspects. In this paper, a coupled electromagnetic-thermal analytical model is established and verified by the experiments, including workpiece emissivity calibration. After the establishment of analytical model, phenomenological study on the effect of magnetic flux concentrator, current parameters were conducted. The FEM simulation helps to further understand the induction heating process, and to quantify the effect of input parameters. Therefore, it is possible to better optimize the process, and to develop new techniques.

Introduction

Induction heating is a process that uses an AC electro-magnetic field to induce heat in a workpiece, and has been used to heat workpiece since the early 1920s. Industrial applications of the technology include heat treating and metal melting, single crystal growing, semiconductor wafer processing, high speed sealing and packaging, and curing of organic coatings [1].

Induction heating occurs when an electrical current is induced into a workpiece by electromagnetic induction and generation of heat by Joule effect [2]. For the induction heating system to be efficient and practical, certain relationships of the frequency and intensity of the electromagnetic field that produces the eddy currents and the properties of the workpiece must be satisfied, for the desired heating rate and depth to be achieved.

The recent development of numerical simulations to calculate thermal histories, phase transformations and stress fields of different materials allow us to develop models which can predict the effects of heat treatment on the metallurgical and mechanical properties of machine component. The establishment of analytical model can be very useful to forecast distortions, residual stresses, phase distributions, hardened depth, thickness of carburized layer, etc. [3] However, numerical models of complex heat treatment, such as the induction heating that involves coupled multi-physics problems, are still in progress [4].

In this paper, an analytical model is established using commercial package Cedrat Flux® 10.3, and is verified by the experiments including workpiece emissivity. With the verification of the analytical model, phenomenological study on the effect of magnetic flux concentrator, current parameters were conducted.

Mathematical model

The numerical calculation of induction heating process combines the electromagnetic computation and the thermal computation. Since the electrical and magnetic properties are temperature dependent, the thermal computations are coupled by the latent heat transformations.

The calculation of the electromagnetic field depends on Maxwell's equations, and the following constitutive relations are additional and hold true for a linear isotropic medium.

$$D = \varepsilon\varepsilon_0 E \quad (1)$$

$$B = \mu_r\mu_0 H \quad (2)$$

$$J = \sigma E \quad (3)$$

Where the parameters D, E, B, J, ε, μ_r and σ denote, respectively, electric flux density, electric field intensity, magnetic flux density, conduction current density, the relative permittivity, relative magnetic permeability, and electrical conductivity of the material.

By taking the above constitutive relations into the Maxwell's Equations, the two governing equations can be deduced:

$$\nabla^2 E - \mu\varepsilon\frac{\partial^2 A}{\partial t^2} = \mu\frac{\partial J}{\partial t} + \nabla\frac{\rho}{\varepsilon} \quad (4)$$

$$\nabla^2 H - \mu\varepsilon\frac{\partial^2 H}{\partial t^2} = -\nabla \times J \quad (5)$$

Using the definition of potentials:

$$B = \nabla \times A \quad (6)$$

$$E = -\nabla\emptyset - \frac{\partial A}{\partial t} \quad (7)$$

Taking equations 6 and 7 into the governing equations 4 and 5, the solution equations are as follows:

$$\nabla^2 A - \mu\varepsilon\frac{\partial^2 A}{\partial t^2} = -\mu J \quad (8)$$

$$\nabla^2 \emptyset - \mu\varepsilon\frac{\partial^2 \emptyset}{\partial t^2} = -\frac{\rho}{\varepsilon} \quad (9)$$

In general, the transient heat transfer process in metal can be described by the Fourier equation [5]:

$$c\gamma\frac{\partial T}{\partial t} + \nabla\cdot(-k\nabla T) = Q \quad (10)$$

Where T is temperature, γ is the density of metal, c is the specific heat, k is the thermal conductivity of the metal, and Q is the heat source density induced by eddy currents per unit time in a unit volume.

For most induction heating problems, boundary conditions combine the heat losses due to convection and radiation. The boundary condition can be expressed as:

$$-k\frac{\partial T}{\partial n} = \alpha(T_s - T_a) + C_s(T_s{}^4 - T_a{}^4) + Q_s \quad (11)$$

where $\partial T/\partial n$ is the temperature gradient in a direction normal to the surface at point under consideration, α is the convection surface heat transfer coefficient, C_s is the radiation heat loss coefficient, Q_s is the surface loss and n denotes the normal to the boundary surface.

The Establishment of Model

The simulation of the induction heating process of is based on the software Flux® 10.3 from Cedrat Inc. Flux® 10.3 is a finite element software application for electromagnetic and thermal physics simulations. The geometry domain of the 2D electromagnetic-thermal field computation is in the Fig. 1.

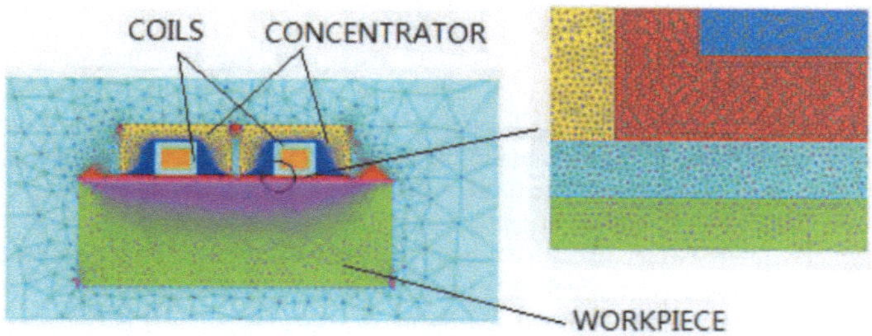

Figure 1. Geometry of the study domain

The workpiece material used in this study is AISI 1045 steel, its nominal chemical composition is shown in Table 1, and the thermal and electro-magnetic properties of the AISI 1045 steel are shown in Fig. 2. The volumetric heat capacity, thermal conductivity, and resistivity are temperature dependent [6]. The measurement of magnetisation curves of AISI 1045 steel is conducted using the Vibrating Sample Magnetometer LAKESHORE-730T (VSM). Apart from the workpiece material properties, we used default properties values for air and water provided by the simulation package. Other materials include inductor (Copper Coil: Relative magnetic permeability=1, Resistivity=0.02E-6 Ωm) and concentrator (Ferrotron 559H from Fluxtrol Inc.: Relative magnetic permeability=17 [7]).

Table 1: Nominal chemical composition of AISI 1045 steel (weight %)

C	Si	Mn	S	P	Cr	Mo	Ni	Al	Cu	Fe
0.45	0.25	0.65	0.025	0.008	0.4	0.1	0.4	0.01	0.17	Bal.

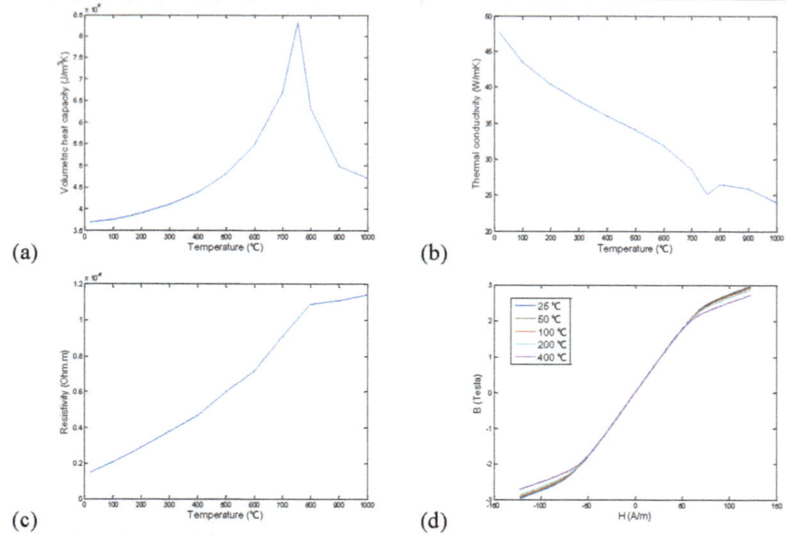

Figure 2. Material properties of AISI 1045 steel: (a) Volumetric specific heat; (b) Thermal conductivity; (c) Resistivity; (d) Magnetisation curves

Experimental Verification

The experimental verification is consisted of two parts, the first part is the emissivity calibration. The reason for conducting the emissivity calibration is due to the nature of the induction heating, the accuracy of thermal coupler may be disturbed by the strong magnetic field, therefore an infrared camera is used to monitor the workpiece temperature. The schematic diagram of emissivity calibration is shown in Fig. 3 (a). A workpiece is first heated inside an oven, and the cooling process is recorded by a thermal coupler and an infrared camera simultaneously. The result from the thermal coupler is then used to calibrate the results from the infrared camera to get the value for emissivity (Fig.3 (b)), which is 0.15 in our case. The resulted emissivity is used for simulation model and later experiments.

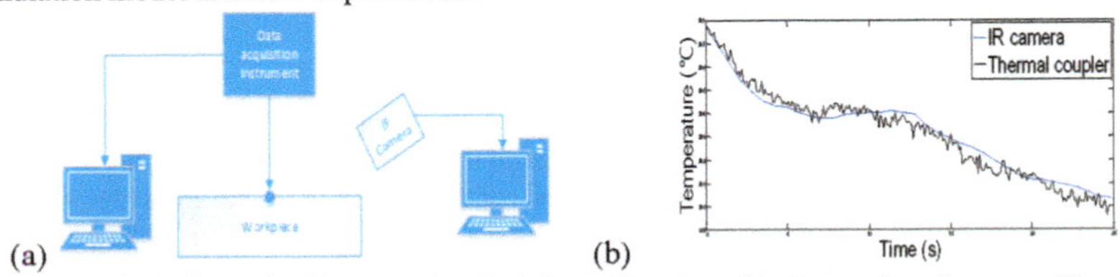

Figure 3. (a) Schematic diagram of emissivity calibration; (b) Comparison between IR camera (emissivity=0.15) results and thermal coupler results

During the second part, the surface temperature distributions of the workpiece during the induction heating process were recorded using NEC infrared camera R300. The schematic diagram of the experimental setup is shown in Fig.4. Two physical experiments were conducted, the current frequency and intensity are 37.6 kHz, 429A and 27.8 kHz, 683A respectively. For each experiment, the temperature profiles over 5 seconds period at points A (on the edge of the concentrator) and B (mid-point between edge of the concentrator and edge of the workpiece) were recorded, and used to compare with the simulation results (Fig.5). A good agreement was obtained between the simulation and experimental temperature profile at point A and B for both experiments, and therefore verified the accuracy of the analytical model.

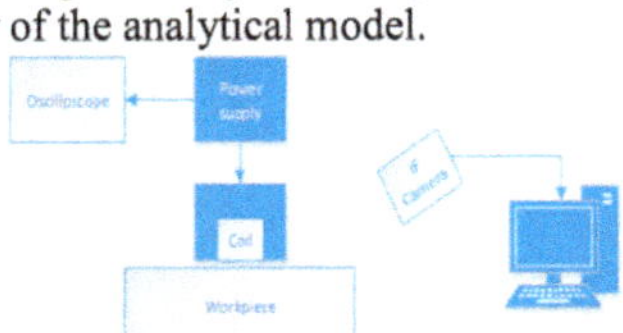

Fig.4. Schematic diagram of the experimental setup

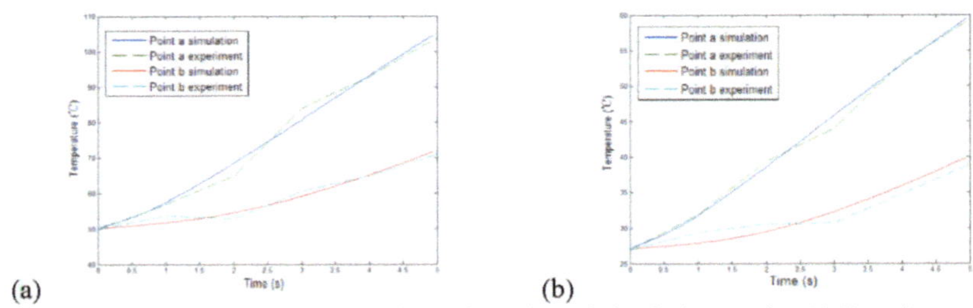

Figure 5. Comparison between experimental results and simulation results: (a) Experiment 1; (b) Experiment 2

The Phenomenological Study

The Effect of Magnetic flux concentrator

The using of magnetic flux concentrators during the induction heating can help the induction heating process to achieve increased productivity by focusing heating on critical areas only, therefore greatly reducing the heating time. In addition, more power is transferred into the selected area of the workpiece with the same coil current. Precision heating occurs exactly where it should, improving workpiece quality, production output and process efficiency [3]. The magnetic flux concentrator will be studied in four aspects: the effect of using a concentrator, concentrator width-x, concentrator height-y (Fig.6 (a)), and its relative permeability. The flux density is measured along the workpiece surface profile (Fig.6 (a)).

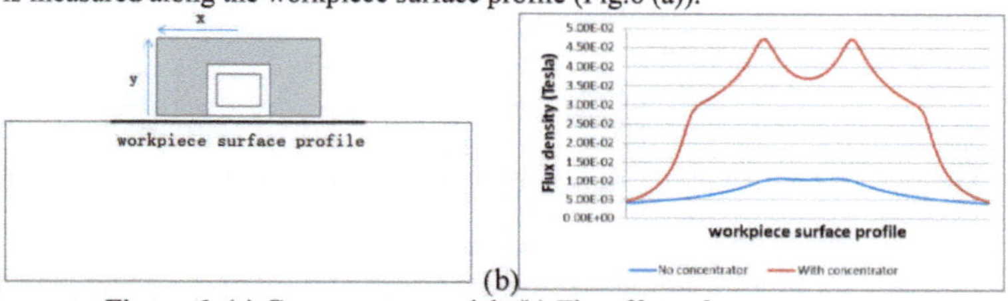

Figure 6. (a) Concentrator model; (b) The effect of concentrator

By simulating the concentrator with different dimensions, (Fig.7): the original condition, width increased by 2mm, width decreased by 2mm, height increased by 2mm, and height decreased by 2mm. It clearly indicated that the change of concentrator height will have higher impact on the surface of the workpiece than the concentrator width.

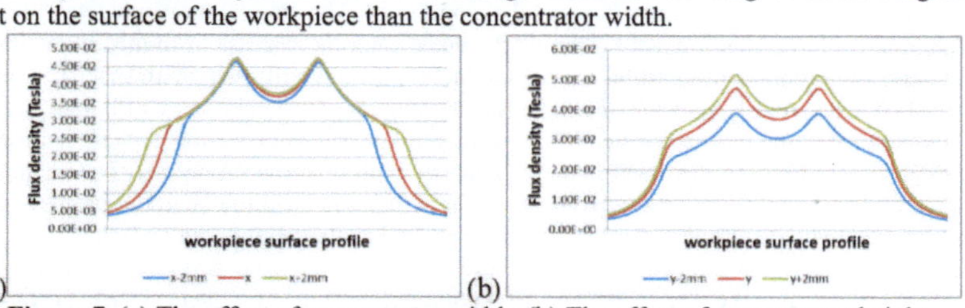

Figure 7. (a) The effect of concentrator width; (b) The effect of concentrator height

For the concentrator relative permeability and following simulations, the point directly below the coil at the center of the workpiece are measured. The simulations of concentrator relative permeability from 10 to 500 were also conducted. As the relative permeability of the concentrator increases, the flux density will also increases, but the increasing rate gradually decreases as the relative permeability increases as showed in Fig.8.

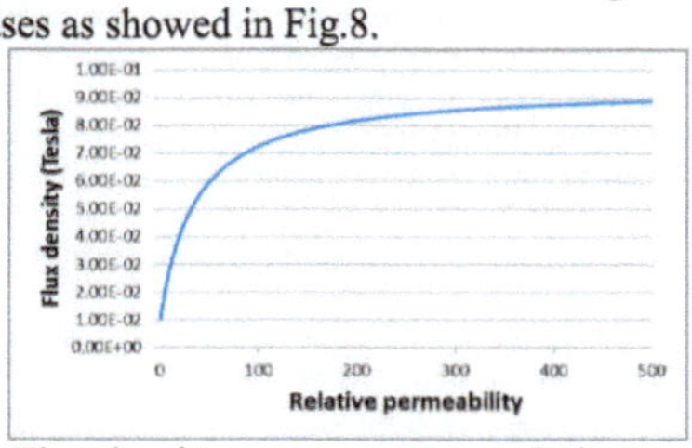

Figure 8. Flux density vs. concentrator relative permeability.

The Effect of Current Parameters

The simulations of frequency from 100 Hz to 100 KHz are conducted. From Fig.9 (a), it indicates that as the frequency increases, the flux density decreases. The intensities of the current are also simulated from 100A to 10000A. Flux density on the workpiece increases with current intensity in a linear relationship (Fig.9 (b)).

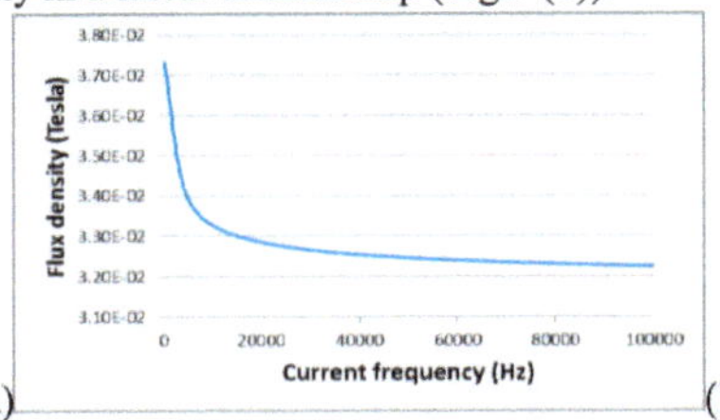

(a)

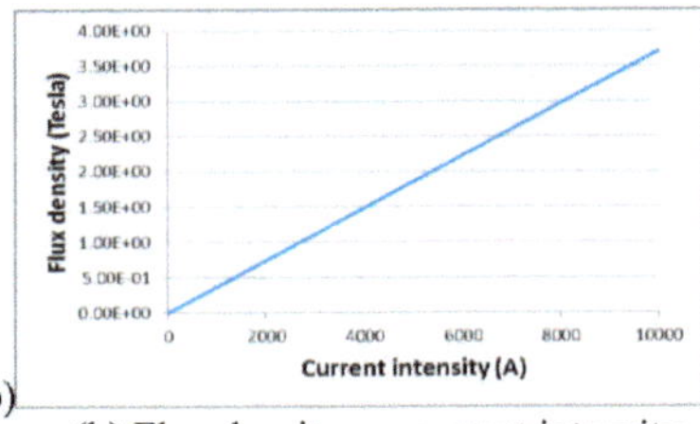

(b)

Figure 9. (a) Flux density vs. current frequency; (b) Flux density vs. current intensity

Conclusion

In this paper, we have presented a 2D electromagnetic-thermal analytical model with accurate material properties. The analytical model is built according to the physical experiment setup. The validity of the analytical model is verified by experiments, including workpiece emissivity calibration. A fairly good agreement was obtained between the simulation and experiment. Besides the establishing of the model, the phenomenological study on the effect of magnetic flux concentrator and current parameters were also conducted. The simulation results clearly indicate the parameters have huge impact on the workpiece during the process of induction heating, that there exist an optimal configuration for each induction heating processes with specific workpiece material. When carefully selected, the induction heating process can be more efficient.

Induction heating is a unique heating method, it induces heat inside the workpiece directly by Joule effect according to the electromagnetic and thermal material properties. The advantage of coupled electromagnetic-thermal analytical model is that with accurate material properties input, it can be used to predicating the parameters like current density and flux field within the workpiece model, which can be very difficult to measure directly. And from this, we can develop new techniques that can take the advantage of induction heating's uniqueness, which cannot be achieved by traditional heating methods.

References

[1] Naar, R. and F. Bay, 2013, Numerical optimization for induction heat treatment processes. Applied Mathematical Modelling, 37(4): pp. 2074-2085

[2] Cho, K.-H., 2012, Coupled electro-magneto-thermal model for induction heating process of a moving billet. International Journal of Thermal Sciences, 60: pp. 195-204.

[3] Okman, O., Z. Dursunkaya, and A.E. Tekkaya, 2009, Generalized transient temperature behavior in induction heated workpieces. Journal of Materials Processing Technology, 209(18-19): pp. 5932-5939.

[4] Naar, R. and F. Bay, 2013, Numerical optimization for induction heat treatment processes. Applied Mathematical Modelling, 37(4):pp.2074-2085

[5] V. Rudnev, D. Loveless, R. Cook, M. Black, Handbook of Induction Heating, Inductoheat, Inc, 2003.

[6] Magnabosco, I., 2006, Induction heat treatment of an ISO C45 steel bar: Experimental and numerical analysis, Computational Materials Science, 35(2): pp. 98-106.

[7] Resource Guide for Induction Technology, 2006, Fluxtrol, Inc., Auburn Hills, MI.

2nd World Congress on Integrated Computational Materials Engineering
Edited by: Mei Li, Carelyn Campbell, Katsuyo Thornton, Elizabeth Holm, and Peter Gumbsch
TMS (The Minerals, Metals & Materials Society), 2013

MODELLING THE PROCESS CHAIN OF COLD ROLLED DUAL PHASE STEEL FOR AUTOMOTIVE APPLICATION

A. Ramazani, U. Prahl

Department of Ferrous Metallurgy, RWTH Aachen University, Germany

Keywords: Dual-phase steel, phase field modeling, micromechanical modeling, geometrically necessary dislocation (GND)

Abstract

This project aims to develop a virtual process chain for the production of components out of cold-rolled dual-phase (DP) steel. The simulation chain starts with cold-rolled strip. During intercritical annealing process all relevant steps like recrystallization, austenite formation and grain growth, ferrite and martensite transformation including bainite fractions and quasi-tempering during hot dip coating and coiling are taken into account. Concerning the final mechanical properties transformation induced micro eigenstresses are described as well as strain partitioning on microscale during cold forming. This multi-scale and process-spanning approach enables the local properties in the part for varying composition and processing conditions. Thus, it can be used for the knowledge driven design and optimization of tailored material and process. To describe all the steps along the process chain, various simulation programs have been linked. By comparison of simulation and experimental results the predictability of this approach can be shown an in a later stage the integrative simulation approach will be further developed towards application for material and process design.

Introduction

The reduction of car body weight in combination with good crash behaviour is a major aim of automotive industry. Main reasons for using high strength multiphase steels are the improvement of the passenger safety, the reduction of fuel consumption and exhaust emission. Multiphase steels combine high strength and high formability, which is a properties combination needed for deep drawing parts in the automotive industry.

The mechanical properties of dual-phase steels are provided by a microstructure which consists of two phases with different hardness, i.e. hard martensitic islands in a soft ferrite matrix. Thereby, not only phase fractions but also distribution and morphology of both phases are responsible for the mechanical properties. [1,2] The formation of dual-phase microstructure from cold-rolled strip occurs during intercritical annealing which involves different metallurgical phenomena, like recrystallization and phase transformations. [3,4] Each of these processes contributes to the establishment of the final microstructure. Thus, the control of the microstructure evolution during the whole processing of dual-phase steels is therefore the key for the processing of steels with predetermined mechanical properties. Hereby, the numerical investigation which is an alternative to expensive experiments can be very helpful.

Nowadays, computational materials science is a powerful approach for understanding physical mechanisms and their interactions during industrial processing allowing the prediction of microstructure as well as of mechanical properties. [5,6] It can be applied for the improvement of process parameters and optimization of chemical composition.

Up to now, different mechanisms along the annealing route of cold-rolled strips in dual-phase processing, as there are recrystallization, austenite formation, grain growth, formation of ferrite and martensite, have been investigated separately. However, the current trend in computational materials science moves towards ICME (Integrative Computational Materials Engineering) where individual approaches on the different scales are combined in order to describe a process chain completely. [7,8] Such combination of models de-scribing single process steps allows the prediction of final product properties from the knowledge of chemical composition and processing conditions. Setting up through-process model is expected to lead to significant improvement in the prediction of microstructure development and therefore, to enable robust production of dual-phase steels with predefined mechanical properties [9-12]

Real materials such as multi-phase steels exhibit on the microscopic scale (microstructure) a heterogeneous material behaviour with constituents of varying strength and deformability. A micromechanical approach based on representative volume elements enables to describe these heterogeneities with the help of continuum mechanical quantities and by this to link mechanical properties at different scales. Thereby it is the goal to derive effective material properties for the component from microstructure quantities and to quantify the effect of microstructural features.

The aim of this contribution is the formulation of a through-process model on a microstructural scale for the production of dual-phase steels from cold-rolled strip by means of Phase-Field approach. Simulation results of the microstructure evolution during intercritical annealing and quenching and of final flow curve will be shown and compared to experimental results.

Experimental Process Simulation

In order to investigate the microstructure evolution during intercritical annealing, tests were performed on Baehr Dilatometer DIL-805A/D. For these experiments, the samples with 1 mm thickness, 4 mm width and 8 mm length were cut parallel to the rolling direction of the ingot. The dilatometric tests have been carried out with a dilatometer type DIL-805A/D fabricated by Baehr Thermoanalyse GmbH. Experiments were realised by a low-pressure environment of the order of 10^{-5} MPa in order to protect the samples from oxidation. During experiments temperature was recorded by a thermocouple. According to this heat treatment, the samples were heated in two-steps to the intercritical temperature of 800 °C, held 60 s at 800 °C and afterward cooled down slowly with 1 °C/s to 600 °C. Along the intercritical heat treatment samples were quenched at various stages in a helium atmosphere. Thus, the microstructure at the chosen temperatures was frozen and afterwards studied by means of light optical microscopy. The corresponding average values of the ferrite fraction were obtained by an image analyser of four nital etched micrographs per sample at magnification 1000x.

Phase-Field Modelling of Microstructure Evolution during Processing

Due to the fact that small variations in chemical composition and industrial process parameters are known to have a strong influence on strength and formability of dual-phase steels, precise control of the microstructure evolution during full processing route is required for achievement of defined mechanical properties. Nowadays it is reasonable to apply physical based numerical investigation approaches, which has reached a level where a property-driven improvement of the microstructure becomes possible.

The Phase-Field approach is developed for the modelling of the phase transformations in multi-component systems [13-15]. For simulation work in this paper the commercial software package

MICRESS® is used which allows the calculation of phase fractions during solid-solid transformations in multicomponent steels as well as the description of the corresponding microstructure evolution. [16]

Therefore, a through-process model for the micro-structure evolution during processing of dual-phase steels from cold-rolled strip was developed and coupled with the available RVE-FE approach for the prediction of mechanical properties within a continuum mechanical framework. In order to understand the relevant features of the transformations occurring during intercritical annealing, all metallurgical processes on the microstructural scale, i.e. recrystallization, austenite formation and decomposition were separately investigated before being combined into one integrative process model. 2D- and 3D-modelling of the microstructure evolution is realised by means of Multi-Component-Multi-Phase-Field method using the commercial software MICRESS® (Fig. 1). It was shown that both 2D- and 3D-Phase-Field approaches enable the simulation of the microstructure evolution during whole processing of dual-phase steels by accurate definition of the input model parameters.

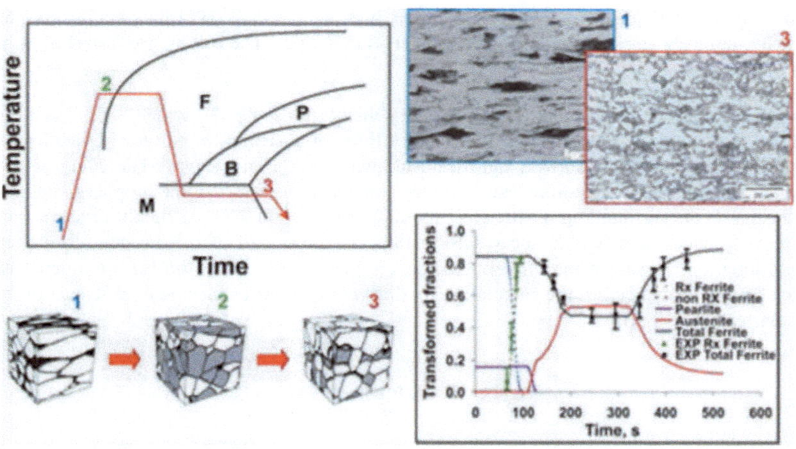

Fig.1: Schematic representation of time-temperature profile for the processing of dual-phase steels (top left); Micrographs before and after heat treatment (top right), Simulated development of microstructure (bottom left); Simulated evolution of transformed fractions (bottom right).

Micromoechanical modelling using representative volume elements

To study the flow behavior of DP steel, micromechanical modelling is a suitable procedure that provides a good description of the deformation at the microlevel. Numerical calculation within the framework of continuum mechanics was performed using FEM. The DP microstructure was embodied in FEM through representative volume element (RVE). 2D RVEs can be created from real microstructures by means of an in-house program (Fig. 2). The minimum RVE size was identified to be 24 μm including at least 20 hard particles [17]. Studying the effect of mesh size with ranging element length shows that no deviation is obtained for elements finer than 0.25 μm. Therefore, quadratic meshes with element size of 0.25 μm and edge length 25 μm were used for 2D modelling. Periodic boundary conditions were imposed on the RVE.

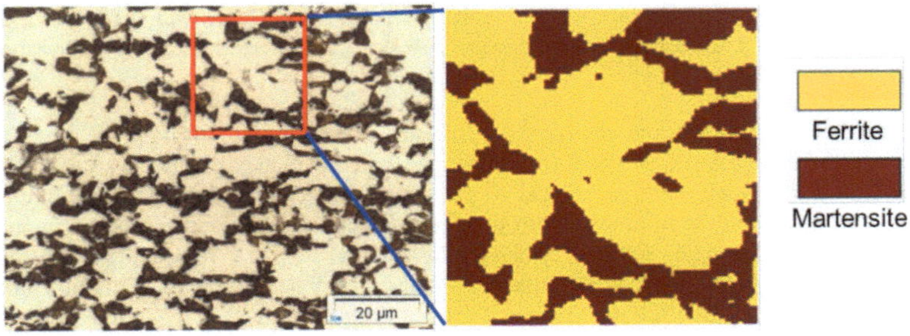

Fig. 2: Selection of 2D RVE for micromechanical modelling of DP steel with 35 % martensite.

The flow curve of individual phases for ferrite and martensite at RT was quantified based on the dislocation-based strain-hardening model [16,18]. For the later introduction of transformation-induced geometrical dislocations for each phase, high-temperature (HT) flow curves are needed at each martensite start (Ms) and finishing (Mf) temperature. The full approach has been given in a former study [18].

Transformation-induced geometrically necessary dislocations (GND), which are created during cooling from intercritical annealing temperature through austenite-to-martensite transformation due to volume expansion, show a significant influence on the initial work-hardening behaviour of DP steels. During cooling austenite-to-martensite transformation imposes volumetric expansion on the surrounding ferritic yielding the formation of GND. As the different variants of martensite cancel each other's shear deformation, the shear part of the austenite-to-martensite trans-formation is ignored and this effect is be assumed to be a pure isotropic volumetric effect which can be introduced in the calculation by different thermal expansion coefficients of ferrite and marten-site [18-20].

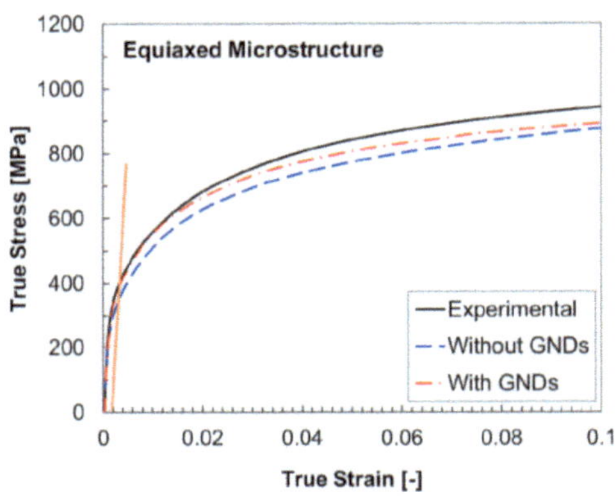

Fig. 3: Comparison between the experimental and predicted flow curves.

After numerically cooling down the RVE from Ms to RT, uniaxial loading in the rolling direction (X-direction) was applied to study the effect of GNDs on the stress and strain distribution in the micro-structure, the macroscopic mechanical properties, and the initial flow behaviour of DP steel. Fig. 7 shows the flow curves of DP steel derived from RVE calculations in comparison to experimental results. From these curves one can observe that calculations including GNDs are capable of predicting the high initial work-hardening rate of DP steel much better than calculations excluding GNDs. Therefore, the initial flow behaviour calculation of DP steel strongly depends on whether GNDs are taken into account. Table 1 shows the YSs determined in this way. In this table, the YSs from experiments and two predictions from numerical tensile tests are compared. GND-based modelling gives an important improvement in the calculation of YSs for the investigated DP steel with various morphologies.

Table 1: Comparison between experimental and predicted yield stresses of studied DP steel with and without considering GNDs.

	Yield stress (MPa)
Experimental	414
Predicted without considering GNDs	369
Predicted considering GNDs	394

Conclusions

Simulations of microstructure evolution during processing of dual-phase steels from cold-rolled strips have been realized by means of Phase-Field Modelling approach. It allows to describe all metallurgical phenomena occurring on a microstructural scale during intercritical annealing, i.e. recrystallization, austenite formation, ferrite formation as a function of chemical composition, starting microstructure and process parameters. The accurate definition of input model parameters yields both 2D- and 3D-approaches to simulate the microstructure evolution successfully. The comparison of 2D- and 3D-simulated results with experimental data demonstrates an overall agreement of the predicted evolution of phase fraction and grain size distribution. 2D RVE FE simulations reveal that, due to accommodation of the volume expansion associated with the austenite-to-martensite phase transformation, GNDs have been introduced, yielding a work-hardened ferrite skeleton after the cooling process, which in turn governs the initial flow behaviour of the compound. Deviations between experimental and simulation results decrease when GNDs are taken into account in the calculations of the flow curve. GND-based modelling yields better prediction of the YSs and initial work-hardening behaviour of DP steel with different martensite morphologies than do experimental results.

Acknowledgements

This paper is written based on the researches that were conducted under Project Nos. MC2.07293 & MC5.06257 in the framework of the Research Programs of the Materials Innovation Institute M2i (www.m2i.nl). Also the authors would like to thank the German Research foundation DFG for the financial support within the Cluster of Excellence Integrative Production Technology for High-Wage Countries".

References

[1] Bag, A.; Ray, K. K.; Dwarakadasa, E. S.: Metall. & Mat. Trans. 30A (1999), pp. 1193-11202

[2] Tomita, Y.: J. of Mat. Sci. 25 (1990), pp. 5179-5184

[3] Militzer, M.; Poole, W.J.: AHSSS Proceedings (2004), pp. 219-229

[4] Huang, J.: Microstructure Evolution during Processing of Dual Phase and TRIP Steels. PhD Thesis, University of British Columbia (2004)

[5] Bäker, M.: Numerische Methoden in der Materialwissenschaft. Braunschweig (2002)

[6] Raabe, D.; Roters, F.; Barlat, F.; Chen, L.-Q.: Continuum Scale Simulation of Engineering Materials. Weinheim Wiley-VCH Verlag (2003)

[7] National Research Council: Integrated Computational Materials Engineering: A Transformational Discipline for Improved Competitiveness and National Security. National Academic Press, Washington D. C. (2008)

[8] G. Gottstein.: Integral Materials Modeling. Weinheim Wiley-VCH-Verlag (2007)

[9] Rudnizki J., Böttger B., Prahl U., Bleck W.: Metall. & Mat. Trans. A (2011), pp. 2516-2525

[10] Thiessen, R.G.: Physically-Based Modelling of Material Responce to Welding. PhD Thesis, TU Delft (2006)

[11] Mecozzi, M.G.: Phase-Field Modelling of the Austenite to Ferrite Transformation in Steels. Ph.D Thesis, TU Delft (2007)

[12] Militzer, M.: Current Opinion in Solid State and Mat. Sci. 15 (2011) 3, pp. 106-115.

[13] Steinbach, I.; Pezzolla, F.; Nestler, B.; Seeßelberg, M.; Prieler, R.; Schmitz, G. J.; Rezende, J. L. L.: Physica D. 94 (1996), pp.135-147

[14] J. Eiken, B. Böttger, I. Steinbach: Phys. Rev. E 73 (2006)

[15] Steinbach, I.: Modelling and Simulation in Mat. Sci. and Eng. 17, 073001-31, (2009)

[16] MICRESS – The Microstructure Evolution Simulation Software: htto://micress.de

[17] Ramazani, A., Mukherjee, K., Prahl, U., Bleck, W., Comp. Mat. Sci. 52 (2012), pp. 46-54.

[18] Ramazani, A., Mukherjee, K., Schwedt, A., Goravanch, P., Prahl, U., Bleck, W.: Int. J. of Plast. 43 (2013), pp. 128–152.

[19] Ramazani, A., Mukherjee, K., Prahl, U., Bleck, W., Metall. & Mat. Trans. A, 43 (2012), pp. 3850-3869.

[20] Liedl, U., Traint, S., Werner, E.A., Comp. Mat. Sci. 25 (2002), pp. 122-128.

2nd World Congress on Integrated Computational Materials Engineering
Edited by: Mei Li, Carelyn Campbell, Katsuyo Thornton, Elizabeth Holm, and Peter Gumbsch
TMS (The Minerals, Metals & Materials Society), 2013

GEOMETRIC ANALYSIS OF CASTING COMPONENTS

Zhibin Quan [1], Zhiqiang Gao[1], Qigui Wang[2], Yunxia Sun[1], Xin Chen [1], Yucong Wang[2]

[1]Institute of Computer Science and Engineering, Southeast University; 2 Si Pai Lou, Nanjing, Jiangsu, 210086, China
[2]General Motors Holdings LLC,823 Joslyn Ave. Pontiac, MI 48340; Pontiac, PA 483730472, USA

Keywords: Geometric Analysis, Casting Components, Virtual Casting Design.

Abstract

Automated manufacturability analysis, aiming to considerate manufacturability aspects at initial stages of product design, is important for the designers in modern manufacturing industry to design manufacturable components. It mainly assists designers to economize the cost and time without compromising the quality and functional requirements of manufacturable parts. This paper deals with feature recognition for automated manufacturability analysis of new casting product. A new region growing method and an octree-based wall thickness computation approach are developed to extract manufacturing features (such as cylinder holes and wall thickness), from a CAD model which is represented by STL format.

Introduction

With the increasingly fierce global competition, to bring competitively priced, well-designed and well-manufactured products to market in a timely fashion is challenging in the manufacturing industry. Since the cost of changes in design phase after initiation of the product development cycle escalates steeply with time, the ability to make essential changes during the design phase translates into significant savings. To achieve this goal, increasing research attention is being directed towards the integration of engineering design and manufacturing. These attempts have led to the evolution of design for manufacturability (DFM) methodologies [1]. DFM involves considering design goals and manufacturing constraints simultaneously, in order to identify manufacturing problems while the product is being designed; thereby reducing the lead time for product development and improving product quality [2-4]. In order to make automated manufacturability analysis, some features such as holes and wall thickness must be recognized first in virtual casting design. The hole sizes can be computed if the holes can be recognized in the model. Then we know whether the holes can be cast or not under the existing industrial conditions. If the wall thickness distribution is known, it can infer that where will be shrinkage and porosity in the model [5].

This study is aimed to develop computational graphic methods to make geometric analysis of CAD model which is represented by STL format. The details of the algorithms and two preliminary data sets are given in the next section. An engine block has been used to validate the algorithms. The validation results show that the developed algorithms can be used for actual industrial applications.

Geometric analysis approaches

Geometric analysis for manufacturability check contains: hole recognition and wall thickness computation. The figure 1 shows an example of the recognized features in an engine block.

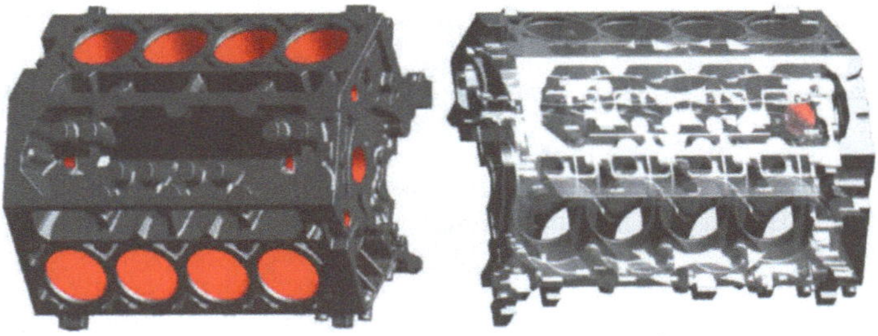

(a) (b)

Figure 1. Geometric analysis results of an engine block. (a) holes displayed by red color, and (b) maximum and minimum wall thickness displayed by red spheres.

Hole Recognition

Topology Re-Construction Freeform surface CAD models represented in STL format is universally supported by all CAD/CAM systems today. The triangles in STL file are discrete and represented by locations of their three points. So we don't know the triangles which share a point or share an edge. Therefore it is necessary to reconstruct their topology with three points in each triangle from location to location, in order to recognize triangles which constitute a hole.

Let $G = (V, E)$ be an undirected graph which is associated with the STL file that represents a CAD model M, where $V = \{v_i | v_i = (x, y_i, z_i), i = 1,2, \cdots, N\}$ is the set of vertices and $E = \{e_{\{ij\}} | e_{\{ij\}} = (v_i, v_j), i, j = 1,2, \cdots, N\}$ is the set of edges. The relation between V and E is

$$e_{\{ij\}} = \begin{cases} 1, & if\ v_i, v_j\ share\ an\ edge \\ 0, & \text{otherwise} \end{cases},$$

where $\sum_{i=1}^{N} \sum_{j=1}^{N} e_{\{ij\}} = M$denotes the number of edges. Let $= \{t_{\{i,j,k\}} | t_{\{i,j,k\}} = (v_i, v_j, v_k)\}$, be the set of triangles. The relationship between T and V is

$$t_{\{i,j,k\}} = \begin{cases} 1, & if\ the\ triangle\ is\ formed\ by\ v_i, v_j, v_k \\ 0, & \text{otherwise} \end{cases},$$

where $\sum_{i=1}^{N} \sum_{j=1}^{N} \sum_{k=1}^{N} t_{\{i,j,k\}} = K$ denotes the number of triangles. Let $n_i = (x, y_i, z_i)$, $i = 1,2, \cdots, K$ be the normal vectors of these triangles.

Let $G' = (V', E')$ be the dual graph of G. The elements of $V'$are triangles. E'represents the adjacent relations of triangles. $V = \{v_i' |, i = 1,2, \cdots, K\}$ is the triangle set and $E = \{e_{\{ij\}}' | e_{\{ij\}}' = (v_i', v_j')\}$ is the edge set. The relation between E' and V' is

$$e_{ij}' = \begin{cases} 1, & if\ v_i', v_j'\ are\ adjacent \\ 0, & \text{otherwise} \end{cases}.$$

In order to recover the topology information, the set E and E' must be assigned values, so that the graph G and G' can be constructed.

To recognize the hole, cylinders must be recognized first. The Gauss map [3] provides a mapping from every point on a curve or a surface to a corresponding point on a unit sphere. Every point on a curve, a surface, or a triangle has its own unit normal. The Gauss mapping is to move that normal to the center of sphere. The point on the sphere where all the normal points are located is called Gauss point. In an ideal condition, the Gauss points of one cylinder and the center of sphere form one plane called Gauss plane [3] (Figure 2(a)(b)).

(a) (b) (c)

Figure 2. (a) Gauss map of a cylinder; (b) a cylinder that has been recognized (blue regions represent cylinder); (c) Region growing.

Seed Selection Each triangle v_o' in STL file, must have three adjacent triangles, v_1', v_2', v_3' as displayed in figure 2(c). Then the triangle v_o' is considered as an active seed triangle, if v_o' satisfies the following two conditions:

Seed selection condition one (also called Gauss condition): the distance $d(n_o, P^{\{i,j,o\}})$ between n_o and the Gauss plane $P^{\{i,j,o\}}$ which is constituted by three points n_i, n_j and $(0,0,0)$ is less than a given threshold $\delta_{P^{\{i,j,o\}}}$, i.e.

$$d(n_o, P^{\{i,j,o\}}) < \delta_{P^{\{i,j,o\}}};$$

Seed selection condition two: let v_i', v_j' be the triangles which satisfy Gauss condition, and v_i, $i = 1,2,\cdots,5$ be the points which constitute the three triangles. Then all v_i must satisfy the following inequality

$$d(v_i, C^{\{i,j,o\}}) < \delta_{C^{\{i,j,o\}}},$$

where $C^{\{i,j,o\}}$ is a cylindrical surface which has minimum fitting error. $\delta_{C^{\{i,j,o\}}}$ is a given threshold.

Region Growing Let v_0' be a active seed triangle, and triangles v_i', v_j' are adjacent to v_0', which make v_0' be active. Let R be an initial growth region formed by v_0', v_i', v_j', P^R be the Gauss plane formed by R, and C_R be the cylindrical surface formed by R, which has minimum fitting error. Let v_k' be one of the triangles which are adjacent to R, i.e. $v_k' \in b(R)$, as displayed in figure 2(c), v_k be one of the three points which form the triangle v_k', but not in R, and v_{centroid} be the centroid of v_k'. v_k' will be added into R, if it satisfies,

Region growing condition one:

$$d(n_k, P^R) < \delta_{P^R};$$

Region growing condition two:

$$d(v_k, C^R) < \delta_{C^R},$$

where δ_{C^R}, δ_{P^R} are given thresholds.

Filter and Measure When all the cylinders have been recognized, the holes will be determined by filtering those which are not cylindrical holes. As we know, the normal vectors of triangles which make up a cylinder have two ways of emission: divergent and aggregate. The aggregate emission of the normal vectors of the cylinder indicates that the cylinder is the hole. After cylindrical holes are recognized, we use the method of ray triangle intersection to calculate the diameter and other parameters of each hole.

Octree-Based Wall Thickness Calculation

A ray triangle intersection method is commonly used to calculate wall thickness of an object. As it is time consuming and inefficient to check a ray with each triangle, it is necessary to filter the triangles that don't intersect with the ray. This can be done by using octree-based algorithm. All triangles are put into different nodes of octree according to their bounding boxes. If a ray intersects with the node, it may intersect with the triangles in the node without consideration of other triangles which are not in.

Definition of Wall Thickness Two generic definitions of the wall thickness are presented here: interior wall thickness for points inside an object, exterior wall thickness for points on the object surface. Interior wall thickness at point P inside an object is defined as twice the minimum distance of P from the nearest surface of the object (Figure 3(a)). Its value can be obtained by growing a sphere or by firing rays from point along all directions towards the object surface. The shortest ray length (distance between point and surface) gives the wall thickness at that point. The definition is mainly for the maximum wall thickness calculation.

Exterior wall thickness at point P on the object surface is defined as the distance between P and Q (Figure 3(b)). A ray is shot from P on the surface in a direction opposite to the local outwards surface normal, to which intersects the immediately opposite surface of the object. The definition is mainly for the minimum wall thickness calculation and the definition is effective only when the surfaces are parallel.

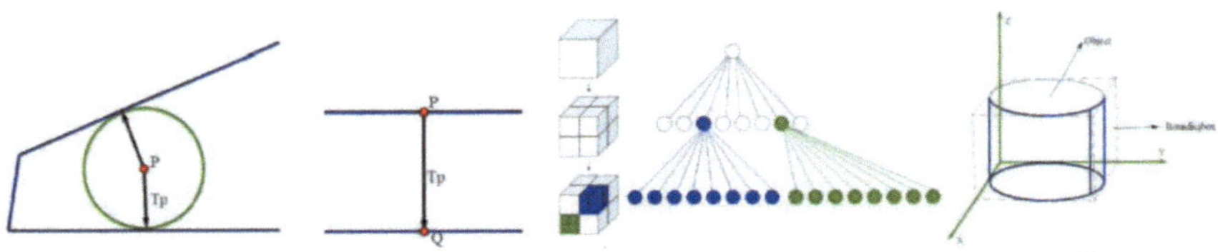

(a) (b) (c) (d)

Figure 3. (a) Interior wall thickness at point P inside an object; (b) Exterior wall thickness at point P on the object surface; (c) Recursive subdivision of a cube into octants in left, the corresponding octree in right; (d) Axis-aligned bounding box of a cylinder.

Octree Data Processing An octree is a tree data structure in which each internal node has exactly eight children except for the leaf node. As shown in figure 3(c), each node in an octree subdivides the space into eight octants and the node corresponds to a cube in space.

In order to add triangles to an octree, each member in the triangle set T must be compared with nodes in the octree. For convenience, a bounding box for each triangle is constructed and is used to be compared with nodes instead. Besides, the bounding box will offer facilities for the following ray-triangle intersection test.

A bounding box is a smallest cuboid, or a rectangle in 2D, containing the object. Figure 3(d) shows the axis aligned bounding box. The edges of the box are parallel to coordinate axes. It can be constructed by getting the maximum and the minimum values in X, Y and Z directions.

For data pre-processing, all triangles in the STL file are added to the corresponding nodes of the octree. During the process, the algorithm compares the bounding box (A) of the triangle with the bounding box (B) of the octree node. If B contains A, it continues to compare A with the octree node's children node until the depth of the current node is more than or equal to the octree's maximum depth or none of the current node's eight children contains A. Then the triangle is added to the corresponding nodes. After all triangles are added, an unbalanced octree is completed.

After data pre-processing, the candidate triangles that probably intersect with a ray can be identified by doing a ray-node intersection test. If a node intersects with a ray, its eight children

(t_i') are continued to be tested until the current node is a leaf one. All candidate triangles are obtained by reading triangle list of each intersection node during the test.

Wall Thickness Calculation The wall thickness intersection point must lie on the triangle which is nearest to P and the ray direction is the normal of the triangle. The direction of the ray can be confirmed by searching the nearest triangle of the point P. The intersection distance is the wall thickness. The method is mainly used to get the minimum wall thickness.

First the flag of each node in the octree is set to false by traversing the octree and the leaf node (B') which P lies in is computed. Then all the candidate triangles are obtained by searching the leaf node. For each candidate triangle (t_j), a ray (r) is shot from P with the direction of t_j's normal, then the intersection distance (d_j) between t_j and r is computed. The minimum d_j is the wall thickness d.

If there is no satisfied d, the searching scope is expanded. At the same time the candidate triangles are got by searching the surrounding leaf nodes. The rest may be deduced by analogy until we get d that we need.

The candidate point P that is used to compute the wall thickness may be around the triangle. In order to get the exact value of the thickness, the nearest triangle t is searched. A ray is made from t in a direction opposite to the local outwards surface normal to intersect the opposite surface of the object immediately. The distance between the two surfaces is exactly the wall thickness.

Algorithm Validation

To verify the correctness of those algorithms, two preliminary data sets (collections of NX UniGraphics CAD models) have been built in NX UniGraphics (UG): one for hole recognition as depicted in table 1, models 1 to 10, the other for wall thickness calculation as depicted in table 1, models 11 to 20. The graphic data sets are all files in the format of STL. Moreover, an engine block (Figure. 1) model has also been used in the validation.

In the open and blind hole recognition, each hole in every model has been recognized, and the hole diameter has been computed (the depth is hard to be verified because of practical limits). To verify the correctness of the diameter computation, a chart is shown in figure 4. Let the original diameter be D_o and the computed diameter be D_c. The absolute relative error is $|D_o - D_c|/D_o$. From the chart (Figure. 4), the maximum value of the absolute relative errors for hole diameter calculation is smaller than 0.05. The average absolute relative error is only 0.97%, as depicted in red line of figure 4.

Figure 4 also shows the absolute relative error of wall thickness (maximum and minimum) between the original data of UG models and the computed data of the algorithm. The average absolute relative error of the maximum wall thickness is 5.49% and the average absolute relative error of the minimum wall thickness is 7.29%, as depicted in green and blue lines of Figure 4. The results are acceptable in actual industrial applications.

Conclusions

A new region growing method and an octree-based wall thickness computation are proposed to extract manufacturing features: holes and wall thickness of CAD models in this paper. The algorithm validation shows that the proposed methods can be used for actual industrial applications, which provide good bases for automated manufacturability analysis of casting components.

Table 1. Test models for hole recognition and wall thickness calculation.

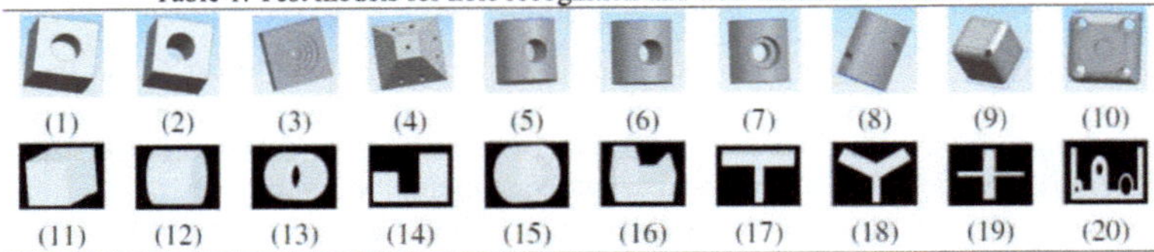

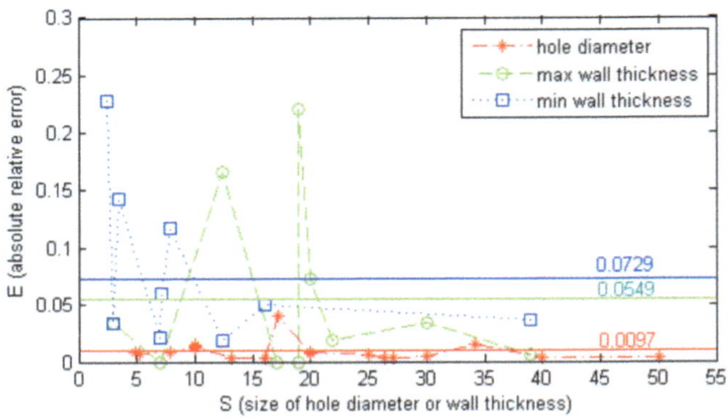

Figure 4. The absolute relative error testing algorithms in the data set.

Acknowledgment

This work is financially supported by Materials Technology Department of GM Global Powertrain Engineering. The support from Dr. Bowang Xiao at GM Powertrain is greatly appreciated.

References

1. Bakerjian R., ed., *Design for Manufacturability, volume 6 of Tool and Manufacturing Enginners Handbook* (Society of Manufacturing Engineers, 1992), 1-125.
2. Babic B., Nesic N. and Miljkovic Z., "A Review of Automated Feature Recognition with Rule-based Pattern Recognition," *Computer in Industry,* 59 (2008), 321-337.
3. Sunil V. B. and Pande S. S., "Automatic Recognition of Features from Freeform Surface CAD Models," *Computer-Aided Design,* 40 (4) (2008), 502-517.
4. Huang Z. and Yip-Hoi D., 2002. "High-level Feature Recognition using Feature Relationship Graphs," *Computer-Aided Design,* 34 (8) (2002), 561-582.
5. Sinha B., "Efficient Wall Thickness Analysis Methods for Optimal Design of Casting Parts," (Paper presented at Engineering Design 2007), 1-4.

2nd World Congress on Integrated Computational Materials Engineering
Edited by: Mei Li, Carelyn Campbell, Katsuyo Thornton, Elizabeth Holm, and Peter Gumbsch
TMS (The Minerals, Metals & Materials Society), 2013

A MICROSTRUCTURE-STRENGTH CALCULATION MODEL FOR PREDICTING TENSILE STRENGTH OF ALSI7MG ALLOY CASTINGS

Yufeng SHI [1], Qingyan XU [1], Rui CHEN [1], Baicheng LIU[1]
Qinfang WU [2], Hongwei YANG [2]

1 Key Laboratory for Advanced Materials Processing of Ministry of Education, Department of Mechanical Engineering, Tsinghua University, Beijing 100084, China
2 Mingzhi Technology Co., Ltd., Suzhou 215217, Jiangsu, China

Key words: secondary dendrite arm spacing, tensile properties, calculation model

Abstract

The mechanical properties of AlSi7Mg alloys are known to be improved by heat treatment. By OM analyses and tension test, the effects of secondary dendrite arm spacing (SDAS) on tensile properties of AlSi7Mg alloys were investigated by using step-shaped castings. Finally, a calculation model of SDAS-tensile properties was developed. Based on the temperature field simulation by FT-STAR, an MCA model was used for simulating the dendrite morphology and SDAS values of AlSi7Mg alloy castings. Using the SDAS values and the calculation model, ultimate tensile strength and elongation would be predicted of AlSi7Mg alloy valve body, which compared well with the experimental data and meet the requirement.

introduction

Currently, countries around the world have increasingly high demand for energy conservation, which prompt the demand for lightweight parts in the automotive, aerospace and other transportation means. As one of the most important light metallic materials, aluminum alloy can reduce the weight of products, which easily achieve the purpose of reducing energy consumption and pollutant emission. Al-Si-Mg cast alloys have drawn widespread concern due to their excellent castability, corrosion resistance, good tensile fatigue properties and especially high strength to weight ratio. AlSi7Mg alloys can obtain increased strength through heat treatment, which have increased use in structural applications[1]. As melting process and melting operation improved, inclusions, intermetallic compounds, oxidation film and hydrogen gas pore in the castings drop rapidly. Meanwhile, the specification of composition have a tight control. Therefore, the cast aluminum alloys have a widespread use and have good competitiveness against the respective wrought materials.

In industrial conditions, empirical equations are fast and easy tools to assess the quality of AlSi7Mg alloy castings. A number of papers have been published about the effects of microstructural parameters, such as secondary dendrite arm spacing (SDAS), the grain size, the size and morphology of eutectic Si particles on the tensile properties of A356 (AlSi7Mg0.3) and A357 (Al-Si7Mg0.6) alloys[2-10].

In the present work, step-shaped castings were produced, in order to investigate the effect of different secondary dendrite arm spacing (λ_2) on the tensile strength (σ_b) and elongation (δ). The experiments included a series of solid solution and aging treatments of AlSi7Mg step-shaped castings, a large number of tensile test and specimens for microstructure characterization. The results have demonstrated that the decreasing of secondary dendrite arm spacing can increase the tensile properties. Finally, empirical equations between tensile strength and elongation of the AlSi7Mg alloys and secondary dendrite arm spacing were obtained. The calculated results by the equations were close to experimental values.

Experimental procedure

At first, excellent melting process of aluminum alloy was performed to minimize porosity and other inclusions and, hence, to reduce deviations in tensile properties. The hydrogen level in the melt was maintained at a low level by using a rotary lance degasser and bubbling high purity Ar gas. Chemical composition of the AlSi7Mg alloy was shown in Table 1.

Table 1. Chemical composition detection of the AlSi7Mg alloy (wt%)

Si	Mg	Fe	Cu	Ti	Al
7.13	0.36	0.149	0.007	0.116	Balance

The alloy was poured into the mould of step-shaped castings with sand casting process. The pouring temperature was 730±5℃, and pouring time was 5 s. The dimensions of the castings were approximately 250 mm in length, 120 mm in width and varied in height of 30, 17, 11, 7, 4 mm (Figure 1). By using a step-shaped casting, the solidification time continued to reduce from thick step to thin one. Samples for the tensile and microstructural characterization were both machined from the middle section of each step, as indicated by the circles in Figure 1.

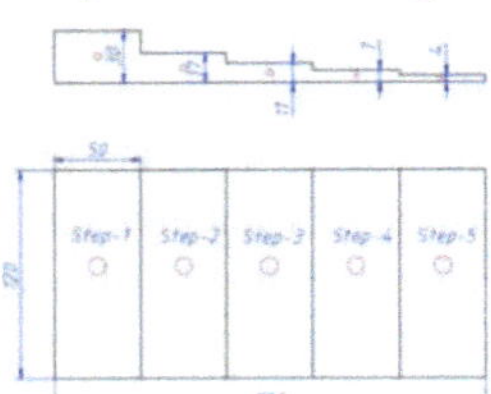

Fig.1 Schematic of the step-shaped casting

The microstructural characterization of the 5 different positions, shown in Figure 2, consisted of primary α-Al dendrites separated by the eutectic Si particles to form a cell pattern, periodically repeated across the metallographic surface. It was found that the primary α-Al dendrites became denser while step thickness increased.

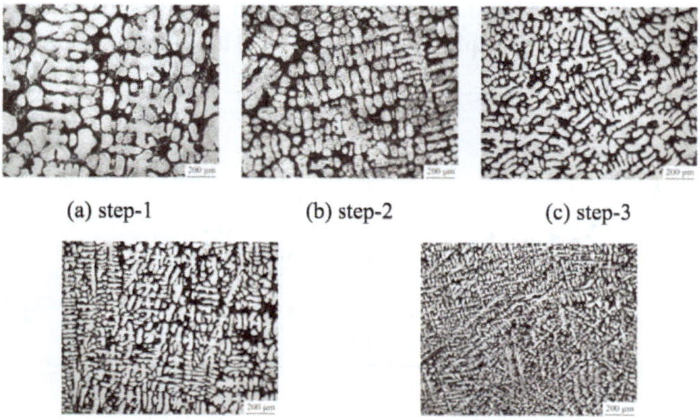

(a) step-1 (b) step-2 (c) step-3

(d) step-4 (e) step-5

Fig.2 Optical micrographs at the 5 different positions of the step-shaped casting

Mechanical testing of the step-shaped casting was performed through tensile tests. The specimen was made according to ASTM B577M specification [11], as shown in Figure 3. The specimen had a gauge length L_0 = 25 mm and a cross-sectional area S_0 =19.2 mm^2. Tensile tests were carried out using a servo-hydraulic testing machine WDW3020 at room temperature.

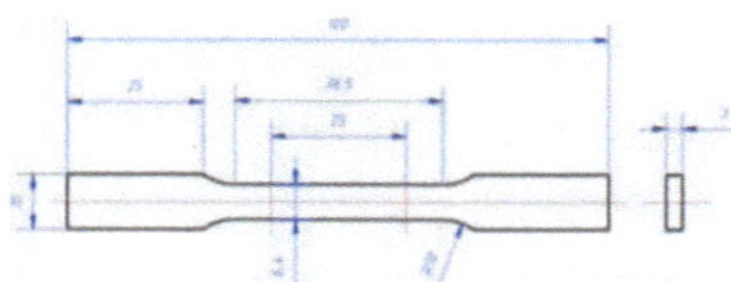

Fig.3 Shape and dimensions of the tensile specimen

Stress-strain curves were obtained by attaching a knife-edge extensometer (25 mm) to the specimen gage length. The experimental data were stored by a testing machine data logger, while the proof strength ($\sigma_{0.2}$), ultimate tensile strength (UTS), and elongation to failure (El) were measured. Figure 4 showed the variation of UTS, El of the 5 different steps with T6 heat treatment. The stress-strain curves indicated that both UTS and El increased with decreasing step thickness. A thin step (step-5) led to an improvement of almost all the mechanical properties (tensile strength, as well as elongation to failure). That is because the thin step had a high solidification rate, which reduced SDAS and had better microstructural characteristics.

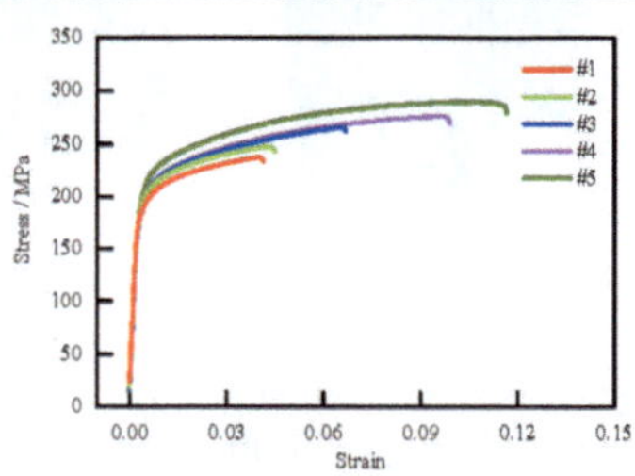

Fig.4 Typical stress-strain curve of the 5 steps

The effect of SDAS on the ultimate tensile strength and elongation was evaluated. The results are reported in the plots through statistical techniques (Figure 5a and 5b).

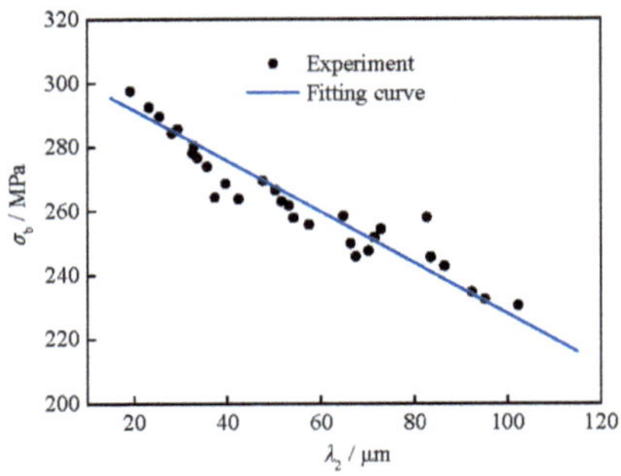

(a) UTS-SDAS relationship

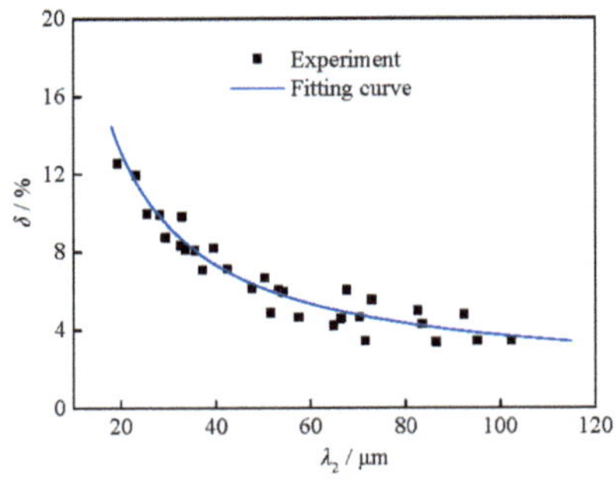

(b) El-SDAS relationship

Fig.5 Experimental data of mechanical properties and SDAS

A regression analysis was performed by using an Origin regression function. However, it was found that UTS (σ_b) and El (δ) had good relationship with SDAS (λ_2). The calculation model is shown as Equation (1) and (2). The σ_b-λ_2 relationship is straight and the δ-λ_2 equation is logarithmic, which is similar to reference [2] and [3].

$$\sigma_b = 307.5 - 0.79\lambda_2 \quad (1)$$

$$\delta = 1.13 + 211.1\lambda_2^{-0.96} \quad (2)$$

Simulation and experimental validation

The equations were exploited to predict the quality for the AlSi7Mg alloy castings. A 3-D valve was shown in Firgure 6, which was casted by sand casting. The alloy used for the valve was an AlSi7Mg alloy of composition (wt%) 7.2 Si, 0.35 Mg, 0.14 Fe, 0.01 Cu and 0.12 Ti. Three positions (P1, P2 and P3) were chosen to predict their mechanical properties (Firgure 6).

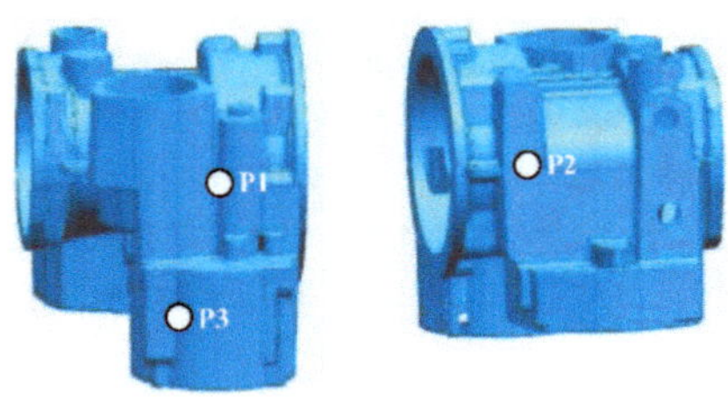

Fig.6 3-D shape of the AlSi7Mg alloy valve

The pouring temperature was 730±5℃, and pouring velocity is 0.29 m/s. The filling and solidification process were simulated using FT-STAR software, developed by Tsinghua University, China. Simulated filling time was about 10 s, compared well with the experimental time 9.8s.

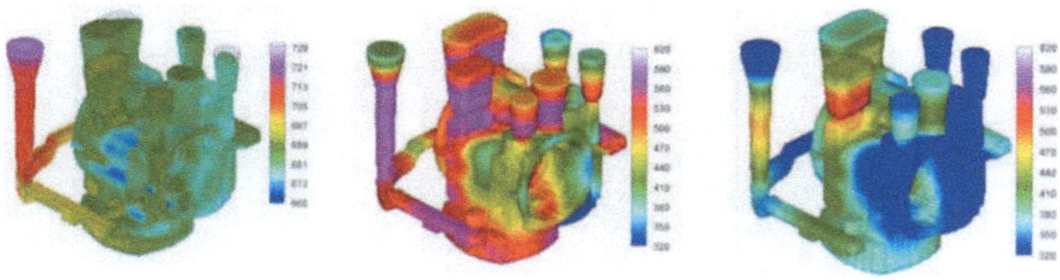

(a) t=9.98s, filled 99% (b) t=329s, solidified 70% (c) t=743s, solidified 100%

Fig.7 Simulation of mold filling and solidification process of the valve

The thickness of the 3 positions was P1>P2>P3. It was observed that P1 displayed a highest solidification time, and the predicted solidification time had a consistent trend with the experimental data.

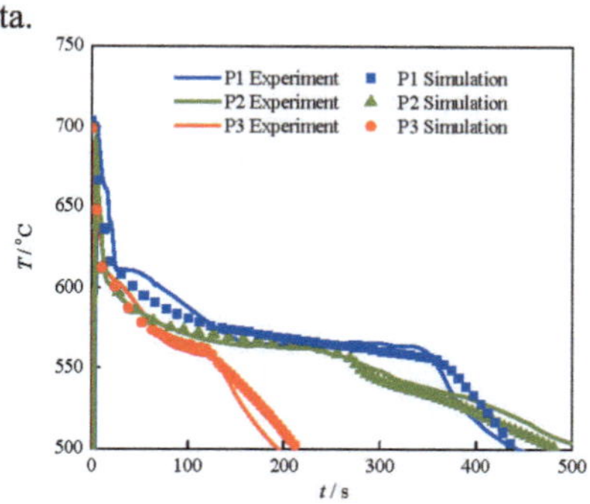

Fig.8 Comparison between simulated solidification path and experimental results

Metallographic specimens were cut from the P1, P2 and P3 regions of the valve, as indicated in Figure 6, and prepared for quantitative measurement of SDAS. Grain morphology of P1, P2 and P3 was predicted by MCA model[12]. Since complex dendrite morphology was characterized by different color for the simulation and experiment, as shown in Figure 8, the predicted dendrites had a good approximation with the experimental microstructural features.

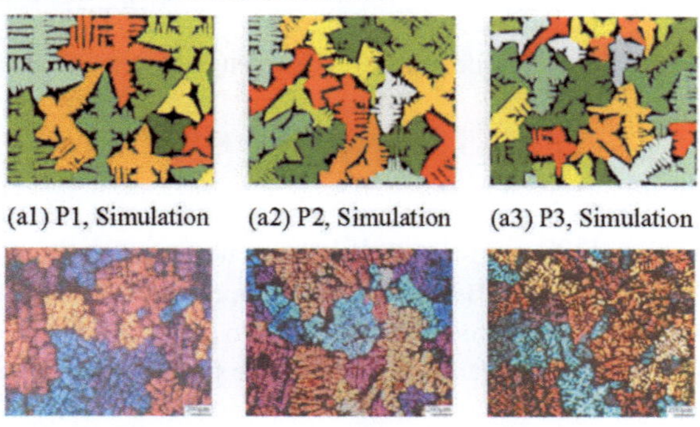

(a1) P1, Simulation (a2) P2, Simulation (a3) P3, Simulation

(b1) P1, Experiment (b2) P2, Experiment (b3) P3, Experiment

Fig.9 Predicted (a1-a3) and experimental microstructures (b1-b3) of P1, P2 and P3

The predicted SDAS had good agreement with experimental data, as shown in Figure 10. The local SDAS values of P1, P2 and P3 varied significantly in different zones of the cast. However, the corresponding mechanical properties also vary significantly.

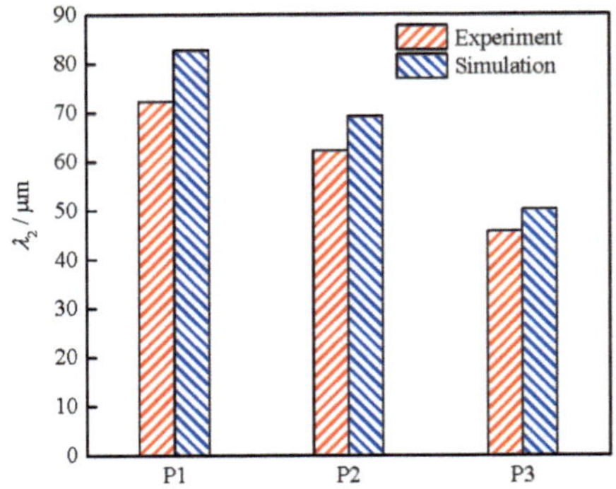

Fig.10 Comparison of SDAS between predicted and experimental data

It is known that SDAS can affect the tensile properties of the valve. As shown in Figure 11, using the equations (1) and (2), the calculated UTS and El met the user's requirement (UTS>207 MPa, El>3%,), and compared well with the experimental data.

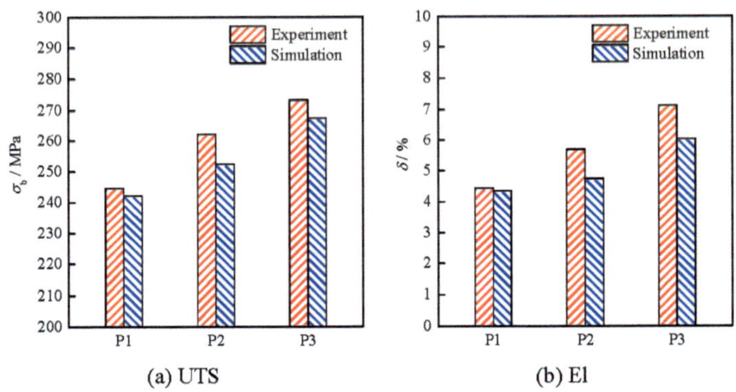

Fig.11 Comparison of tensile properties between predicted and experimental data

conclusion

A good relationship between the mechanical properties and the SDAS values was developed for the widely used AlSi7Mg alloys, while using condition was sand casting and T6 heat treatment. Based on the experimental SDAS and tensile property data, the regression equations would be performed to predict UTS and El of AlSi7Mg alloys, binding the MCA model, which had a role to fast detect the product quality of the sand castings.

Acknowledgments

This research is funded by National Basic Research Program of China (No.

2005CB724105, 2011CB706801), National Natural Science Foundation of China (No. 10477010), National High Technology Research and Development Program of China (No.2007AA04Z141) and Important National Science & Technology Specific Projects (No.2009ZX04006-041-04).

References

[1] N.D. Alexopoulos and S.G. Pantelakis, "Quality evaluation of A357 cast aluminum alloy specimens subjected to different artificial aging treatment," *Materials and Design*, 25(2004), 419-430.

[2] L. Ceschini, Alessandro Morri, Andrea Morri, et al, "Correlation between ultimate tensile strength and solidification microstructure for the sand cast A357 aluminium alloy," *Materials and Design*, 30(2009), 4525-4531.

[3] S.C. Sun, B. Yuan and M.P. Liu, "Effects of moulding sands and wall thickness on microstructure and mechanical properties of Sr-modified A356 aluminum casting alloy," *Trans. Nonferrous Met. Soc. China*, 22(2012), 1884-1890.

[4] Q.G. Wang, "Microstructural effects on the tensile and fracture behavior of aluminum casting alloys A356/357," *Metall Mater Trans A*, 34(2003), 2887-2899.

[5] S.G. Shabestari and H. Moemeni, "Effect of copper and solidification conditions on the microstructure and mechanical properties of Al-Si-Mg alloys," *J Mater Process Technol*, 153-154(2004), 193-198.

[6] S.G. Shabestari and F. Shahari, "Influence of modification, solidification conditions and heat treatment on the microstructure and mechanical properties of A356 aluminum alloy," *J Mater Sci*, 39(2004), 2023-2032.

[7] Q.G. Wang, D. Apelian and D.A. Lados, "Fatigue behavior of A356/357 aluminum cast alloys. Part II. Effect of microstructural constituents," *J Lights Met*, 1(2001), 85-97.

[8] Q.G. Wang and C.H. Cáceres, "The fracture mode in Al–Si–Mg casting alloys," *Mater Sci Eng A*, 241(1998), 72-82.

[9] Z. Lia, A.M. Samuel, F.H. Samuel, C. Ravindran, S. Valtierra and H.W. Doty, "Parameters controlling the performance of AA319-type alloys. Part I. Tensile properties," *Mater Sci Eng A*, 367(2004), 96-110.

[10] E. Ogris, A. Wahlen, H. Luchinger, P.J. Uggowitzer, "On the silicon spheroidization in Al-Si alloys," J Light Met, 2(2002), 263-269.

[11] ASTM B577M, "Standard test methods of tension testing wrought and cast aluminum- and magnesium-alloy products. Annual book of ASTM standards, vol.02.02," West Conshohocken, PA: ASTM International, 2001, 419-429.

[12] Y.F. Shi, Q.Y. Xu, B.C. Liu, "Modeling and simulation of dendritic growth in solidification of Al-Si-Mg ternary alloys," Proceedings of the 13th International Conference on Modeling of Casting, Welding and Advanced Soldification Process, Austria, 2012.

2nd World Congress on Integrated Computational Materials Engineering
Edited by: Mei Li, Carelyn Campbell, Katsuyo Thornton, Elizabeth Holm, and Peter Gumbsch
TMS (The Minerals, Metals & Materials Society), 2013

VALIDATION OF HIGH STRENGTH CAST Al-Zn-Mg-Cu ALUMINUM FOR USE IN MANUFACTURING PROCESS DESIGN

M.D. David, R.D. Foley, J.A. Griffin, C. Monroe

The University of Alabama at Birmingham
1150 10th Avenue South Room 254; Birmingham, AL, 35294, USA

Keywords: CALPHAD, Intermetallic Phases, 7xxx, SEM, X-ray diffraction

Abstract

The wrought 7xxx series aluminum alloys are some of the highest strength aluminum alloys. Historically, cast alloys with similar compositions have not been produced because their long solidification ranges decrease castability and the intermetallics that form during solidification reduce ductility. To develop cast Al-Zn-Mg-Cu alloys, an understanding of the intermetallic formation is needed. Six experimental Al-Zn-Mg-Cu alloys were cast under 1 MPa (10 atmospheres) of pressure. Zinc concentrations of 8 and 12wt% and Zn: Mg ratios of 1.5 and 5.5 were produced. Copper was held constant at about 0.9wt%. Thermal analysis, x-ray diffraction, and scanning electron microscopy (SEM) were used to determine the solidus, liquids and time at solidus, and to identify the amount, composition and crystal structure of the as-cast intermetallics in the alloys. Thermodynamic modeling, via a CALPHAD method, was also used to predict these parameters. SEM and x-ray diffraction traces revealed that the intermetallics in the alloys were η and T. The volume fraction of these phases increased with zinc content and decreased with Zn:Mg ratio. The results predicted by calculation showed excellent agreement with the identification and amount of phases present. Calculation also accurately predicted the composition of the T phase while there was greater difference between the predicted and measured composition of the η phase. A comparison between the calculated and measured liquidus and solidus and time at solidus showed that the values followed similar trends and the closest predictions were of the liquidus temperatures.

Introduction

The wrought 7xxx series aluminum alloys are some of the highest strength aluminum alloys. Historically, cast alloys with similar compositions have not been produced because their long solidification ranges decrease castability and intermetallics form during solidification that reduce ductility. The increased need for high strength, lightweight materials is driving the development of new cast aluminum alloys. Developments in casting and simulation techniques make the production of these cast alloys possible. An understanding of the intermetallic formation is critical since the intermetallics are not broken up during processing.

Recent studies have determined the equilibrium secondary phases that form in 7xxx series aluminum alloys [1-6]. The intermetallics that can occur are: η, T, and S. The η phase ranges from $MgZn_2$ to CuMgAl and is generally described as $Mg(Al,Cu,Zn)_2$. The T phase can vary from $Mg_3Zn_3Al_2$ to $CuMg_4Al_6$, and may be identified as $Mg_3(Al,Cu,Zn)_5$. The S phase is identified as Al_2CuMg and has a narrow composition range [7]. The formation of these phases is strongly dependent on the alloying elements. A high Zn:Mg ratio favors the precipitation of η. As this ratio decreases, the volume fraction of T over η significantly increases [1]. The amount of S phase increases with Cu content [8].

The overall properties of Al-Zn-Mg-Cu alloys are controlled by the resulting microstructure. Prediction of phase equilibrium and properties of these multicomponent alloys is complex compared to binary and ternary systems. Thermodynamic modeling and simulation using software packages such as JMatPro has been widely applied to various complex alloys including aluminum alloys [1,3,8]. These software packages, which are based on the CALPHAD (Calculation of Phase Diagram) method, use theoretical models and thermodynamic database to predict phase stability and optimize properties.

The objective of this research is to validate the reliability of thermodynamic calculations and simulations comparing its predictions to the microstructural analysis of the experimental alloys.

Experimental Procedure

Six Al-Zn-Mg-Cu wedge castings were produced with the compositions shown in Table 1 [9]. The wedge ranged in thickness from 2.5 to 5 cm over a 20 cm length. Zinc concentrations of 8 and 12wt% and Zn:Mg ratios of 1.5 to 5.5 were tested. Copper was held approximately constant at about 0.6-1%. Twenty-one pound heats of each alloy were melted in a resistance furnace and pouring was done between 730-760°C (1350-1400°F). The castings were solidified using a pressure vessel pressurized at 1 MPa (10 atm). Thermal analysis cups and spectrometer samples were poured from each heat. The solidus and liquidus temperatures were determined from the cooling curves.

Table I. Chemical Compositions of Alloys Produced

Alloy	Zn (wt%)	Mg (wt%)	Cu (wt%)
8Zn-5Mg	7.9	4.9	0.9
8Zn-2Mg	8.1	2.1	0.9
8Zn-1.4Mg	8.2	1.4	0.9
13Zn-7Mg	13.1	7.3	1.0
13Zn-3Mg	12.5	3.4	0.8
12Zn-2Mg	12.4	2.0	.63

JMatPro Version 6.2.1 was used to simulate the solidification of the experimental aluminum alloys. Thermodynamic analysis was performed using the actual composition of each alloy shown in Table 1 with the assumption that the amounts of additional elements were negligible. The starting temperature was at 730°C with a step temperature of 1°C. Final solidification was assumed to occur at 0.03 wt% liquid cutoff. The secondary phases present and their corresponding amount and composition were determined. The solidus and liquidus temperatures for all chemistries were also noted. Using a cooling rate of 1.5°C/s, the time spent at the solidus temperature was estimated on the calculated cooling curve.

The wedge castings were sectioned and samples were removed approximately 4.8 cm from the thinnest portion of the wedge to validate the calculated results. To identify the phases found, x-ray diffraction (XRD) was performed on the samples using a Siemens D-500 Diffractometer with Cu radiation and a graphite monochrometer. Scans were conducted at 40kV and 30mA between 2θ of 30° and 90° using a step size of 0.02° and a dwell time of 29 seconds at each step.

The as-cast microstructure was characterized with an FEI FEG-650 SEM Scanning Electron Microscope (SEM) equipped with an EDAX TEAM SDD 30mm^2 energy dispersive x-ray detector (EDS). The accelerating voltage used was 15 kV. EDS scans were performed on the phases in the samples to obtain their approximate compositions. Standardless analysis was used to calculate the compositions. Image analysis was used to quantify the volume fraction of intermetallic phases that formed in the alloys. Measurement was done at 2000X magnification on backscatter electron SEM images using ImagePro Plus image analysis software. Between 50 and 200 images were acquired from each sample to ensure at a coefficient of variation of 5% or less.

Results and Discussion

The intermetallic phases identified by x-ray diffraction and predicted are listed in Table II. The results agreed for all alloys with the exception of the 8Zn-2Mg alloy. Both η and T phase were predicted to form in this alloy but only η was identified by x-ray diffraction. These results are expected as the amount of T phase predicted was very low at 0.6 volume % and would not have been resolved with x-ray diffraction.

Representative backscatter electron SEM images of the as-cast microstructure of the alloy with the lowest and highest alloying additions (8.2Zn-1.4Mg and 13.1Zn-7.3Mg) are shown in Figure 1. All chemistries exhibited some larger intermetallics in both eutectic and non-eutectic forms at the grain boundaries. Adjacent to the grain boundaries there was a precipitate-free zone followed by small second phase platelets in most samples.

The measured and calculated quantity of the intermetallics for each alloy is given in Table II. The measured values were determined by image analysis and the small platelets phases were excluded since accurate measurements could not be performed. The amount of secondary phases decreased with increasing Zn:Mg ratio and increased with zinc content. The experimental data was lower than the values predicted and Figure 2 shows the correlation between the two data sets (error bars are 95% confidence intervals). The linear relationship between them shows an offset between the experimental and calculated values of about 1 vol%. This could be due in part to the exclusion of the small platelet phases in the measurement results.

Table II. Identification and Volume Percent of Intermetallic Phases Measured Experimentally and Predicted by JMatPro

Alloy	Zn:Mg	Phases Present		Total Vol% of Phases	
		X-Ray Diffraction	Calculated	Experimental	Calculated
8Zn-5Mg	1.61	T	T	6.25	7.71
8Zn-2Mg	3.86	η	η T	1.32	3.20
8Zn-1.4Mg	5.86	η	η	0.78	1.84
13Zn-7Mg	1.79	T	T	13.58	15.00
13Zn-3Mg	3.68	η T	η T	5.22	6.53
12Zn-2Mg	6.20	η	η	1.95	3.25

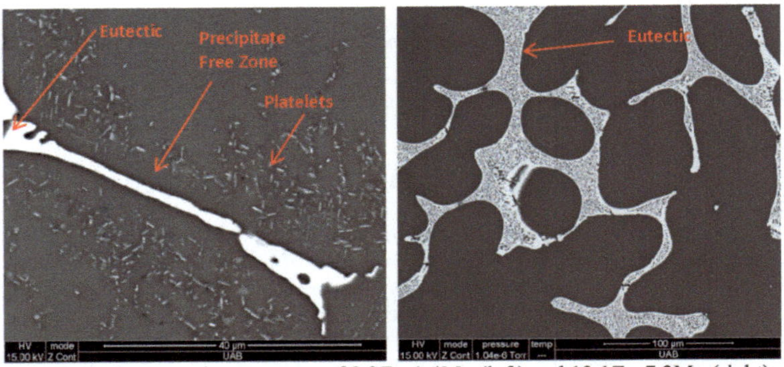

Figure 1. As-cast microstructure of 8.2Zn-1.4Mg (left) and 13.1Zn-7.3Mg (right).

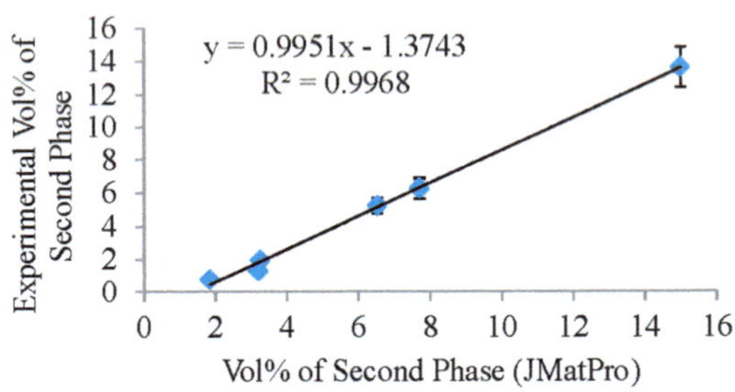

Figure 2. Experimental vs theoretical vol% of second phase in each alloy

The large intermetallics consisted of Al, Zn, Mg, and Cu. The compositions of the phases were determined using EDS spectra and the experimental (italics) and calculated results are shown in Table III. Because of emission volume issues and the standardless analysis, some error in the experimental results is expected. The error will be larger when the second phase is smaller. A difference of about 3 wt% or less would be acceptable because of these limitations and because of local variations in composition.

The measured composition of the T phase was very close to the predicted values and the results were closer at higher alloying levels where the size of the second phase was larger. There was a significant difference in the predicted and measured compositions of the η phase. The measured aluminum, and copper concentration were higher than predicted while the zinc concentration was lower. The measured and predicted magnesium concentrations were close. The greatest difference between the predicted and measured value occurred in the zinc concentration and this ranged from 6 to 16 wt% with larger differences at smaller alloying levels. The difference in copper ranged from 1 to 8% and in aluminum from 2 to 6 wt%.

The calculated and measured liquidus, solidus and time at solidus are summarized in Table IV. Representative cooling curves have been published [9]. The difference between the calculated and experimental liquidus temperatures ranged from 0 to 8 °C. The calculated solidus temperatures, T_s, were dependent on the solidification cut-off set in JMatPro. For this set of alloys, the difference between the experimental and calculated solidus ranged from 5 to 16°C.

The calculated time at the solidus was significantly dependent on cooling rate. The deviation between the predicted and experimental results could be due to the variation between the solidification conditions during the actual casting process and the thermodynamic models used in JMatPro.

Table III. Chemical Composition of Predicted and Identified Phases

Alloy		Phase	Wt% of Elements in Each Phase			
			Al	Zn	Mg	Cu
8Zn-5Mg	Calculated	T	28	40	26	7
	Measured (EDS)	*T*	*25*	*42*	*24*	*9*
8Zn-2Mg	Calculated	η	11	57	18	15
	Measured (EDS)	η	14	46	21	19
8Zn-1.4Mg	Calculated	η	8	59	17	15
	Measured (EDS)	*η*	*14*	*43*	*20*	*23*
13Zn-7Mg	Calculated	T	27	44	26	3
	Measured (EDS)	*T*	*27*	*44*	*26*	*4*
13Zn-3Mg	Calculated	T	18	54	23	5
	Measured (EDS)	*T*	*18*	*53*	*21*	*7*
12Zn-2Mg	Calculated	η	6	71	17	7
	Measured (EDS)	*η*	*8*	*65*	*19*	*8*

Table IV. Comparison of Experimental Thermal Data for Aluminum Alloys and Data Predicted with Solidification Cut-off of 0.03wt% Liquid and Cooling Rate of 1.5°C/s

Alloy	Liquidus (T_L), °C		Solidus (T_s), °C		Time at Solidus, s	
	Calculated	Measured	Calculated	Measured	Calculated	Measured
8Zn-5Mg	620	618	484	472	46	30
8Zn-2Mg	634	629	477	467	23	10
8Zn-Mg	637	637	473	457	27	2
13Zn-7Mg	598	590	478	473	83	89
13Zn-3Mg	620	626	477	466	45	26
12Zn-2Mg	627	624	476	463	38	16

Summary and Conclusions

Using x-ray diffraction and SEM, the secondary phases in the aluminum alloys were identified and their amount and composition were measured. SEM and x-ray diffraction traces revealed that the main intermetallics in the alloys examined were η and T. The volume fraction of these phases increased with zinc content and decreased with Zn:Mg ratio. The results predicted showed excellent agreement with the identification and amount of phases present. Calculation also accurately predicted the composition of the T phase while there was greater difference between the predicted and measured composition of the η phase. A comparison between the calculated and measured liquidus and solidus and time at solidus showed that the measured and predicted values followed the same trends and the closest predictions were of the liquids temperatures. The differences between the theoretical and experimental thermal data could be due in part to the deviation of actual casting conditions to theoretical models and calculations incorporated in JMatPro.

Acknowledgements

"Research was sponsored by the US Army Contracting Command Joint Munitions & Lethality Contracting Center, Picatinny, NJ on behalf of Benet Laboratories, and was accomplished under Cooperative Agreement Number W15QKN-10-2-0001. The views and conclusions contained in this document are those of the authors and should not be interpreted as representing the official policies, either expressed or implied, of Benet Laboratories or the U.S. Government. The U.S. Government is authorized to reproduce and distribute reprints for Government purposes notwithstanding any copyright notation heron."

References

1. J.-J. Yu, X.-M. Li, and X.-Q. Yu, "Thermodynamic Analyse on Equilibrium Precipitation Phases and Composition Design of Al-Zn-Mg-Cu Alloys," *Journal of Shanghai Jiaotong University (Science)*, 17 (3) (2012), 286-290.

2. S. T. Lim, I. S. Eun, and S. W. Nam, "Control of Equilibrium Phases (M, T, S) in the Modified Aluminum Alloy 7175 for Thick Forging Applications," *Mater. Trans.*, 44 (1) (2003), 181-187.

3. J-J. Yu and X-M. Li, "Modelling of the Precipitated Phases and Properties of Al-Zn-Mg-Cu Alloys," *Journal of Phase Equilibria and Diffusion*, 32 (4) (2011), 350-360.

4. C. Mondal and A. K. Mukhopadhyay, "On the Nature of T($Al_2Mg_3Zn_3$) and S(Al_2CuMg) Phases Present in As-Cast and Annealed 7055 Aluminum Alloy," *Mater. Sci. Eng. A*, 391 (2005), 367-376.

5. Y. li, P. Li, G. Zhao, X. Liu, and J. Cui, "The Constituents in Al-10Zn-2.5Mg-2.5Cu Aluminum Alloy," *Mater. Sci. Eng. A*, 397 (2005), 204-208.

6. X.-M. Li and M. J. Starink, "Effect of Compositional Variations on Characteristics of Coarse Intermetallic Particles in Overaged 7000 Aluminium Alloys," *Mater. Sci. Tech.*, 17 (2001), 1324-1328.

7. John E. Hatch, ed., Aluminum: Properties and Physical Metallurgy (Ohio: American Society for Metals, 1983), 51-52, 79.

8. N. Saunders, "The Application of Calculated Phase Equilibrium to Multi-Component Aluminum Alloys," *J. JILM*, 51 (2001), 141-150.

9. E. Druschitz, R.D. Foley, J.A. Griffin, "High Strength Cast Al-Zn-Cu-Mg Aluminum", Transactions of the American Foundry Society, Paper 13-1568, 2013.

2nd World Congress on Integrated Computational Materials Engineering
Edited by: Mei Li, Carelyn Campbell, Katsuyo Thornton, Elizabeth Holm, and Peter Gumbsch
TMS (The Minerals, Metals & Materials Society), 2013

THE SIMULATION AS PREDICTION TOOL TO DETERMINE THE METHOD OF RISER CALCULATION MORE EFFICIENT

Suarez Lisca, L. H., Coello Machado, N. I.

Universidad Central "Marta Abreu" de Las Villas
Carretera a Camajuaní km 5 ½; Santa Clara, Villa Clara, 54830, Cuba

Keywords: Simulation casting, quality, risers design

Abstract

The riser must be adequate to satisfy the liquid and solidification shrinkage requirements of the casting. In addition, the riser itself will be solidifying, so the total shrinkage requirement to be met will be for the riser/casting combination. The total feeding requirement will depend on the specific alloy, the amount of superheat, the casting geometry, and the molding medium. The shape of a casting will affect the size of the riser needed to meet its feed requirements for the obvious reason that the longer the casting takes to solidify, the longer the riser must maintain a reservoir of liquid metal. A variety of methods have been devised to calculate the riser size (shape factor method, geometric method, the modulus method) needed to ensure that liquid feed metal will be available for as long as the solidifying casting requires. In this research has been calculated the riser geometry by different methods for a piece type wheel and the simulation has been used to determine which of the methods it is more efficient.

Introduction

The consequences of a low casting yield are well known in the foundry industry: lower profits due to increased production costs and decreased capacity. Additional melted metal and heats, as well as the increased labor and materials costs required for production, are the primary reasons for the increased costs. Furthermore, it is recognized that a higher casting yield has the side benefit of lowering casting cleaning costs. Essentially, when a foundry achieves the highest possible yield, it can operate at maximum capacity, maximizing its revenues[1]. Casting is a very simple process; apparently, melting, pouring and made the mold is only, but each one has other processes more complex. Green sand casting process involves many process parameters which affect the quality of the casting produced. The rigorous control of the scrap, the use of molding sands appropriate, the correct production of the molds, the good design of the gates and risers systems, among other, they are some of the processes that it is necessary to keep in mind to achieve a maximum quality of the cast pieces. If to that said previously it is added that the optimization of anyone of them, choosing in those cases where it exists more than a way for their determination the best and more economic, it guarantees that the process is efficient and it can guarantee himself the time that delays a product in arriving again at the market is smaller. The main objective of this paper is improvement the efficiency of the sand casting process; to guarantee that it has been divided in three steps: first, to determine the main factors that influence in the final quality of the product, second, to carry out an analysis using the simulation for determines the action of the main factor and to optimize the results obtained in the simulation.

A survey applied experts in foundry and that reflects the main parameters that affect the final quality of the cast piece (see figure 1), provides an evaluation and a vision of those that more affect the productive process, one of them is the pouring process and the design the gate and riser system, the results can see in the figure 2. In these graphics two ways of analysis of the results are presented, the average of points confers by the experts in each variable and the frequency with which a certain valuation appears for the certain ones as more influential.

	Expert 1	Expert 2	Expert 3	Expert 4	Expert 5
Design of Foundry technology					
Personal qualification	9	5	9	8	4
Method applied to obtain the solution	9	1	9	6	4
Preparación de la mezcla de moldeo y machos					
The use of the right sand	9	10	8	9	8
The use of the right clay	9	10	9	9	8
Additives	9	1	9	9	8
Right control in laboratory	10	5	9	8	6
Personal qualification	8	6	7	7	9
Preparacion de modelos y cajas de macho					
Quality of the flask	8	1	8	4	8
Quality of the tools for molding	8	5	7	6	7
Molding method	8	5	7	7	7
Personal qualification	9	5	6	8	8

Figure 1 Part of the survey applied to experts

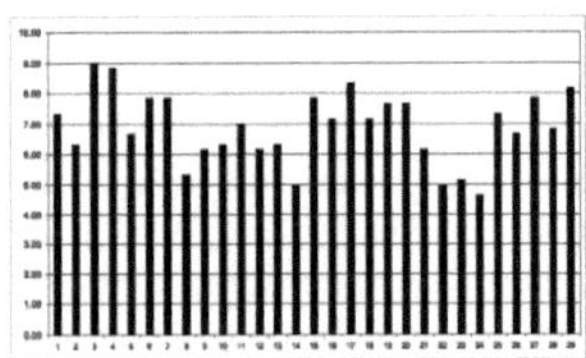

Figure 2 Average of points confer by the experts in each variable

The gating and risering systems as one of the main factors for their effect in the quality of the casting is identified, a brief study is pointed out on the same ones and it is determined for wheel of AISI 1045 which is the effective feeding distance depending on the geometric characteristics of the cast.

Effect of the riser geometry on the final quality

The risers are elements that compose the feeding system, they are metal reservoirs which feed molten metal into the casting as it freezes and shrinks and the main principle for designing the riser size is the directional solidification (DS), as a consequence of the requirement of the shrink phenomena and of contraction in the liquid; therefore, any quantitative method to be used in the sizing of the feeding system with intentions of controlling their quality, it should be based in two approaches of metallurgical nature that are:

1. The value of the solidification shrinks and of the shrink in the liquid state, magnitudes that depend on the composition of the metal or alloy; pouring temperature and the mold strength.
2. Considerations of some metallurgists aspects:
    - The solidification front.
    - The space distribution of the temperature,
    - The gating system pressure,
    - The alloy of the molten metal,

- The number of risers that is necessary,

The good design and the optimum quantity of risers to use guarantee furthermore a good sanity of the piece, the elimination of an additional metal that bears the current generation or the additional expense of fossil fuels. Although no simple principle exists for the design of feeding systems, extensive studies have enabled a quantitative approach to be developed to supplement the improvised techniques traditionally employed. These studies, mainly carried out in the field of steel castings, have coupled theory with the systematic collection of experimental data. Notable contributions include those of Heuvers [2], Chvorinov, Briggs, Caine, Wlodawer and groups at M.I.T. and the United States Naval Research Laboratory and others how Beckermann in the field of feeding-distance rules for steel cast [3]. The size and shape of each feeder head must satisfy two requirements. Firstly, the head must freeze sufficiently slowly to ensure that liquid metal will be available throughout the freezing of the section to be fed, so enabling directional solidification to continue from the casting into the head. Secondly, the head must be capable of supplying a sufficient volume of liquid to compensate for liquid and solidification shrinkage. Any resulting pipe must remain clear of the casting. These two requirements may be referred to respectively as the *freezing time* and *volume feed capacity* criteria. The shape of a feeder head is not, however, determined solely by the need to maximize freezing time. Other factors include the timing of the demand for feed metal, affecting the shape of the shrinkage cavity in the head, and the permissible area of junction with the casting: this should be as small as possible to minimize fettling costs. To check the utility of a method or other, we take a piece of steel, type wheel, and the risers are calculated. In the table 1 the results of the calculations are shown, using the different methods.

Table 1 Data of the risering dimension calculate for different ways

Method	Dr (mm)	Hr (mm)	Neck (mm)	Volume (dm^3)	Weight (kg)
Moduli method	112.5	112.5	20	1.118	8.72
Heuvers method	84	130	15	0.72	5.62
Feeding distance method	73.21	73.21	15	0.308	2.40
Proposed method	117	117	20	1.260	9.8

According to the results, it can say that the difference between methods is significant, for that reason in future works will be make an experiment to check which is the most effective and more economical of them. Maximizing casting yield, which is defined as the weight of a casting divided by the weight of the metal poured to produce the casting (*i.e.*, including metal that solidifies in the risers, gating, downsprue, *etc.*), is an important consideration in the steel casting industry [1]. An increase in casting yield decreases production costs; with increased yield, production of the same number of castings requires less melted metal and fewer heats, as well as reduced labor and material costs required for production. Also, higher yield usually has the side benefit of lower casting cleaning costs. One effective way to improve casting yield is through riser optimization, where "optimized" means (1) the riser has the minimum possible volume to provide sufficient feed metal to the casting, without the riser pipe extending into the casting; and (2) the smallest number of risers are used, while still ensuring that the risers are close enough to each other to produce a sufficiently sound casting. Computer simulation of the casting process is becoming an indispensable tool in the effort to increase casting yield. Through the use of simulation, foundries are able to evaluate modifications to casting designs without having to actually produce the casting, thus saving time, material resources, and manpower. However, computer simulation must be applied on a case-by-case basis, and its effective use requires expertise as well as accurate data for many process variables [4].

Due to these limitations, risering rules are still widely used in the steel casting industry. Risering rules dictate riser size and placement by determining (1) the riser size necessary to supply adequate feed metal to a casting section, and (2) the feeding distance (FD), which is the maximum distance over which a riser can supply feed metal to produce a sound casting. According to Beckermann [1] different survey indicates that simulation is used for less than 10 pct of the tonnage of steel castings produced, and that risering rules (or rules-based software) are used to rig about 80 pct of the tonnage produced.[1,2] Due to the prevalence of rules-based rigging in the steel casting industry, any attempt to increase casting yield in a general sense must begin with these rules. Even if simulation is used, risering rules are still useful to develop a reasonable starting point for simulation, which will shorten the iterative optimization cycle.

Different ways to calculate the geometry of the riser

Casting is an important industrial process for manufacturing near net-shaped products such as engine blocks, crankshafts, and turbine blades. Among various shape casting methods, sand casting is very common, and more than 80% of casting parts is produced by this method [5]. Generally, production of cast parts has two design stages: product design and process design. The product design is performed by design engineers based on the expected service requirements. In this context, topology optimization (Harzheim and Graf 2005, 2006) is common to achieve the optimal design. The process design stage then is performed by manufacturing engineers to produce sound castings that satisfy the a priori defined criteria for filling and solidification defects, mechanical properties, dimensional tolerance, residual stresses, and production cost. Riser design is one of the main parts of the casting process design. When molten metal enters into the mold cavity, its heat is transferred through the mold wall. In the case of pure metals and eutectics (narrow bound solidifications), the solidification proceeds layer by layer (like onion shells) starting from the mold wall and proceeding inward. The moving isothermal interface between the liquid and the solid region is called the solidification front. As the front solidifies, it contracts in volume (density of solid is usually larger than the liquid) and draws molten metal from the adjacent liquid layer. When the solidification front reaches the central hot spot, there is no more liquid metal left, and a void—shrinkage cavity—is formed. This is avoided by attaching an additional part to the casting, which is called riser or feeder. The riser should be designed so that the controlled progressive directional solidification is established and the volumetric shrinkage is compensated. Therefore, the optimal design of feeding system is a key to produce sound castings. The riser design not only controls remained defects in the casting, but also affects the cooling rate and so mechanical properties and thermal stresses. Furthermore, it has a considerable effect on the production cost. As mechanical properties of castings are considerably affected by process conditions, particulary riser design.

Determination of the FD

A great deal of effort has been expended to develop rules for determining riser FDs in steel castings. Many researchers have developed empirical relations for determining feeding distances in carbon and low-alloy (C&LA) steels. These rules are typically based on experimental casting trials performed in the 1950s by Bishop, Myskowski, and Pellini at the Naval Research Laboratory (NRL), as well as on similar casting trials conducted by the Steel Founders' Society of America (SFSA). An extensive review of empirical FD relations for C&LA steels is provided in previous work by Beckermann [6].

Simulation in casting process

The replacement of physical experiments with software simulations is increasingly common in many sections of the industry today. Some numerical experiments are carried out in order that optimal tooling and process parameters are selected to get products right first time-avoiding time-consuming and costly physical experimentation. Other studies aim to obtain a deeper understanding of the effect of varying process parameters (sensitivity studies) towards optimizing a process. However, numerical experiments that were based on the DOE method are rare in the open literature. The fact that workers are only just starting to consider such a combination for casting related simulations is apparent from a recent paper where the DOE method is applied to numerical simulations of aluminum permanent mold casting. The aim of their investigation was determinate what is the best method for optimal riser design in sand casting process. In summary, whilst DOE methods and the use of computer simulations are no longer new to the manufacturing industry, instances of combining the two for achieving significant increases in productivity during a problem solving exercise are relatively scarce and the effectiveness of this strategy therefore remains to be investigated. Simulation studies, when used in several areas of investigation, are quite useful to study the behavior of some phenomena in which different virtual situations are generated by the researcher using some specialized software. Robustness studies are rather common in statistic research; many of them are used to observe the behavior of an estimator under several hypothetical situations that could happen in practice [7].

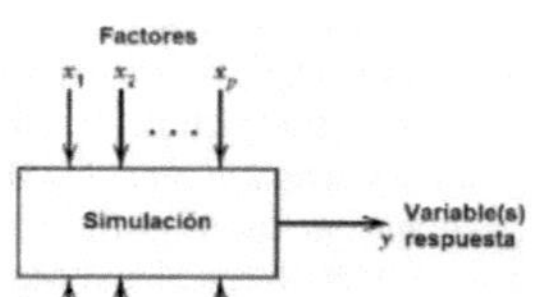

Figure 3 General scheme of a simulation study. Adapted from a scheme presented in [7]

Simulation studies, when used in several areas of investigation, are quite useful to study the behavior of some phenomena in which different virtual situations are generated by the researcher using some specialized software. Robustness studies are rather common in statistic research; many of them are used to observe the behavior of an estimator under several hypothetical situations that could happen in practice.

The risers design process optimization

In occasions with an experiment is not enough to obtain the looked for answers or the achieved levels of improvements are not enough, for what is necessary to experience in a sequential way until finding the wanted level of improvement. In the case of this experiment, a high number of variables that intervene as entrance variables exist. As one of the objectives of the work it is to determine the good conditions of the process decides to use the methodology of the answer surface, since the same one allows to find conditions of good operation of a process to improve the quality of a product. Response surface methodology, or RSM, is a collection of mathematical and statistical techniques that are useful for the modeling and analysis of problems in which a response of interest is influenced by several variables and the objective is to optimize this response. In the design of risers interview different variables, riser geometry (high, diameter), temperature of the melt, type of molding sands and overheat temperature. The answer or result waited in this case would be the action riser design about the sanity of the piece, that which influences directly on the final quality of the piece and the process.

Conclusion

1. The foundry process has a great complexity due to the great quantity of variables that intervene in the process, subjective many of them.
2. Through the survey carried out experts it can be proven that one of the main parameters that affects the quality of the pieces is the filling of the same ones, influencing the geometry and the calculate method of the risers.
3. Don't have a bibliography that demonstrates which is the calculate method of risers more effective neither economic.
4. Through the consulted literature it is evidenced that the simulation like prediction tool is a way to decrease costs, the time that delays a product in arriving again at the market and the quality of the casting.
5. For the optimization of the process the methodology RSM is the better way to guarantee the optimum point of the risers design process.

References

1. Beckermann, C., O. Shouzhu, and K.D. Carlson, *Feeding and risering of high-alloy steel castings.* Metallurgical and materials transactions B, 2005. **36B**(97).
2. Wloadawer, R., *Directional Solidification of Steel Castings* ed. P. Press. 1966.
3. Beckermann, C., et al., *Development of New Feeding-Distance Rules Using Casting Simulation: Part I. Methodology.* Metallurgical and Materials transactions B, 2002. **33B**.
4. Tavakoli, R. and P. Davami, *Optimal riser design in sand casting process by topology optimization with SIMP method I: Poisson approximation of nonlinear heat transfer equation.* Struct Multidisc Optim, 2008. **36**: p. 193-202.
5. Gunasegaram, D.R. and B.J. Smith. *MAGMASOFT helps assure quality in a progressive Australian iron foundry.* in *32nd National Convention of the Australian Foundry Institute.* 2001. Fremantle, Australia.
6. Coello Machado, N.I. and L.H. Suárez Lisca. *Casting parts quality improvement using forecast and control process parameters.* in *5th International doctoral students workshop on logistics.* 2012. Magdeburg, Germany: Prof. Dr. -Ing. habil. Prof. E. h. Dr. h. c. mult. Michael Schenk.
7. MONTGOMERY, D., *Design and analysis of experiments. Fifth Edition.* 1997, New York: John Wiley & Sons, Inc.

2nd World Congress on Integrated Computational Materials Engineering
Edited by: Mei Li, Carelyn Campbell, Katsuyo Thornton, Elizabeth Holm, and Peter Gumbsch
TMS (The Minerals, Metals & Materials Society), 2013

MULTI-OBJECTIVE OPTIMIZATION OF MICROSTRUCTURE IN WROUGHT MAGNESIUM ALLOYS

B. Radhakrishnan, S.B. Gorti and S. Simunovic

Oak Ridge National Laboratory
Oak Ridge, TN 37831, USA

Keywords: twin spacing, texture, crystal plasticity, multi-objective optimization

Abstract

The microstructural features that govern the mechanical properties of wrought magnesium alloys include grain size, crystallographic texture, and twinning. Several processes based on shear deformation have been developed that promote grain refinement, weakening of the basal texture, as well as the shift of the peak intensity away from the center of the basal pole figure - features that promote room temperature ductility in Mg alloys. At ORNL, we are currently exploring the concept of introducing nano-twins within sub-micron grains as a possible mechanism for simultaneously improving strength and ductility by exploiting a potential dislocation glide along the twin-matrix interface – a mechanism that was originally proposed for face-centered cubic materials. Specifically, we have developed an integrated modeling and optimization framework in order to identify the combinations of grain size, texture and twin spacing that can maximize strength-ductility combinations. A micromechanical model that relates microstructure to material strength is coupled with a failure model that relates ductility to a critical shear strain and a critical hydrostatic stress. The micro-mechanical model is combined with an optimization tool based on genetic algorithm. A multi-objective optimization technique is used to explore the strength-ductility space in a systematic fashion and identify optimum combinations of the microstructural parameters that will simultaneously maximize the strength-ductility in the alloy.

Introduction

The poor room temperature formability of magnesium compared to steel and aluminum alloys is attributed to the hexagonal close packed (HCP) crystal structure that allows easy dislocation slip only along the close packed basal plane [1]. Room temperature deformation is largely accommodated by the operation of basal slip and twinning on $(10\bar{1}2)$ planes, commonly known as "tension" or "extension" twins. However, the operation of tension twins during sheet forming operations results in the reorientation of the grains such that the c –axis (the long axis of the HCP unit cell) is parallel to the sheet thickness and produces the so-called "basal" texture. The sheet cannot accommodate further thickness reduction because both the basal slip and the tension twin mechanisms are essentially shut down. Typically improved room temperature ductility in wrought magnesium alloys is obtained by alloying with rare-earth (RE) elements that results in significant grain refinement and the formation of weak textures that promote the activation of dislocation glide in the prism and pyramidal planes [2]. The ease of formation of compression twins along the $(10\bar{1}1)$ planes is considered to be one of the key mechanisms of texture weakening in RE containing alloys. However, there is a recent push for developing RE-free magnesium alloys because of the uncertainty in the global RE market.

In metallic polycrystals there is generally an inverse relationship between strength and ductility. In engineering materials, the conventional approach for simultaneously increasing the strength and ductility is to refine the grain size. However, below a limiting grain size where significant dislocation activity within the grain cannot be supported, the strength again has an inverse relationship with ductility. Recent research has shown that the introduction of nanotwins within ultra-fine grains can lead to significant enhancement in strength without loss of ductility [3]. The strength level is dictated by the mean spacing between the twins and the ductility is determined by the ability of the twin-matrix interface to allow dislocations to glide. There has been recent activity in trying to extend the above nano-twinning concept for FCC copper to hexagonal close packed (HCP) magnesium alloys. Unfortunately, the main challenge in HCP alloys is that below a certain grain size (roughly 3-4 μm for Mg), twinning as a major deformation mode practically ceases, and deformation occurs almost entirely by dislocation slip [4]. Although alloying is a potential solution, finding the right solute to promote twinning has been a challenge. Recent experiments using ball milling have provided crucial evidence for mechanical alloying of magnesium with up to 10% titanium that leads to twinning inside nanoscale grains [5]. It is significant to note that the twins produced in the Mg-Ti alloy are the compression twins that are produced in the RE-containing alloys. The twin plane, $(10\bar{1}1)$ also coincides with the pyramidal slip planes and therefore there is a potential for dislocation glide along the twin plane.

In recent years, several shear deformation processes have been used to reduce the grain size and weaken the basal texture in wrought magnesium alloys in order to improve the room temperature ductility. The shear textures that result from these processes also shift the peak of the basal pole by a rotation about the transverse direction (TD). Therefore, the microstructural variables that are used in the multi-objective optimization include the texture spread, texture rotation and the twin spacing. The objective is to find the optimum combination of the microstructural properties that maximizes the strength and ductility in the presence of the twin-matrix glide mechanism described above.

Approach

The computational approach consists of coupling a crystal plasticity based finite element (CPFE) code with an optimization routine. The crystal plasticity theory is used to formulate an elasto-viscoplastic constitutive model based on anisotropic elasticity and dislocation slip on the basal, pyramidal and prismatic slip systems that operate in HCP magnesium. The grains are assumed to contain a collection of compression twins along the $(10\bar{1}1)$ planes that are produced by prior deformation processing as described in the previous section. The objective of the optimization simulations is to find improved combinations of strength and ductility for such a microstructure assuming that the mechanical response is governed essentially by dislocation slip on the usual slip planes. However, the slip systems are now classified into two different types. Slip systems that are inclined to the $(10\bar{1}1)$ planes are considered hard modes because of the nanoscale confinement of the matrix regions between the twins, and the slip systems parallel to the $(10\bar{1}1)$ planes are considered soft modes. The critical resolved shear stresses for initiating slip along the hard modes are obtained from the Hall-Petch data for magnesium and the appropriate twin spacing. The critical resolved shear stress for the soft modes are obtained from the Hall-Petch data and the appropriate grain size. The constitutive model is implemented in a finite element framework with grain level discretization. The details of the CPFE model are described elsewhere [6]. In addition to computing the stress-strain behavior of the polycrystal, a failure model is built into the analysis in order to define the strength-ductility combination for a specific microstructure. The failure criterion is based on a critical value of the tensile hydrostatic stress and shear strain. The critical shear strain concept is based on Jerusalam et al [7] for face centered cubic copper. However, the implementation of the criterion leads to a monotonic increase in ductility with reduced twin spacing. Since the operating hydrostatic tensile stress also scales inversely with twin spacing and the material may fail by interphase separation above a critical stress, the current study includes the additional stress criterion. The critical stress and shear strain values are not known presently. However, they can be calibrated using experimental data as and when they become available. Currently, there is an ongoing experimental effort at ORNL to develop Mg-Al-Zn-Ti alloys using severe plastic deformation approaches [8].

The variables used in the optimization technique are the spread and the rotation of the basal poles describing the crystallographic texture and the twin spacing. The texture is varied from strong basal where all the grains are along the [0001] direction to a weak texture where the orientation of each grain deviates from [0001] by a specified angle that can be systematically varied. The rotation of the basal pole about the transverse direction is also varied systematically from 0 to 45 degrees. The optimization task proceeds as follows. An initial population based on a random selection of the texture-spread parameter, texture-rotation parameter and the twin spacing is chosen. The population is used to spawn separate CPFE runs. In the current study, the initial population consisted of 10 individuals. Each individual was run using a computational domain of size 20x20x20 cubic elements. Each element is considered a "grain" that has a specific crystallographic orientation assigned using the generated grain orientations and twins with given spacing. At the end of the run, the mean strength and mean ductility values are calculated for the population. Subsequently, individuals whose strength-ductility product exceeds the mean strength-mean ductility product of the population are retained for the next generation run. The ones that fail this criterion are reassigned by randomly picking from one of the successful individuals. The second generation proceeds after a random reassignment of the texture spread, texture rotation and the twin spacing in the vicinity of the original values. In the current simulations a $\pm$ 20% variation in the values was used to reassign the texture and the twin spacing. The runs are continued in a similar fashion for several generations. The results shown in the current study are after 10 generations.

Results and Discussion

Table 1 shows the typical texture spread, texture rotation and twin spacing values that are randomly selected in a given population and their strength and ductility levels.

Table 1. Typical Results for Runs Belonging to a Generation

Gen No	Texture Spread	Twin Spacing	Texture Rotation	Strength (Mpa)	Ductility
1	2.40E-01	5.38E+01	1.32E+01	2.31E+03	8.11E-02
2	5.08E-01	6.39E+01	2.91E+01	2.20E+03	8.88E-02
3	4.65E-01	6.32E+01	3.01E+01	2.22E+03	8.88E-02
4	1.89E-01	5.47E+01	1.09E+01	2.30E+03	8.11E-02
5	2.01E-01	5.79E+01	1.20E+01	2.29E+03	9.60E-02
6	2.17E-01	5.94E+01	1.24E+01	2.26E+03	9.36E-02
7	1.83E-01	5.65E+01	1.13E+01	2.30E+03	9.10E-02
8	2.01E-01	6.45E+01	1.14E+01	2.18E+03	8.88E-02
9	1.90E-01	5.67E+01	1.35E+01	2.30E+03	9.10E-02
10	2.12E-01	6.30E+01	1.23E+01	2.21E+03	9.12E-02

A typical (0002) pole figure showing a weak texture with spread and weak texture due to TD rotation are shown in Fig. 1.

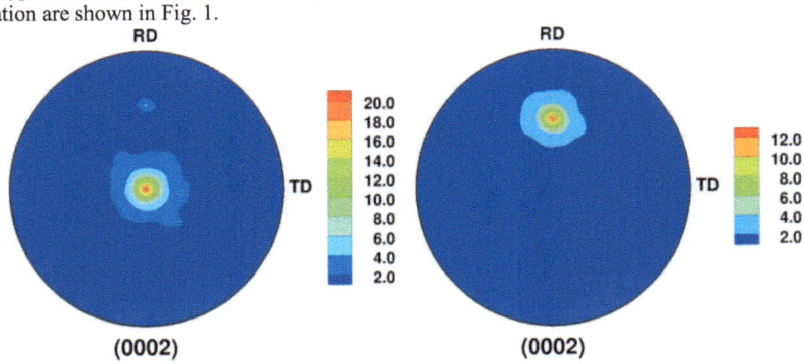

Fig. 2. Typical examples of textures with spread (left) and TD rotation with spread (right) in the form of (0002) pole figures

In magnesium alloys, the weakening of the texture due to texture spread and texture rotation about TD can result in the activation of non-basal slip systems. The stress-strain response for the 10 individuals after the generation 1 and generation 10 are shown in Figs 3. Figure 3 shows that the strength level and the strain at failure vary with the texture spread, texture rotation, and the twin spacing. It also shows that the optimization process resulted in the identification of the texture parameters and the twin spacing that optimizes the strength and ductility of the alloy.

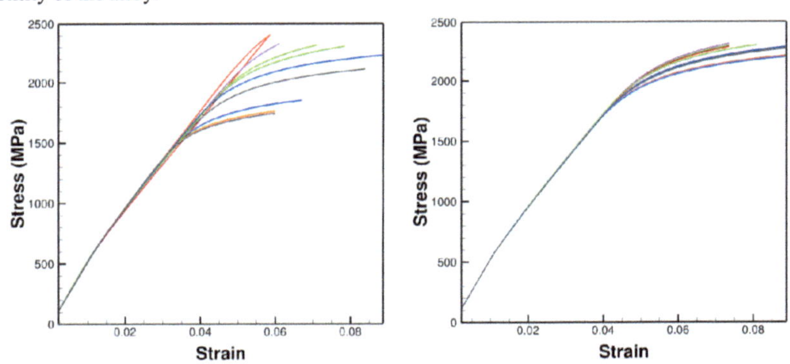

Fig. 3. Stress-strain curves after generation 1 (left) and generation 10 (right)

By comparing the curves after generation 1 and generation 10, we can see that individuals with high strength and low ductility and individuals with low strength and high ductility are eliminated from the population. The individuals that survive the generations are those with an intermediate strength and significantly higher ductility.

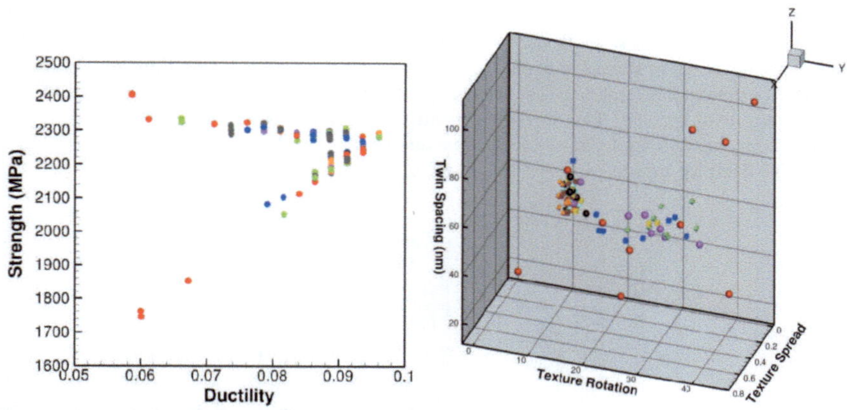

Fig. 4. Strength-ductility combinations for each run as a function of generations (left) and clustering of the microstructural variables at the optimum level (right)

Fig. 4 shows the strength-ductility combinations of the runs as a function of the generation. It is clear that as the generations progress individuals with better combinations are identified. At the end of generation 10 individuals with a strength level of roughly 2800 MPa and a ductility level of about 9.5% are identified. Fig. 6 also shows that the microstructural variables, twin spacing, texture spread, and texture rotation all cluster around the optimum values at the end of generation 10. We now examine the results in the parameter space to identify the effects of individual parameters on the strength and ductility. For this purpose, we present three-dimensional plots with strength or ductility as the dependent variable.

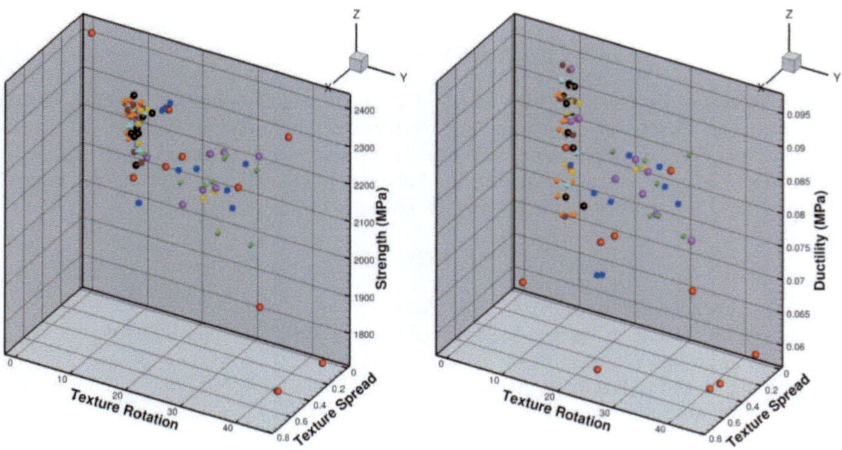

Fig. 5. Strength and ductility as a function of the texture rotation and texture spread

Fig. 5. Shows the strength and ductility of the individuals as a fuction of the texture rotation and texture apread. The high strength individuals appear to be clusterd around a texture rotation of about 10° and a texture spread of about 0.1. However, a range of ductility values are obtaind for roughly the same texture rotation and texture spread variables. Therefore, ductility is controlled more by the twin spacing. This is shown in Fig.6. where it is seen that the variation of twin spacing from about 40 nm to 50 nm results in a dramatic increase of ductility from about 6.5% to 9.5%. The ductility appears to fall on either side of this range.

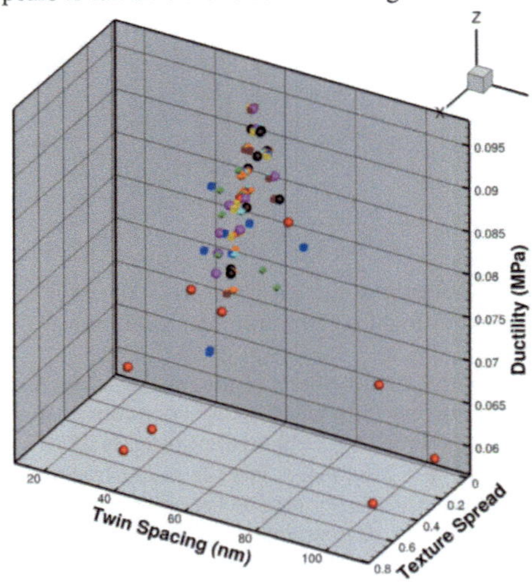

Fig. 6. The effect of twin spacing on the ductility (left) and the clustering of the microstructural parameters

Acknowledgments

Research sponsored by the Laboratory Directed Research and Development program at Oak Ridge National Laboratory, managed by UT-Battelle, LLC, under contract DE-AC05-00OR22725 for the U.S. Department of Energy. This research used resources of the Center for Computational Sciences at Oak Ridge National Laboratory, which is supported by the Office of Science of the U.S. Department of Energy under contract DE-AC05-00OR22725.

References

1. M.R. Barnett, et al. JOM, 61 (2009), pp. 19-24.
2. N. Stanford, and M. R. Barnett, Mater Sci and Eng. A., 496 (2008) pp. 399-408.
3. K. Lu, L. Lu, and S. Suresh, Science, 324 (2009) pp. 349-352.
4. Q. Yang, Q. and A. K. Ghosh, Acta materialia, 54 (2006) pp. 5159-5170.
5. K.L.Wu et al., Scripta Materialia, 64 (2011) pp. 213-216.
6. S.B. Gorti and B. Radhakrishnan, this Proceedings.
7. A. Jérusalem, M. Dao, S. Suresh and R. Radovitzky, Acta mater, 56 (2008) pp. 4647-4657.
8. Laboratory Directed Research and Development (LDRD) Program, ORNL 2012.

[2nd] World Congress on Integrated Computational Materials Engineering
Edited by: Mei Li, Carelyn Campbell, Katsuyo Thornton, Elizabeth Holm, and Peter Gumbsch
TMS (The Minerals, Metals & Materials Society), 2013

A COMPUTATIONAL FRAMEWORK FOR INTEGRATED PROCESS DESIGN OF HIGH PERFORMANCE PARTS

Kuldeep Agarwal[1], Rajiv Shivpuri[2]

[1]Department of Automotive and Manufacturing Engineering Technology,
Minnesota State University Mankato, 205 Trafton Center East, Mankato, MN 56001
[2]Integrated Systems Engineering, The Ohio State University
210 Baker Systems, 1971 Neil Avenue, Columbus,OH 43210

Keywords: Design, Process, Anomaly, Forging

Abstract

High-performance failure-critical parts such as aeroengine disks are manufactured in a sequence of processing steps such as solidification, deformation, heat treatment and finishing. The fatigue and failure performance of these parts are often governed by the material and processing state. Uncertainties in the material structure, defects and anomalies play a major role in the uncertainties in performance. This paper includes a hybrid computational framework for integrated materials and process design with achieved through model decomposition based on Bayesian probabilistic formulation. Efficacy of this approach is demonstrated by applying it to the selection of forging parameters for maximizing life of titanium disk with hard alpha anomaly.

Introduction

With increased societal reliance on complex engineered systems, their performance and failure involves significant human and environmental risk, and life cycle costs. The higher the complexity of the product and the larger the number of associated components, the higher is their probability of failure or malfunction. Failure may consist of multiple performance attributes.

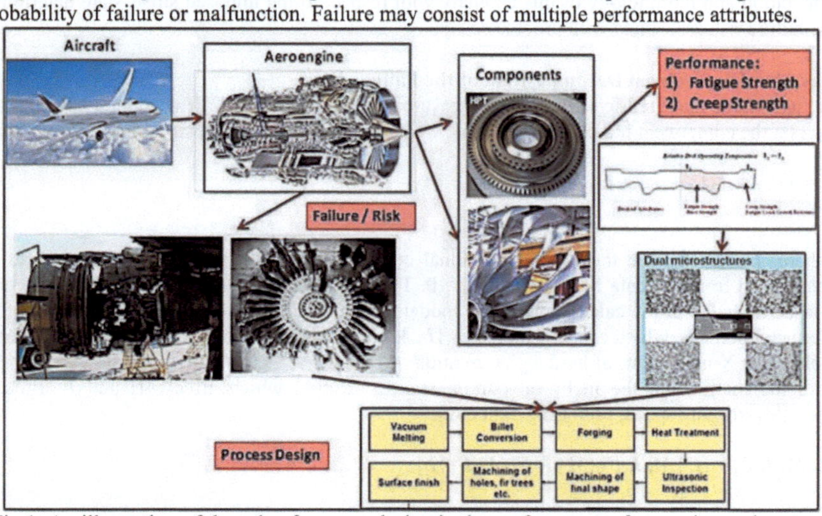

Fig 1: An illustration of the role of process design in the performance of an engineered system.

While a majority of these failures can be related to pilot error and adverse weather conditions, a significant amount of these failures are related to the failure of aircraft subcomponents such as aeroengines, landing gears, airframe etc. For example, on July 19, 1989 the DC-10 aircraft of United Airlines Flight 232 suffered an uncontained failure of its number 2 engine which caused severing of the control systems and to the crash of the aircraft during landing at Sioux City, Iowa, killing 110 of its 285 passengers and one of its 11 crew members. Accident analysis by the NTSB related this engine failure to a casting related defect in the forged disk of the engine [1, 2]. Fig. 1 shows the role played by the manufacturing process (process design) in the performance and failure of the complex aircraft system. In this, the performance of the complex system (e.g. aircraft), is determined by the performance of the sub-systems (e.g. engine), which is in turn is dependent on the performance of a component of the sub-system (such as a turbine disk), and its interactions with other components (such as blades assembled in the disc). Finally the performance (creep, fatigue, fracture etc.) of the component or a part is wholly dependent on its process of manufacture (melting, billet forging, disk forging, heat treatment, machining and inspection). Hence the performance of the complex engineered system is dependent on the manufacturing process parameters such as in forging or casting design.

In this paper, we illustrate the ICME systems approach to the material and processing based design of engineered system. As an alternative to the deterministic design approach we presents a hybrid approach consisting of probabilistic Bayesian hierarchical decomposition together with multi body computational model to assess the risk associated with an embedded discrete anomaly, and to reduce this risk through intelligent design of material processing. To determine the life of the system we first decompose the life of the disk into the multi-level description consisting of processes, processing design parameters, material and structural states and the mechanical performance attributes. The linking between the scales is obtained through acyclic Bayesian network of the decomposed state. The process design parameters (such as temperatures, velocities, preform design etc.) and the material state (anomaly motion and rotation during forge processing) is linked to the failure state via multibody computational modeling of the embedded inhomogeneity in a self-consistent titanium matrix. The location, orientation and morphology of this anomaly with respect to the imposed stress state determines the failure risk associated with process design.

Bayesian Hierarchical Decomposition of the Failure Risk

Let us assume that the life of the disk is measured and is denoted as $\boldsymbol{L}$. If we do not have the life data on the component but only on coupons, we will make the following simplifying assumption:

$$L = \alpha_1 L_1 + \alpha_2 L_2 + \ldots.. + (1 - \alpha_1 - \alpha_2 - \ldots) L_m \quad (1)$$

Where, L_1, ... L_m are the life of individual coupons; α*'s* are parameters which need to be determined from the data $\Sigma\, \alpha_i = 1$ and $\alpha_i > 0$. The role of material and its processing state can also be included in the calculation of life model through the joint probability distribution of all these components, which can be denoted by $[L, M, I, F]$ where $\boldsymbol{L}$ life; $\boldsymbol{M}$ defect state for the hard alpha (like X-movement of hard alpha, rotation of hard alpha etc.); ***I:*** internal state variables for both the matrix and the inclusion (strain, strain rate etc.) which affect ***M;*** and ***F***: Process variables (forging speed, preform height etc.).

$$[L, M, I, F] = [L \mid M, I, F]\,[M \mid I, F]\,[I \mid F]\,[F] \ldots\ldots\ldots\ldots\ldots\ldots \quad (2)$$

The aim is to find this joint distribution based on the observed data $\boldsymbol{D}$. Let us assume that the defect state is completely characterized by the location and rotation of hard alpha during the forging process. Also, let us assume that these parameters are only affected by the forging velocity and billet temperature. Thus, our posterior model can be written as:

$$[\boldsymbol{L}, \boldsymbol{M}, \boldsymbol{I}, \boldsymbol{F} \mid \boldsymbol{D}] \propto [\boldsymbol{D} \mid \boldsymbol{L}, \boldsymbol{M}, \boldsymbol{I}, \boldsymbol{F}]\,[\boldsymbol{L}, \boldsymbol{M}, \boldsymbol{I}, \boldsymbol{F}] \quad (3)$$

Where: $\boldsymbol{L}$ represents the life of the disk; $\boldsymbol{M}$ has two components representing the X-movement (X) and rotation (θ) for the hard alpha; $\boldsymbol{I}$ has two components I_M and I_H representing Strain (ε) and Strain rate ($\dot{\varepsilon}$) for the matrix and hard alpha respectively.

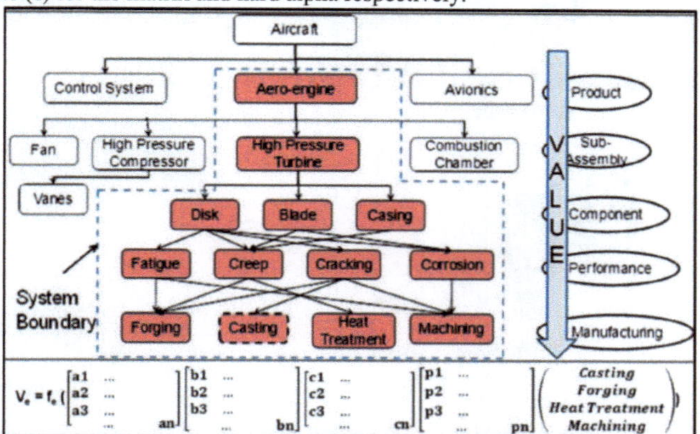

Fig 2: Hierarchical decomposition of the system into individual elements and the estimation of failure risk as a value (Ve). The challenge is to obtain the matrices that link process to design.

The Bayesian idea for hierarchical modeling is that given an ordering of variables in a conditional distribution, we can assume some conditional independence to simplify the model. Using the hierarchical approach the material processing design parameters can be linked to the disk life through the material structural state as shown in Fig 2. This multilevel causal design has material processing at the lowest level, material parameters at middle levels and life (performance) at the highest level. The value proposition (disk failure) is linked to the primitive variables of process design. In the forging process, these design primitives include preform design, forging temperature, forging speeds, contact times, post forging cooling etc. These design primitives have significant impact on the material structure and defects, which in turn affect the desired mechanical performance indicators.

Failure of the Titanium Rotor Disk due to Hard Alpha Anomaly

Analysis of years of aircraft accident data showed that the leading cause of engine related accidents for turbofan engines is the uncontained failure of a safety critical part such as an aeroengine disk [1,2]. Examining the processing history of the turbine disc, it seen that it is manufactured in a sequence of steps starting with vacuum melted (VIM/VAR) ingot that is converted using hot cogging, hot forged, heat treated, ultrasonically inspected for defects, machined to final shape, finish machined, surface finished and finally inspected before shipping. Due to the reactive nature of titanium to oxygen and nitrogen in the atmosphere, several defects or anomalies are generated during the melt processing and they evolve during subsequent deformation processing. These include macrosegregation of oxygen, beta stabilizing elements, and nitrogen-stabilized hard alpha particles (LDIs) and high-density inclusions (HDIs). The occurrence of HDI in the final product is rather rare with improved melting practices such as hearth or scull melting. The so-called low-density inclusions (LDI) - hard-alpha or alpha-I inclusions- originate in a number of different ways, but the principal source with high nitrogen contents is in the original manufacture of titanium sponge. This alpha stabilized phase (hard alpha HA) is very brittle and after mechanical working normally contains cracks and may retain some porosity from the original sponge structure, Fig 3 (left). Flow stresses for hard alpha with different nitrogen contents and that of the titanium matrix are included in Fig 3 (right).

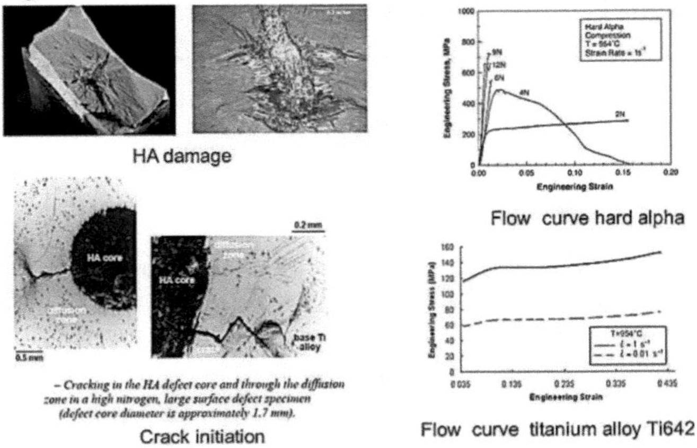

Fig 3: HA damage and cracking (left) and flow stress of HA and the Ti64 matrix [1].

Computational Model for Embedded Inhomogenieties in Titanium Matrix

While there are several approaches available to estimate the effect of embedded inclusions in a deforming matrix such as those of Eshelby, Mori-Tanaka and self-consistent method with some of these extended to rate sensitive plasticity [3,4], our objective was to accurately determine the micromechanics of the large anomalyin the evolving polycrystalline matrix (α/β titanium alloy). With this objective, a computational model for the HA anomaly embedded in the titanium matrix was developed using the multibody lagrangian description available in the FORGE software. Details of this description can be found elsewhere [5,6]. To model this defect inside the titanium matrix, we use multi-body approach of 3-D FEM in which the titanium matrix (red), the hard alpha anomaly (yellow) and the surrounding high nitrogen diffused zone (green) are treated as three separate bodies with mutual interactions. The interface between these bodies is treated as symmetric Lagrangian contact using "sliding surfaces approach." The plastic behavior of each of the three bodies is based on the constitutive behavior reported in FAA report [1]. The results of this modeling are included in Fig 4. Also in this figure are the experimental results and the results of the computational model based on DEFORM software as reported in the FAA report. It is seen that the multibody description accurately predicts the location and deformation of the hard inclusion as well as the softer diffusion zone surrounding it.

This validated model is then used in the parametric computational experiments to evaluate the effect of forging processing parameters on the micromechanics of the embedded hard inclusion. While deformation parameters in both the pancake and finish forging were varied, the results of this latter study are included in Fig 5. The processing parameters have a greater influence on the location of the defect and lesser on its orientation.

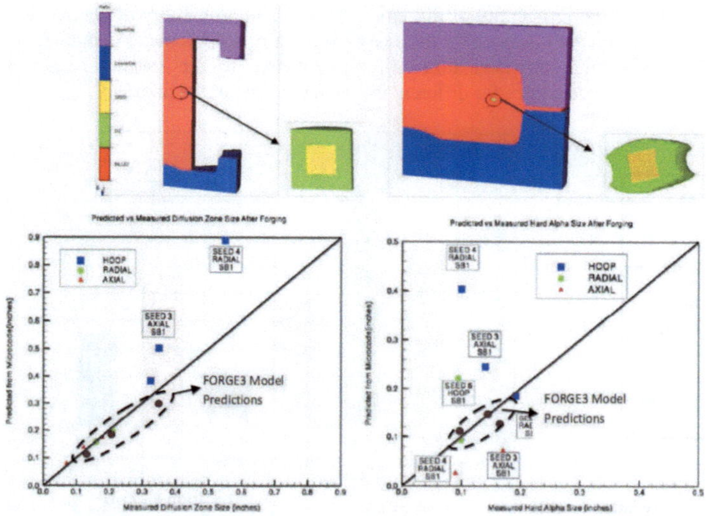

Fig 4: Validation of the embedded anomaly model using multibody slide-line approach.

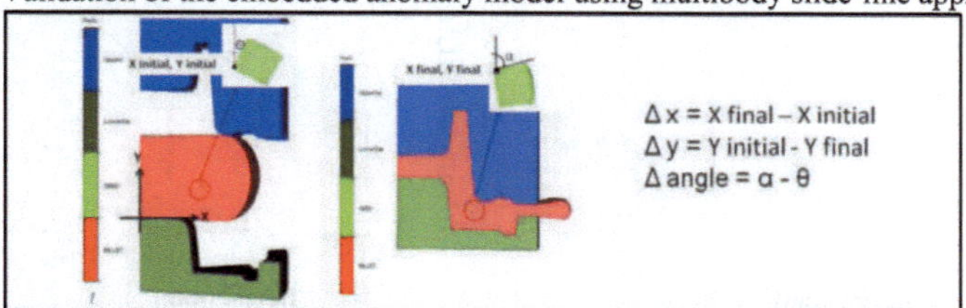

% N	Temp (°C)	Die Velocity	Δ x	Δ y	Δ angle	Deformation
2	850	5	15.4	70.5	68.5	0.4 – 0.6
2	850	70	7.3	60.21	67.5	0.7 – 0.9
2	950	5	16.28	74.88	69.3	0.2 – 0.4
2	950	70	8.1	62.43	67.1	0.5 – 0.6

Fig 5: Computational study of the relating primitive forging parameters to the inclusion kinetics.

Estimating the Processing Dependent Failure Risk

In this case study, four (4) different locations of the hard alpha in the initial billet and two different chemical compositions (2%N, 4%N) were chosen and the design was run for these locations. It is assumed that a vacuum melted billet will have only one of these defects and will have only one of the chemical compositions. The design however can be extended for various different cases. Although more complicated models can be built, for sake of simplicity "normal" models are assumed in this work. The life data used came from the FAA report [1.2] based on their experimentation and is given Fig 6. These data points also have the process conditions which were used to generate the other data. The hard alpha is assumed to be of the same size and shape in all the locations. This is based on inspection results and studies by previous researchers. The hard alpha is assumed to be a cylinder with diameter of 5.08 mm and height of 5.08 mm. These locations are identified as Loc1, Loc2, Loc3 and Loc4 in Fig 6. All the hard alphas are assumed to have an equal probability of formation = 0.003 and probability of detection = 0.5. A Bayesian network estimated was used for the design of the forging process to maximize life in all the four cases. Note that the normalized life is sensitive to the forging height, forging temperature and forging velocities for different nitrogen contents of the anomaly.

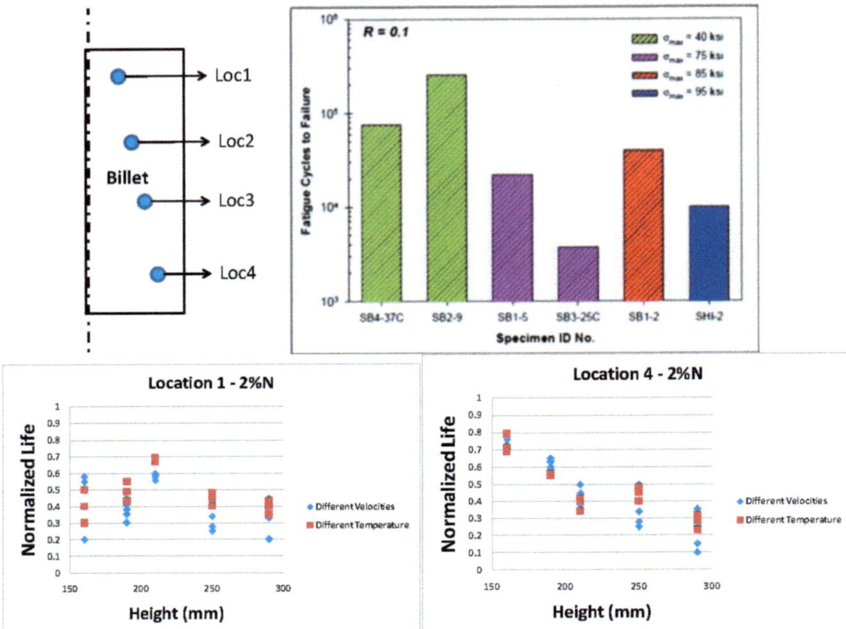

Fig6: Results of fatigue test reported in FAA report [ ] and the predicted normalized life factors.

Conclusions and Acknowledgement

In this paper a hybrid approach is presented that enables tailoring process design parameters to maximize inclusion sensitive life of an aeroengine disk. This hybrid approach consists of Bayesian decomposition of the multiscale system into hierarchical states, and a multi-body based computational framework for inclusion sensitive design. The approach is shown to adequately link the processing and material parameters to the performance attributes of the manufactured product. The authors acknowledge support from Transvalor, FORGE3 software provider, and NSF for their grant CMMI-1000330.

References

[1] Turbine Rotor Material Design, FAA Final Report DOT/FAA/AR-00/64, 2000.
[2] Turbine Rotor Material Design – Phase II, Final Report DOT/FAA/AR-07/13, 2008.
[3] Segurado J et al., Int. J. of Plasticity 28 (2012) 124-140
[4] Semiatin et al., Metall. Mater, Trans. A 33A, Aug. 2002: 2719-2727
[5] Mura T., Micromechanics of Defects in Solids, 2nd Ed., Martinus Nijhoff Publishers, 1991.
[6] Barboza J, Fourment L, EUROMECH 435- Friction and Wear in Meal Forming, 18-20 June 2002, Valenciennes, France.

2nd World Congress on Integrated Computational Materials Engineering (ICME)

Materials Data for ICME

2nd World Congress on Integrated Computational Materials Engineering
Edited by: Mei Li, Carelyn Campbell, Katsuyo Thornton, Elizabeth Holm, and Peter Gumbsch
TMS (The Minerals, Metals & Materials Society), 2013

CONSIDERATION OF ECOSYSTEM FOR ICME

Weiju Ren

Oak Ridge National Laboratory
Materials Science and Technology Division
MS-6069, Building 4515, Oak Ridge, Tennessee, 37831, USA

Keywords: database, cyber infrastructure, integrated computation, ICME

Abstract

As the Integrated Computational Materials Engineering (ICME) emerges as a hot topic, computation, experimentation, and digital database are identified as its three major components. Efforts are being actively made from various aspects to bring ICME to reality. However, many factors that would affect ICME development still remain vague. This paper is an attempt to discuss the needs for establishing a database centered ecosystem to facilitate ICME development.

Introduction

As the Integrated Computational Materials Engineering (ICME) emerges as a hot topic, computation, experimentation, and digital database are identified as its three major components. Similar to that for the Materials Genome Initiative (MGI), to bring the three components together for a successful development, integration is apparently considered as the key.

In all human efforts on materials engineering, experimentation has been employed throughout the history and computation for several decades up to date. Although data were involved all the time in experimentation and computation, database is for the first time separately identified as a major component in the game. For materials engineering, the concept of ICME has undoubtedly heralded a new and exciting era that promises accelerated development. To increase chances of success, efforts are worth making to discuss environment conducive to the integration. This paper is intended to outline an ecosystem that would facilitate development of ICME through considering major activities that should typically be conducted for a complete ICME process, the way these activities should be integrated, the role database systems should play, and some challenges and possible solutions for establishing an ICME ecosystem.

A Promise for Accelerated Materials Engineering Development

Based on words, the "Integrated Computational Materials Engineering" can be understood as materials engineering through an integrated computational approach. Engineering with materials should include materials development/selection, manufacture processing, and product design, with the goal to control the materials-process-structures-properties-design-performance chain and satisfy product performance requirements. Engineering with materials can be traced back millions of years in human history. The early progress of human civilization is often classified in terms of materials, e.g. the Stone Age (lasted roughly 3.4 million years), Copper Age, Bronze Age, and Iron Age. During these long periods of time, very slow progress was made in engineering with materials, mostly through the trial-and-error approach and accidental discoveries. This slow rate was changed when human civilization entered the era of the Industrial

Revolution. With many new tools and technologies, materials engineering was greatly enhanced and progress became much faster through a combination of knowledge-based logical reasoning and trial-and-error experimentation. Ironically, as more was learned about materials, to engineer with materials became increasingly complicated, often involving handling of data and relations between data in massive quantity and in multiple time and length scales that extends beyond capacity of the human brain, and further acceleration of materials engineering became a difficult struggle, leaving trial-and-error experimentation coupled with knowledge-based logical reasoning the major approach for development.

With arrival of the Information Age, rapid increase in computational power has provided a great boost to human capacity for logical reasoning and data processing. In the past few decades, significant progress has been achieved in computational modeling and simulation, with various algorithms and software tools developed for materials engineering. Accomplishments from computer aided engineering with materials, particularly visualization of materials characterization in a wide range of length and time scales from atoms to engineering products, frequently generate excitement in the materials community. However, when it comes to accurately controlling the materials-process-structures-properties-design-performance chain and satisfying product performance requirements, the excitement often quickly cools down to a difficult reality, in which the majority of development in materials engineering still heavily relies on the traditional approaches, and the old analogy of bread recipe evolution still remains largely valid for new materials development. This undesirable reality begs the question of what must be further developed to accelerate the progress in materials engineering.

In the past few years, ICME has emerged as a promising answer to the question. A consensus is gradually reached that with integration of multidisciplinary and multiscale computational processes backed up by experimentation, materials development and selection, manufacture processing, and product design can be effectively optimized to gain a much improved control over the materials-process-structures-properties-design-performance chain for product fabrication. Because most of the optimization is expected to be done in a virtual environment at high speed, the entire process of materials engineering should be greatly accelerated.

Activities and Ecosystem Conducive to ICME

In the past few years, many attempts were made to develop ICME out of its infancy, with some case studies demonstrating remarkable results. [1] It is generally believed that because a large amount of computational codes and software tools that have evolved over the past several decades are easily available, adequate integration of these codes and tools for product trade-off studies and multidisciplinary optimizations can accelerate materials engineering.

For initial ICME development, leveraging the codes and tools that have proven valid in actual engineering applications over decades is obviously a practical and efficient approach because their modeling and simulation results can be considered trustworthy. However, when these codes and tools are required to provide predictive simulations beyond their repeatedly validated engineering application regions even if it is still within their technical domain, their reliability can often be seriously challenged. Moreover, for further ICME development, many new codes and software tools must be developed from scratch and be trusted before sufficient engineering application experiences can be accumulated. In its discussion on ICME development, the National Research Council (NRC) attributed the current industrial reluctance in accepting ICME to "the slow conversion of science-based computational tools to engineering tools, …" [2] As a matter of fact, the root of the reluctance also often lies in a lack of confidence in the reliability of modeling and simulation for engineering applications, particularly when the engineering applications are required beyond the experienced regions, or when unfamiliar codes and tools are introduced. For accelerated materials engineering development, it is apparently impractical to build confidence in computational codes and software tools through long-term evolutionary revisions in actual engineering applications. The confidence must be built using well-established methodologies and workflows for verification and validation (V&V), which must be clearly identified as important ICME activities. A workflow for modeling, simulation, and experimentation for validation can be depicted by Figure 1.

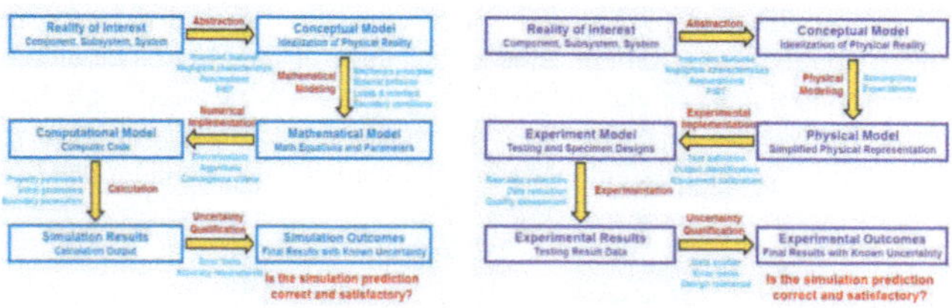

a) Modeling and simulation process with verification and uncertainty quantification.

b) Experimentation process for generation of validation data.

Figure 1: A desirable workflow for modeling, simulation, experimentation for validation.

To validate modeling and simulation, the simulation outcomes are compared with the experimental outcomes and revised until satisfaction is reached, as shown in Figure 2.

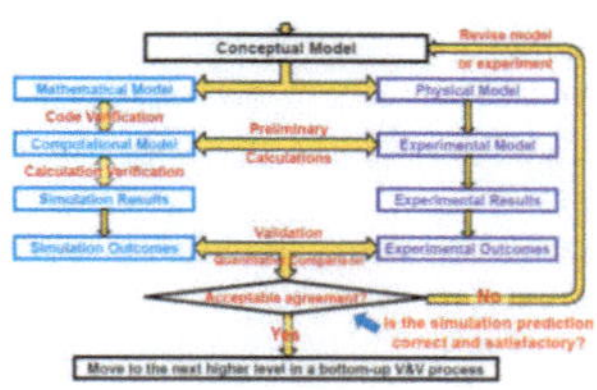

Figure 2: Verification and validation for modeling and simulation.

The process would use and generate a lot of data, which must be collected and managed as evidence for reliability of the simulation. It would also involve data quality assessment (DQA) and uncertainty quantification (UQ). All the data involved up to this point are only limited to one specific level of the modeling and simulation process. If an ICME project covers multiple length and time scales, e.g. the ideal case from atomic to full product level with components, the data involved can become enormous.

Although methodologies for DQA, UQ, and V&V are currently still under development with opinions and approaches different from the desirable ones discussed above, it is commonly agreeable that 1) If computational results are to be trusted for manufacturing of products, DQA, UQ, and V&V must be conducted, and 2) Despite different methods, V&V requires systematic collection of evidence for correctness of modeling and reliability of simulation. Apparently, these activities would involve considerable data generation, collection, management, and use. In addition to these activities that support the computation, the experimentation component of ICME also involves extremely data-intensive activities including digitalized materials microstructural analysis, materials testing and characterization, materials development and modification, materials selection and optimization, and materials processing.

As it was noted by NRC, "ICME can be defined as the integration of materials information, captured in computational tools, with engineering product performance analysis and manufacturing-process simulation." [2] It is clear that all the activities discussed above must heavily rely on dataflow to become effective. If integration is the key for ICME, apparently it is the dataflow that relates and integrates various ICME activities. It should also be pointed out that as an integral part of ICME, knowledge (i.e. the know-how) provided by experts in all these activities is often applied and communicated in the form of dataflow. Therefore, for long-term effective ICME development, an ecosystem as shown in Figure 3 should be established with a network of database systems as a tangible framework for integration, through which activities and communications between different disciplines of the ICME community can be conducted.

Figure 3: Ecosystem that effectively provides integration of various ICME activities.

Build the Ecosystem for Integration

To create the ICME ecosystem will obviously require a cyber infrastructure to be established to provide a tangible framework, and an ICME community to be organized to overcome barriers such as copyrights, proprietary information protections, and conflict of interests to enable data exchange. Sharing data between organizations is obviously an ideal time- and cost-efficient approach. In the past few years, the Oak Ridge National Laboratory (ORNL) has developed a digital database system dubbed the Gen IV Materials Handbook for international materials data collaboration among nine signatories/countries, through which experiences have been accumulated and are discussed below for consideration of ICME ecosystem development. [3]

The fact that ICME would involve massive dataflow from one activity to another for integration requires the data to be managed in digital form that can be easily transferred and processed. Because computational simulation is mostly based on logical reasoning, relations between different data become very useful and must be traced with accuracy. Therefore, a desirable database system for ICME must be able to manage not only digital data, but also relations between the data. Traceability to background information is often found crucial not only for trustworthiness of the data but also for correctly interpreting and analyzing the data.

To effectively integrate ICME activities, the database system must enable easy extraction of desired data for analysis, presentation, and transfer. This capability, particularly in very large scales, can only be provided in a database system that manages data by its most basic elements, i.e. the individual data point, so that any specific data groups can be called out for automated reformatting, processing, and transfer. To develop such a system, an anatomy analysis must be conducted on the body of the information of interest to delineate its fine structures and define its most basic data elements along with their relations. The analysis results are used as the basis for custom design of database schema for construction of the database management systems (DBMS) with various virtual data containers. Once the basic data elements are uploaded into the virtual containers, the data can be conveniently manipulated because each virtual container is a part of the database system with specific functionalities for operation, and any groups of data can be easily called out and transferred to other ICME activities, e.g. to external modeling software for simulation, and the simulation result data can also be uploaded back into the database system.

To establish an ecosystem shared by the ICME community, it is crucial that data collections can be systematically processed, compared, and analyzed using software tools. Unfortunately, materials data generated by different organizations usually do not have the same schemas and formats. Analysis and comparison of such data collections often require significant effort in reformatting before they can be meaningfully studied. In the Gen IV database system, this hurdle is successfully overcome using a master record layout, which includes attributes for all collaborating countries, and each country inputs data into their own attributes, leaving the rest blank. When displayed, the DBMS presents only the attributes filled with data. As a result, each country has their data displayed in their own schemas. Because all the data are now loaded in the same master schema, comparison and analysis of the data become an easy task. The records and attributes of interest can be selected and called out to generate a table for comparison and analysis. Data gaps are also clearly identified by empty cells in the table. The called out data can be further transferred to other ICME activities for integration operations. Many techniques developed for effective digital data management are summarized in Reference [4].

To share data from different sources, standardization is a great hurdle. Instead of rigorous standardization, the hurdle can be effectively reduced using standards-compliant schemas and ontologies that enable database interoperability by developing data exchange mechanisms, thus allowing database systems to largely keep their original terminologies, schemas, and formats. Recently, the Gen IV Materials Handbook and the MatDB (which is the materials database system of the European Union) successfully demonstrated preliminary database interoperability by developing mechanisms for exchange of tensile test data. [5, 6] Counterparts of tensile data attributes in the two database systems were identified, and data were transferred from one database system to the other through intermediate XML files with mutually agreed standards-compliant schema. Although mapping attributes to intermediate standards-compliant schema may still be considered as a standardization operation, it only involves the two interoperating database systems in a much less rigorous fashion and requires minimum revision in these database systems. As the interoperability is expanded to more database systems, mapping attributes to standards-compliant data formats will become an evolutionary standardization process, which can naturally grow as the ICME ecosystem develops.

It can be expected that to build an ICME network of database systems, one of the major challenges is to develop business models for its financial and technical sustainability. Before the great added value of a well-structured digital database system over a traditional compilation of data documents is commonly understood and popularly desired/used for integrating ICME activities, it is difficult to attract investment and expertise to ensure a healthy growth of the network for ICME ecosystem development. In this regard, it is imperative for the ICME community to develop commonly acceptable methods for monetary quantification of the value of data, database, digital data, digital database, and digital database system. With clear and commonly accepted price tags, many barriers for data exchange would collapse, paving the path for effective ICME development. To accumulate and preserve a wealth of digital data for the future, the ICME community should promote mandatory requirements for digital data plan and submittal as an integral part of materials R&D proposals and publications.

Summary

To accelerate materials engineering development, the concept of ICME has undoubtedly heralded a new and exciting era. To increase chances of success, an ecosystem conducive to integration of various activities for computational materials engineering should be established. In short-term the integration can be achieved by coordination and incorporation of legendary experiment results and data, computational codes, and simulation software packages that have been proven trustworthy through many years of application to actual manufacturing of products. However, for long-term ICME growth, expanding their computational predictions into unproven regions, as well as development of new experiments, models, and simulations cannot be avoided, which would necessitate integration of a wide range of activities such as DQA, UQ, V&V, and so on. A review of a complete ICME process suggests that it is the dataflow (including information and knowledge application in the form of data) that actually integrates different ICME activities. To facilitate the integration, an effective network of digital database system must be established as a tangible framework along with the organization of an ICME community that interacts and communicates through the network to provide an ecosystem conducive to ICME development. Some experiences from international materials data collaboration for nuclear development at ORNL are discussed for consideration of the ICME ecosystem development.

References

[1] "Proceedings of the 1st World Congress on Integrated Computational Materials Engineering (ICME)," Sponsored by TMS, Wiley, A John Wiley and Sons, Inc., Publication, ISBN 978-0-47094-319-9, edited by John Allison, Peter Collins, and George Spanos.

[2] "Integrated Computational Materials Engineering: A Transformation Discipline for Improved Competitiveness and National Security," Committee on Integrated Computational Materials Engineering, National Research Council, http://www.nap.edu/catalog/12199, 2008.

[3] "Gen IV Materials Handbook Functionalities and Operation (3A) – Handbook Version 3.0 – ", ORNL/TM-2011/248_3A, U. S. Department of Energy Generation IV Nuclear Energy Systems Program, U. S. Department of Energy, February 15, 2012, Weiju Ren.

[4] "Effective Materials Property Information Management for the 21st Century", invited publication by the Journal of Pressure Vessel Technology, Vol. 133, Issue 4, pp. 044002, August, 2011, Weiju Ren, David Cebon, and Steve Arnold.

[5] "Interoperability of Material Databases 2011 Progress Report," International Nuclear Energy Research Initiative, Project Number: 2010-005-E, Program Area: Gen IV, Oak Ridge National Laboratory, October 2011, Hans-Helmut Over, Weiju Ren, and Peter Hähner.

[6] "Interoperability of Material Databases 2012 Progress Report," International Nuclear Energy Research Initiative, Project Number: 2010-005-E, Program Area: Gen IV, Oak Ridge National Laboratory, December 2012, Lianshan Lin and Weiju Ren.

Acknowledgments

This work is sponsored by the U.S. Department of Energy, Office of Nuclear Energy Science and Technology under contract DE-AC05-00OR22725 with Oak Ridge National Laboratory, managed by UT-Battelle, LLC.

2nd World Congress on Integrated Computational Materials Engineering
Edited by: Mei Li, Carelyn Campbell, Katsuyo Thornton, Elizabeth Holm, and Peter Gumbsch
TMS (The Minerals, Metals & Materials Society), 2013

CROSS-SCALE, CROSS-DOMAIN MODEL VALIDATION BASED ON GENERALIZED HIDDEN MARKOV MODEL AND GENERALIZED INTERVAL BAYES' RULE

Yan Wang[1], David L. McDowell[1], Aaron E. Tallman[1]

[1]Georgia Institute of Technology
801 Ferst Drive NW, Atlanta, GA 20332-0405

Keywords: Irradiation, Uncertainty Quantification, Imprecise Probability, Hidden Markov Model

Abstract

Reliable simulation protocols supporting integrated computational materials engineering (ICME) requires uncertainty to be quantified. In general, two types of uncertainties are recognized. Aleatory uncertainty is inherent randomness, whereas epistemic uncertainty is due to lack of knowledge. Aleatory and epistemic uncertainties need to be differentiated in validating multiscale models, where measurement data for unconventionally very small or large systems are scarce, or vary greatly in forms and quality (i.e., sources of epistemic uncertainty). In this paper, a recently proposed generalized hidden Markov model (GHMM) is used for cross-scale and cross-domain information fusion under the two types of uncertainties. The dependency relationships among the observable and hidden state variables at multiple scales and physical domains are captured by generalized interval probability. The update of imprecise credence and model validation are based on a generalized interval Bayes' rule (GIBR).

Introduction

Not all uncertainties or errors in simulation can be readily represented by classical probability distributions. In general, two types of uncertainties are recognized. Aleatory uncertainty is inherent randomness, whereas epistemic uncertainty is due to lack of knowledge. Epistemic uncertainty is a result of conflicting information from multiple sources, conflicting beliefs among experts' opinions, lack of data, lack of time for introspection, measurement error, lack of dependency information, truncation errors during numerical treatments, etc. In contrast to aleatory uncertainty, epistemic uncertainty can be reduced.

In modeling and simulation (M&S), epistemic uncertainty manifests from errors associated with the models and input data. For instance, in finite-element analysis, epistemic uncertainty arises from the truncation error involved in linear and nonlinear approximations of strain fields using polynomials, imprecise input parameters in the model, mismatch of structures and geometric configurations between the modeled and the true physical ones, etc. In kinetic Monte Carlo simulation, epistemic uncertainty is largely from the imperfect knowledge and resulting simplification of transition paths and reaction rates, as well as dynamic rates because of external loads, crowding effects, and other unknown correlations. In molecular dynamics simulation, epistemic uncertainty is mainly from the inaccurate potential functions (both forms and parameters), boundary conditions, local approximation of potentials, cut-off distance, high strain rates/short simulation time scales, as well as measurement errors in experimental data during model calibration. In contrast to epistemic uncertainty, various sources such as disturbed boundary and loading conditions, different sampling sizes and running times, inherent variations of material compositions, and other randomness and fluctuation contribute to the aleatory component of uncertainty.

Since aleatory and epistemic uncertainties arise from distinct sources and have very different characteristics, we should represent them differently, preferably in different forms, so that they can be easily differentiated. Aleatory uncertainty is traditionally and predominantly modeled using probability distributions. In contrast, epistemic uncertainty has been modeled in several ways, such as probability, interval, fuzzy set, random set, basic probability assignment, etc. Here we use intervals to quantify epistemic uncertainty.

Interval is as simple as a pair of numbers, i.e. the lower and upper bounds. The reason to choose an interval representation is two-fold. First, an interval is a natural means for human users to communicate information and is simple to use. It has been widely used to represent a range of possible values, an estimate of lower and upper bounds for numerical errors, and the measurement error because of the available precision as the result of instrument calibration. Second, an interval can be regarded as the most suitable way to represent the lack of knowledge. Compared to other forms, specification via an interval has the least assumptions. It only needs lower and upper bounds, without any assumption of distributions between them. Given that epistemic uncertainty arises intrinsically from lack of knowledge, a representation with the least assumptions is most desirable.

Probability has certain limitations in representing epistemic uncertainty. The accuracy of a predictive simulation depends heavily on fundamental understanding of the underlying physical and chemical processes. Lack of perfect knowledge and fundamental insight inevitably renders models imperfect. Any assumption regarding distributions in M&S introduces a bias. The most significant limitation of probability is that it does not differentiate 'total ignorance' from other probability distributions. A problem arises because introducing a uniform or any particular form of distribution implicitly introduces extra information that cannot be justified in the case of zero knowledge. This leads to the Bertrand-style paradoxes. "Knowing the unknown" does not represent total ignorance. Although the Bayesian approach has been proposed to reduce the bias introduced in assuming a distribution, and it serves the purpose well in an ideal situation where we have plentiful data without measurement errors, its limitation remains in the real-world applications where lack of data or imperfect measurement lingers. Therefore, it is desirable to have more general and efficient approaches to incorporate epistemic uncertainty in M&S, with minimal assumptions regarding probability distributions and their parameters. Moreover, it is desired that such approaches be less computationally demanding than the traditional Bayesian learning and update approach.

Basic Elements of Generalized Interval Probability

Interval or imprecise probability, represented by lower and upper probability bounds as $[\underline{p}, \overline{p}]$, is a generalization of classical probability that simultaneously models the two uncertainty components; aleatory uncertainty is modeled by probability whereas epistemic uncertainty by interval. When $\underline{p} = \overline{p}$, the degenerated interval probability becomes the traditional probability. Differing from other forms of imprecise probabilities, such as Dempster-Shafer theory [1,2], coherent lower prevision [3], p-box [4], etc., the recently proposed generalized interval probability [5] provides a simplified probabilistic calculus structure that ensures ease of application.

Generalized Interval Probability

A generalized interval $\mathbf{x} := [\underline{x}, \overline{x}]$ is defined as a pair of numbers. $[\underline{x}, \overline{x}]$ is called *proper* if $\underline{x} \leq \overline{x}$, and *improper* if $\underline{x} \geq \overline{x}$. The introduction of improper intervals greatly simplifies the calculus structure of interval probability and makes it very similar to the one in the classical precise probability theory. The calculation of generalized intervals is based on the Kaucher interval arithmetic [6]. In generalized interval probability theory, the probability measure has the value of generalized interval. Therefore, both [0.2,0.4] and [0.4,0.2] are valid probabilities. The relationship between proper and improper intervals is established by a *dual* operator. For instance, dual[0.2,0.4]=[0.4,0.2] and dual[0.4,0.2]=[0.2,0.4].

In generalized interval probability, *conditional probability* is uniquely defined as $\mathbf{p}(X \mid Y) := \mathbf{p}(XY) / \mathrm{dual}\,\mathbf{p}(Y) = [\underline{p}(XY) / \underline{p}(Y), \overline{p}(XY) / \overline{p}(Y)]$. As a result, the *generalized interval Bayes' rule* states that $\mathbf{p}(E_i \mid A) = \mathbf{p}(A \mid E_i)\mathbf{p}(E_i) \big/ \sum_{j=1}^{n} \mathrm{dual}\,\mathbf{p}(A \mid E_j)\,\mathrm{dual}\,\mathbf{p}(E_j)$ where E_i (i=1,..,n) are mutually disjoint events as the partition of the sample space, and most importantly $\sum_{j=1}^{n} \mathbf{p}(E_j) = 1$, which is called the logic coherent constraint (LCC). With simple algebraic calculation, the probabilistic reasoning based on generalized interval probability is very similar to the traditional one in the classical probability. In contrast, other forms of imprecise probabilities must rely on linear or nonlinear optimization methods to estimate probability lower and upper bounds, which is computationally cumbersome.

Generalized Hidden Markov Model (GHMM)

A GHMM [7] was recently proposed to capture correlations of variables between scales, as illustrated in Fig. 1, where dependencies between random variables at three length scales Ω_x, Ω_y, and Ω_z are captured. The state variables, denoted as x_i, y_j, and z_k, respectively associated with these three scales, are hidden. Their true values have to be predicted and inferred by some measurable or observable quantities X_i, Y_j, and Z_k, respectively, via physical experiments. The correlation relationships are expressed as conditional probabilities. For instance, $\mathbf{p}(x_i|x_{i,1}, x_{i,2})$ captures the dependency within one scale, $\mathbf{p}(y_j|x_{i,1}, x_{i,2})$ captures the one between scales, and $\mathbf{p}(X_i|x_i)$, $\mathbf{p}(Y_j|y_j)$, and $\mathbf{p}(Z_k|z_k)$ are between the observable and hidden variables. It should be noted that these 'scales' may also be constituted by domains (e.g., time, different loosely coupled model environments), so that the conditional probabilities can be expressed not only in terms of disparate length and time scales, but associated sets of models, even multiple models.

Application of GHMM in Model Validation under Uncertainties

Similar to the Bayesian approach in model validation [8, 9], the validation of simulation models under both epistemic and aleatory uncertainties at multiple scales can be performed based on the generalized interval Bayes' rule. Given the experimental measurements Y_1, …, Y_M and Z_1, …, Z_N in separate domains or scales, the interval probability that the model has the parameter x is

$$\mathbf{p}(x \mid Y_1,\ldots,Y_M,Z_1,\ldots,Z_N) = \frac{\mathbf{p}(x)\int\cdots\int\begin{bmatrix}\mathbf{p}(Z_1,\ldots,Z_N \mid z_1,\ldots,z_N)\mathbf{p}(Y_1,\ldots,Y_M \mid y_1,\ldots,y_M)\\ \mathbf{p}(z_1,\ldots,z_N \mid y_1,\ldots,y_M)\mathbf{p}(y_1,\ldots,y_M \mid x)\end{bmatrix} dz_1\cdots dz_N dy_1\cdots dy_M}{\mathrm{dual}\int\cdots\int\begin{bmatrix}\mathbf{p}(Z_1,\ldots,Z_N \mid z_1,\ldots,z_N)\mathbf{p}(Y_1,\ldots,Y_M \mid y_1,\ldots,y_M)\\ \mathbf{p}(z_1,\ldots,z_N \mid y_1,\ldots,y_M)\mathbf{p}(y_1,\ldots,y_M \mid x_1,\ldots,x_L)\mathbf{p}(x)\end{bmatrix} dz_1\cdots dz_N dy_1\cdots dy_M dx} \quad (1)$$

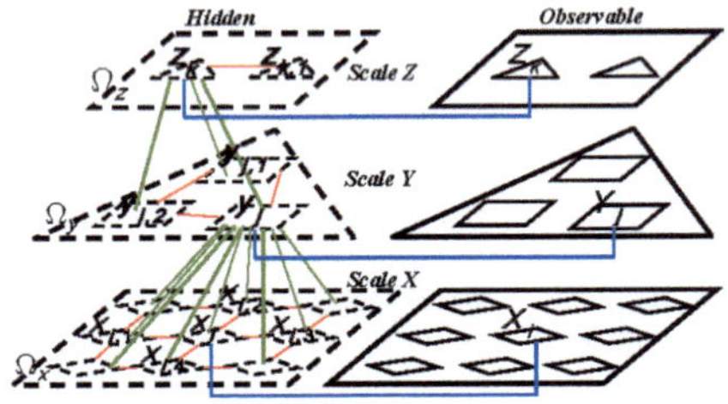

Figure 1. The generalized hidden Markov model captures spatial and scale dependencies.

where $y_1, \ldots, y_M$ and $z_1, \ldots, z_N$ are hidden variables of physical properties at different domains or scales and related to x. In the numerator of Eq.(1), $\mathbf{p}(z_1,\ldots,z_N|y_1,\ldots,y_M)$ and $\mathbf{p}(y_1,\ldots,y_M|x)$ represent the correlations between hidden variables of properties at different scales, whereas $\mathbf{p}(Z_1,\ldots,Z_N|z_1,\ldots,z_N)$ and $\mathbf{p}(Y_1,\ldots,Y_M|y_1,\ldots,y_M)$ capture the correlations between hidden and measurable properties. The key to applying Eq. (1) is to find conditional probability values from experiments and prior experience.

An Example of A Molecular Dynamic Simulation Model for Irradiation on Fe

Here we demonstrate how to validate models using the generic GHMM approach with a molecular dynamics (MD) simulation of point defect generation in Fe crystals that are subject to high energy knock-on atom collisions (irradiation) and resulting cascade events. A MD simulation model of the Fe crystal was constructed in LAMMPS. For each combination of different energy levels and radiation angles, 16 simulation runs are conducted. The probabilities that a stable Frenkel pair are generated, also equivalently known as damage function $v(T)$ with the transfer or recoil energy T, are collected and shown in Fig. 2 with the label 'mid'. Because of the model and numerical errors involved, the probabilities do not necessarily always increase as the energy level increases, as shown in Fig. 2(b) and (c). Therefore, interval probabilities with lower and upper limits are used. The interval probability widths are calculated by the standard deviation of binomial distributions, where error bounds are added to and subtracted from the middle or nominal values of probabilities. Interval probabilities capture the uncertainties and errors involved in the MD simulation model. To ensure the non-decreasing pattern of the lower and upper cumulative distribution functions (CDF) $P(T \leq t)$, a filtering procedure is taken to 'smooth out' the empirical CDF's as follows. If the upper CDF value at a given energy level is less than the one at a lower energy level (immediately on its left in the chart), its value is set to be the same as the one at the immediate lower energy level. Similarly, if the lower CDF value at an energy level is larger than the one at a higher energy level (immediately on its right in the chart), its value is set to be the same as the one at the immediate higher energy level.

During physical experiments, the total displacement cross section σ, as the indicator of the amount of point defects, is not measured directly. Rather, it is based on the measurement of the electrical resistivity change rates $\Delta\rho/n$ that is correlated with σ. Instead of using the traditional empirical analytical relationship between $\Delta\rho/n$ and σ where uncertainty is ignored, we consider the correlation between the two in terms of conditional probability. Some examples of the correspondence between the two quantities from experimental measurement are shown in Table I [10], where the measurement error is listed in terms of the minimum and maximum values of $\Delta\rho/n$. In other words, at a particular level of maximum transmitted energy T_m, one value of σ

corresponds to a range of possible values of $\Delta\rho/n$ and a uniform distribution between the bounds can be assumed. The correlation between the two quantities thus can be represented as conditional probability $P(\Delta\rho/n|\sigma,T_m)$. Similarly, the correspondence between the total displacement cross section σ and the damage function ν is established by conditional probability $P(\sigma,T_m|T)$, instead of using an empirical analytical relation $\sigma = \int_0^{T_m} \nu(T)d\sigma(T)dT$ where $d\sigma(T)$ is the differential recoil energy cross section or the probability of producing a recoil of energy between T and T+dT. The range of energy in the experiments is from 20eV to 130eV.

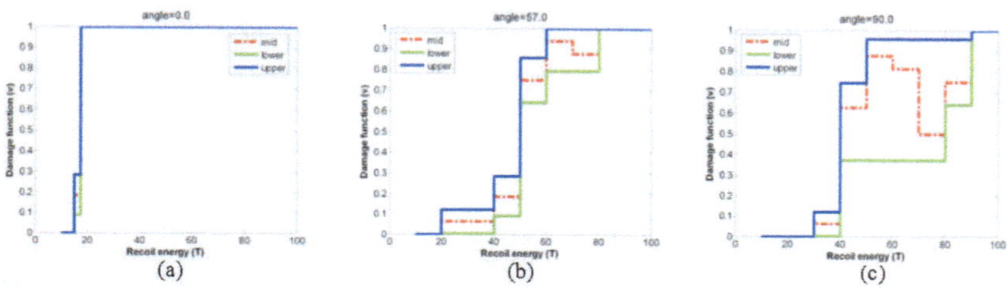

Figure 2. The example cumulative probability distributions of the generation of a stable Frenkel pair at different recoil energy (eV) and angles by MD simulation.

Table I. Examples of correspondence between the displacement cross section σ (barns) and the resistivity change rates $\Delta\rho/n$ (10^{-26} Ωcm/(e$^-$/cm^2)) (with minimum and maximum values) at each level of maximum transmitted energy T_m (eV) in a head-on collision [10]

<100>				<111>			
T_m	σ	$\Delta\rho/n$: *min*	$\Delta\rho/n$: *max*	T_m	σ	$\Delta\rho/n$: *min*	$\Delta\rho/n$: *max*
70	23.1578947	7.3366283	8.21892725	70	29.0789474	9.25116666	9.55835715
100	32.8947368	9.51916799	10.4014669	100	39.2105263	12.8622778	13.1694683

To validate the MD simulation model by the experiments that are conducted at macroscopic scale, we can compare the prior probability of model parameters θ in MD P(θ) and the posterior probability P(θ|Δρ/n), calculated as

$$\mathbf{p}(\theta \mid \Delta\rho / n) = \frac{\mathbf{p}(\theta)\int\cdots\int\left[\mathbf{p}(\Delta\rho / n \mid \sigma, T_m)\mathbf{p}(\sigma, T_m \mid T)\mathbf{p}(T \mid \theta)\right]dTd\sigma dT_m}{\mathrm{dual}\int\cdots\int\left[\mathbf{p}(\Delta\rho / n \mid \sigma, T_m)\mathbf{p}(\sigma, T_m \mid T)\mathbf{p}(T \mid \theta)\mathbf{p}(\theta)\right]dTd\sigma dT_m d\theta} \qquad (2)$$

As an illustration, a simple numerical example is shown as follows. Given a set of MD simulation model parameter θ, the interval probability that a Frenkel pair will be generated at an energy level less than 70 eV at direction <111> is estimated from Fig.2(b) as $\mathbf{P}(T\leq70|\theta)=[0.7923,1]$. From LCC, we know $\mathbf{P}(T>70|\theta)=[0.2077,0]$. Additionally, $\mathbf{P}(T\leq70|\sim\theta)=[0,1]$ and $\mathbf{P}(T>70|\sim\theta)=[1,0]$ since there is a lack of knowledge when the model parameters are different.

From Table I and the assumption of uniform distribution of $\Delta\rho/n$, we have $\mathbf{P}(\Delta\rho/n\leq9.4048|\sigma=29.0789,T_m=70)=[0.5,0.5]$, $\mathbf{P}(\Delta\rho/n>9.4048|\sigma=29.0789,T_m=70)=[0.5,0.5]$, $\mathbf{P}(\Delta\rho/n\leq9.4048|\sigma=39.2105,T_m=100)=[0,0]$, and $\mathbf{P}(\Delta\rho/n>9.4048|\sigma=39.2105,T_m=100)=[1,1]$. We do not have any information about what we did not measure. That is, with total ignorance, we have $\mathbf{P}(\Delta\rho/n\leq9.4048|\sigma\neq29.0789,T_m\neq70)=[0,1]$, $\mathbf{P}(\Delta\rho/n>9.4048|\sigma\neq29.0789,T_m\neq70)=[1,0]$, $\mathbf{P}(\Delta\rho/n\leq9.4048|\sigma\neq39.2105,T_m\neq100)=[0,1]$, and $\mathbf{P}(\Delta\rho/n>9.4048|\sigma\neq39.2105,T_m\neq100)=[1,0]$. Furthermore, $\mathbf{P}(\sigma=29.0789,T_m=70|T\leq70)=[0.01,1]$, $\mathbf{P}(\sigma\neq29.0789,T_m\neq70|T\leq70)=[0.99,0]$,

$\mathbf{P}(\sigma=39.2105, T_m=100|T\leq 70)=[0,0.01]$, $\mathbf{P}(\sigma\neq 39.2105, T_m\neq 100|T\leq 70)=[1,0.99]$, $\mathbf{P}(\sigma=29.0789, T_m=70|T>70)=[0,0.01]$, $\mathbf{P}(\sigma\neq 29.0789, T_m\neq 70|T>70)=[1,0.99]$, $\mathbf{P}(\sigma=39.2105, T_m=100|T>70)=[0.01,1]$, $\mathbf{P}(\sigma\neq 39.2105, T_m\neq 100|T>70)=[0.99,0]$.

Suppose the prior probability is $\mathbf{P}(\theta)=[0.4,0.5]$, the posterior probability after an instance of $\Delta\rho/n\leq 9.4048$ is observed is $\mathbf{P}(\theta|\Delta\rho/n\leq 9.4048)=[1,0.5]$ based on Eq. (2). The value of posterior probability increases compared to the prior. However, because of the lack of knowledge, the width of the interval probability also increases. Interval probability provides the extra information of how significant the epistemic component of uncertainty plays in the model validation. The slight overlap between the two intervals does not invalidate the model. Further experiments are needed, if more robust conclusions are desirable. As more knowledge is obtained, the posterior probability will converge to the classical precise one with the interval width gradually reduced.

Concluding Remarks

The GHMM and generalized interval Bayes' rule improve the robustness of the model validation process, where measurement data have inherent systematic errors and computational models contain intrinsic model errors. During validation, interval-valued posterior probability distributions of model parameters or model predictions are updated with the collected data from multiple scales or domains based on the generalized interval Bayes' rule. Validation is done by comparing the interval posterior probability with the prior. A substantial difference between the two invalidates the model. The epistemic uncertainty component during the inference provides us extra information so that more robust conclusions can be obtained.

Acknowledgements

This research is being performed using funding received from the DOE Office of Nuclear Energy's Nuclear Energy University Programs.

References

[1] Dempster A. (1967) Upper and lower probabilities induced by a multi-valued mapping. *Annals of Mathematical Statistics*, **38**(2):325-339

[2] Shafer G.A. (1990) Mathematical Theory of Evidence. Princeton University Press, Princeton, NJ.

[3] Walley P. (1991) Statistical Reasoning with Imprecise Probabilities. Chapman & Hall, London.

[4] Ferson S., Kreinovich V. Ginzburg L., Myers D.S., and Sentz K. (2003) Constructing probability boxes and Dempster-shafer structures. *Sandia National Laboratories Technical report* SAND2002-4015, Albuquerque, NM.

[5] Wang Y. (2010) Imprecise probabilities based on generalized intervals for system reliability assessment. *International Journal of Reliability & Safety*, **4**(4): 319-342

[6] Kaucher E. (1980) Interval analysis in the extended interval space IR. *Computing Supplementa*, Vol.2, 33-49

[7] Wang Y. (2010) Multiscale uncertainty quantification based on a generalized hidden Markov model. *Journal of Mechanical Design*, **133**(3): 031004

[8] Babuška, I., Nobile, F., and Tempone, R. (2008). A systematic approach to model validation based on Bayesian updates and prediction related rejection criteria. Computer Methods in Applied Mechanics and Engineering, **197**(29), 2517-2539.

[9] Oden, T., Moser, R., and Ghattas, O. (2010). Computer predictions with quantified uncertainty, Part I. SIAM News, **43**(9), 1-3.

[10] Vajda, P. (1977). Anisotropy of electron radiation damage in metal crystals. Reviews of Modern Physics, **49**(3), 481.

2nd World Congress on Integrated Computational Materials Engineering
Edited by: Mei Li, Carelyn Campbell, Katsuyo Thornton, Elizabeth Holm, and Peter Gumbsch
TMS (The Minerals, Metals & Materials Society), 2013

APPLICATION OF STATISTICAL AND MACHINE LEARNING TECHNIQUES FOR CORRELATING PROPERTIES TO COMPOSITION AND MANUFACTURING PROCESSES OF STEELS

P.D Deshpande[1], B. P. Gautham[1],
A. Cecen[2], S. Kalidindi[4],
A. Agrawal[3], A.Choudhary[3]
[1]Tata Research Development and Design Centre, Tata Consultancy Services;
54B Hadapsar Industrial Estate; Pune, Maharashtra, 411013, India

[2]Drexel University; Mechanical Engineering & Mechanics;
Philadelphia, PA 19104, USA

[3]Northwestern University; Dept. of Electrical Engineering and Computer Science;
Evanston, IL 60208, USA

[4]Georgia Institute of Technology; Woodruff School of Mechanical Engineering;
Atlanta, GA, USA

Keywords: Material Informatics, Regression Analysis, Processing-Property Linkages, Artificial Neural Networks

Abstract

Establishing correlations between various properties of alloys and their compositions and manufacturing process parameters is of significant interest to materials engineers. Both physics-based as well as data-driven approaches have been used in pursuit of this. Of various properties of interest, fatigue strength, being an extreme value property, had only a limited amount of success with physics based models. In this paper, we explore a systematic data driven approach, supplemented by physics based understanding, employing various regression methods with dimensionality reduction and machine learning methods applied to the fatigue properties of steels available from the National Institute of Material Science public domain database to arrive at correlations for fatigue strength of steels and present an assessment of the residual errors in each method for comparison. This study is expected to provide insights into the methods studied to make objective selection of appropriate method.

Introduction

Data driven exploration of material science has been making considerable progress in the recent years and has been pursued as "Materials Informatics" [1,2]. Progress in this direction is fuelled by the availability of large amounts of experimental and simulation data along with advances in data analytics. Various statistical techniques along with machine learning methods are currently employed in problems involving complex physical interdependencies, such as bioinformatics, to a great degree of success. These approaches have led to an improved understanding of the field and to establish quantitative relationships. The complex interdependencies between variables of interest can be captured using various data analytic methods [3], including artificial neural networks [4-7].

This paper deals with application of select regression methods with dimensionality reduction and machine learning methods applied to the fatigue properties of steels available from the National Institute of Material Science (NIMS) public domain database [8]. The aim is to arrive at correlations for fatigue strength of steels in terms of composition and processing conditions and present an assessment of the residual errors in each method for comparison. The methods explored include Polynomial regression, Least-square regression with non-linear transformation of variables, Artificial Neural Network Regression and Reduced Error Pruning Tree.

Correlating steel properties to composition and manufacturing processes

In this section, we compare various statistical and machine learning techniques for deriving a correlation between manufacturing process parameters and composition with fatigue strength for steels. All the techniques have been applied on the same data-set.

Data-Set Used for Analysis

Fatigue Dataset for Steel from NIMS MatNavi, one of the largest databases in the world for materials was utilized in the present work. The data used contains the following entities:

- Chemical composition - %C, %Si, %Mn, %P, %S, %Ni, %Cr, Cu %, Mo% (all in wt. %)
- Upstream processing details - ingot size, reduction ratio, non-metallic inclusions
- Heat treatment conditions – temperature, time and other process conditions for normalizing, carburizing-quenching and tempering processes
- Mechanical properties - YS, UTS, %EL, %RA, hardness, Charpy impact value (J/cm^2), rotating bending fatigue strength @ 10^7 cycles.

A total of **437 data records** were collected from a list of available data sheets dealing with carbon and low alloy steels (371 observations), carburizing steels (48 observations) and spring steels (18 observations). This data pertains to various heats of each grade of steel and different processing conditions. In all the studies, the correlations are sought to be made for fatigue strength in terms of composition and process parameters.

Data Pre-processing and Application of Regression Techniques

The systematic data-driven approach utilized in this work comprises of two main steps:

1. Rigorous description of data to provide an efficient basis for exploring possible linkages between the fatigue strength and composition and processing conditions; and
2. Identifying such linkages through established protocols, such as regression or machine learning techniques.

Data pre-processing

Prior to the regression analysis, the data was conditioned for a stable and well defined regression scheme. Firstly, the data was checked for any inherent co-linearity and any co-linear parameter that showed a Pearson product moment coefficient higher than 0.95 with another parameter was eliminated. Secondly, singular value decomposition was performed on the dataset, which is a matrix factorization defined as:

$$D = U \times S \times V \qquad (1)$$

where, D is the data matrix such that every observation is represented by a row and each column is an explanatory variable, U is the matrix of left singular vectors, V is the matrix of right singular vectors and S is the diagonal matrix of singular values. In this case, $A = U \times S$ is a transformation of D where the data is represented by a new set of explanatory variables such that each variable is a known linear combination of the original explanatory parameters. The dimensions of A are also referred to as the Principal Components (PC) of the data. The principal components are used in regression techniques.

K-means Clustering

The entire available data-set was assessed for visible clustering by employing k-means clustering technique. The cluster plot demonstrates inherent clustering in the available data, which agrees with the *a priori* knowledge of the data-set. The distinct clustering in the available data represents 4 clusters according to the grade of steels as depicted in Figure 1. These clusters however do not offer sufficient data-points to create individual meta-models for each cluster and hence, for all methods used, the entire data-set is used to develop correlations.

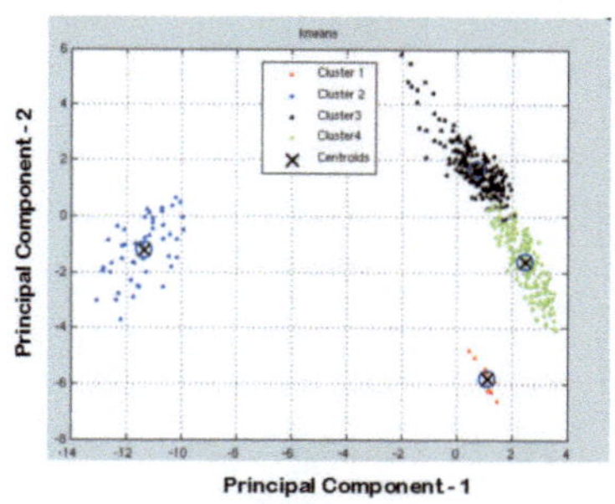

Figure 1: K-means clustering of NIMS normalized data

Polynomial Regression Technique

In the linear regression analyses conducted, the error associated with each data point was expressed as the absolute residual:

$$E = |F - f^p(PC_1, PC_2 \ldots PC_n)| \qquad (3)$$

where, F is fatigue strength and $f^p(PC_1, PC_2 \ldots PC_n)$ denotes a pth-order polynomial function of n principal components (PC_i). The polynomial coefficients are established by minimizing the residual sum of squares (RSS) in the entire dataset. The measure of error and the extracted polynomial fit depend critically on the selection of both p and n. Akaike Information Criterion (AIC) is a relative measure of good fit applied in literature for model selection and depicted in Figure 2. AIC provides only relative quality of a fit. The smaller the values of AIC relatively better the fit. The law of succinctness (Ockham's razor) suggests that out of all viable fits, the least complex one (the one with the lowest number of explanatory terms (R)) should be used. The AIC is defined as:

$$AIC = K \cdot \ln\left(\frac{1}{K}\sum_{k=1}^{K} E_k{}^2\right) + 2 \cdot R \qquad (2)$$

This information is used in the selection of appropriate order of the polynomial. The results of the selected polynomial fit in explaining the entire dataset using leave one out cross validation (CV) are presented in Figure 3.

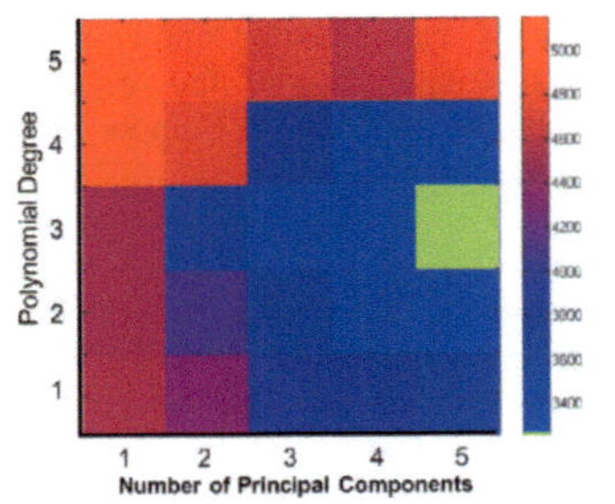

Figure 2: The Akaike Information Criterion for varying polynomial degrees and PCs

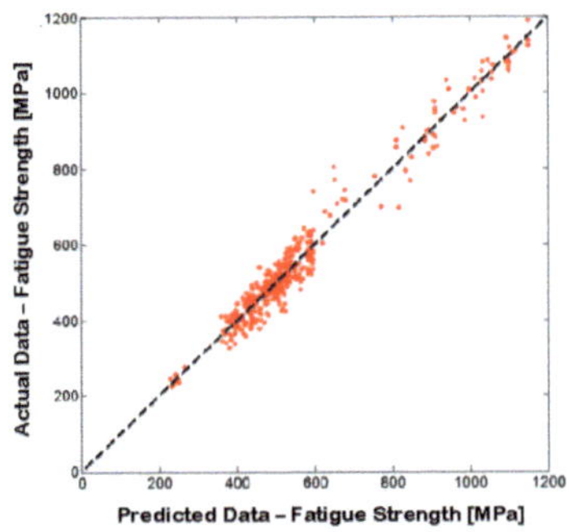

Figure 3: Polynomial regression results [r-squared value = 0.9626]

Regression Post Non-Linear Transformation of Select Input Variables

A non-linear transformation of temperature terms based on the physically known Arrhenius relationship was carried out and the resulting data-set was used for linear regression. The results obtained were marginally improved as presented in Figure 4 with r-squared value of 0.9719. However they do not sufficiently explain the scatter towards the higher order fatigue strength steels and prompt us to apply other non-linear techniques.

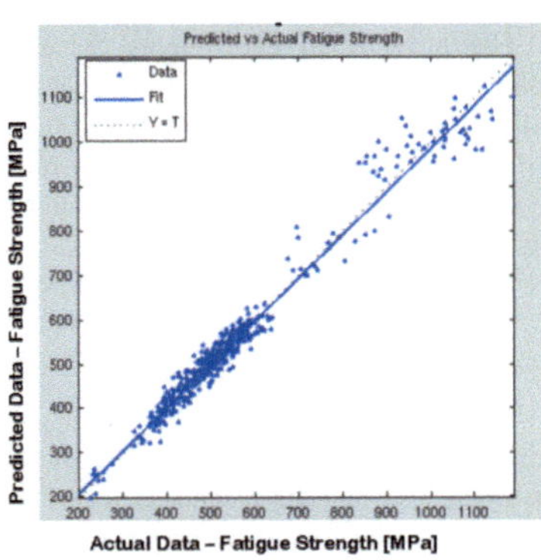

Figure 4: Non-linear model including exponential temperature terms [r- squared = 0.9719]

Artificial Neural Network

An artificial neural network was designed with 25 input and 25 hidden notes and a single output node and trained using Matlab® ANN toolbox. The data-set used for training consists of 70% of records with 15% used for validation and 15% for testing. The results have been encouraging as presented in Figure 5.

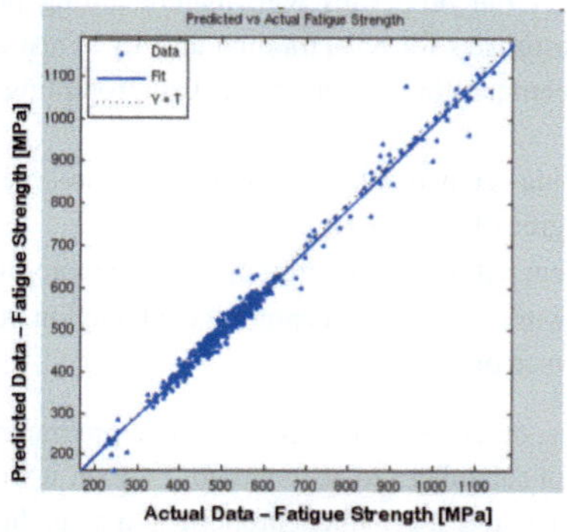

Figure 5: ANN based regression [r-squared = 0.9867]

Decision Tree Based Predictive Modeling

A Reduced Error Pruning Tree (REPTree) is an implementation of a fast decision tree learner. A decision tree consists of internal nodes denoting the different attributes and the branches denoting the possible values of the attributes, while the leaf nodes indicate the final predicted value of the target variable. REPTree builds a decision/regression tree using information gain/variance and prunes it using reduced-error pruning. In general, a decision tree construction begins at the top of the tree (root node) with all of the data. At each node, splits are made according to the information gain criterion, which splits the data into corresponding branches. Computation on remaining nodes continues in the same manner until one of the stopping criterions is met, which include maximum tree depth, minimum number of instances in a leaf node, minimum variance in a node. Leave one out cross validation (CV) was used to evaluate model performance, which is essentially equivalent to n-fold cross validation, where n is the number of instances in the dataset. The results of Decision Tree Predictive Modeling are depicted in Figure 6.

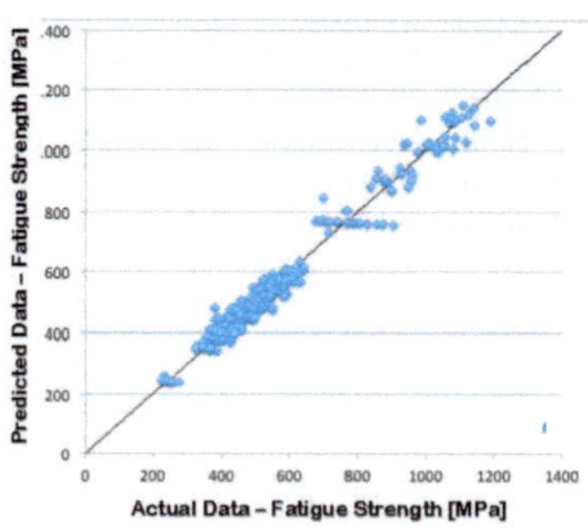

Figure 6: Scatter plot of the REPTree predictions for steel fatigue strength [r-squared = 0.9726]

Summary of our collective findings and future work

Given the exploratory nature of our work, it was encouraging to see good potential in various techniques in correlating material properties with manufacturing processes and compositions using. The r–squared error estimates for 85% training and 15 % test data for ANN regression or cross validation for other techniques have been above 0.96 prompting us to explore further along these directions.

- Despite the limited data-set available in the open literature, predictions were possible with a reasonable degree of accuracy
- The inherent non-linear nature of data prompts us to explore further non-linear methods
- Artificial Neural networks are able to capture the interaction between variables and therefore offer improved predictions

Besides the above techniques, other methods such as genetic programming, ensemble modeling, random forest, etc. will be apt candidates to be explored. Further, it would be worthwhile to explore the use of easily measurable mechanical properties to be included as input variables to predict difficult to measure properties such as the fatigue strength.

References

1. S. R. Kalidindi, S. R. Niezgoda, and A. A. Salem, "Microstructure informatics using higher-order statistics and efficient data-mining protocols", JOM, 63, pp. 34-41, 2011.
2. K. Rajan, "Materials Informatics", Materials Today, 8 (10) (2005), 38-45.
3. B. P. Gautham, R. Kumar, et. al. "More efficient ICME through materials informatics" Proceedings of 1st World Congress on Integrated Computational Materials Engineering, TMS 2011
4. L.A. Dobrzański, , M. Kowalski, J. Madejski, "Methodology of the mechanical properties prediction for the metallurgical products from the engineering steels using the Artificial Intelligence methods", Journal of Materials Processing Technology 2005
5. Sumantra Mandala, P.V. Sivaprasada et. al., "Artificial neural network modeling of composition–process–property correlations in austenitic stainless steels" Journal: Materials and Manufacturing Processes Volume 24, Issue 2, January 2009
6. Mohanty, S. Datta & D. Bhattacharjee, "Composition–Processing–Property Correlation of Cold-Rolled IF Steel Sheets Using Neural Network by Materials and Manufacturing Processes" Volume 24, Issue 1, 2008
7. H. K. D. H Bhadeshia, "Neural Networks in Materials Science" ISIJ (The Iron and Steel Institute of Japan) International, 1999
8. http://tsuge.nims.go.jp/top/fatigue.html

2nd World Congress
on Integrated Computational Materials Engineering (ICME)

Building Blocks for ICME

2nd World Congress on Integrated Computational Materials Engineering
Edited by: Mei Li, Carelyn Campbell, Katsuyo Thornton, Elizabeth Holm, and Peter Gumbsch
TMS (The Minerals, Metals & Materials Society), 2013

Towards an Integrative Simulation of Microstructural Response to Case Hardening of Microalloyed Steels

Patrick Fayek[1], Thomas Petermann[1], Ulrich Prahl[1]

[1]Department of Ferrous Metallurgy, RWTH Aachen University; Intzestr. 1, 52072 Aachen, Germany

Keywords: Microalloying, case hardening, grain growth, phase-field modeling, Zener force, precipitation simulation, integrative simulation

Abstract

Case hardening is more efficient when the process is carried out at elevated temperature. The process is though limited by grain coarsening phenomena which occur when the state of precipitates in the material is not sufficient. The presented approach comprises the simulation of the precipitate volume and size, dependent on chemical composition, temperature and previous processing steps and the subsequent simulation of the microstructural response using phase-field calculations including the actual precipitation state. The model is applied to microalloyed 18CrNiMo6-7 case hardening steels with different microalloying concepts. It is calibrated for one of the concepts and then applied on the other in order to show its feasibility and the absence of fitting factors. Simulation of the grain coarsening behavior in relation to the precipitate development due to i.e. ripening and dissolution is needed for the pre-evaluation of processing schemes and materials design. Therefore, a whole processing route has to be considered, which in general is currently the focus of the Cluster of Excellence "Integrative Production Technology for High-Wage Countries" as part of the research area "Integrative Computational Materials and Process Engineering (ICMPE)" [1]. Aim is to consider all relevant processes by linking various simulation programs. A virtual platform (AixViPMaP) is therefore under development for the communication and exchange of the simulation results [2].

Introduction

Carburization of steel is a treatment in which a surface layer up to several millimeters is enriched with carbon. It is usually carried out for several hours at temperatures around 950 °C and completed by a quenching process, leading to a hardened wear resistant surface layer, while retaining favorable strength and toughness properties within the core [3]. The process can significantly be shortened by acceleration of the carbon diffusion. This can be achieved by increasing the concentration gradient of carbon according to the first of Fick's laws [4], as i.e. realized in double/multi-stage treatments [5], leading to an increased diffusion stream into the material [4], or by increasing the process temperature, since the diffusion coefficient obeys an Arrhenius relation [4]. The latter may indeed be the most effective measure. A reduction of the soaking time of about 50 % is possible when increasing the soaking temperature from 950 °C to 1050 °C assuming a hardening depth of 1 mm [6].

Annealing treatments at these hardening temperatures necessitate the presence of precipitates to inhibit the formation of improper grain sizes. According to the Zener theory, precipitates induce a force which counteracts the driving pressure for grain growth [7]. At 950 °C it is possible to avoid grain coarsening throughout aluminum-nitride precipitates when a sufficiently high amount of aluminum and nitrogen is alloyed. A further elevation of the temperature is limited due to an increase of the solubility product, making higher alloying contents necessary, which itself is limited due to a decline of purity [8]. Elevated temperatures are also accompanied by a coarsening of the precipitate phase, thus decreasing the restraining force against grain coarsening [9].

Niobium- and titanium-carbonitrides exhibit significantly higher solidus temperatures, forming more stable precipitates [10]. The solubility products are lower [11, 12] and they are more stable against coarsening.

Materials and modeling methods

The investigated materials were microalloyed variants of an 18CrNiMo6-7 case carburizing steel. The respective chemical compositions are displayed in Table I. Variant A is an example of the commonly used aluminum microalloying concept for case hardening steels. Variant B in contrast is additionally alloyed with niobium to increase the amount of precipitates especially at elevated temperatures.

Table I. Chemical composition of the investigated steel variants in weight percent.

Element	C	Si	Mn	Cr	Mo	Ni	Al	Ti	Nb	N
Var. A	0.16	0.22	0.56	1.64	0.31	1.54	0.030	0.0024	0.005	0.012
Var. B	0.17	0.20	0.54	1.64	0.32	1.56	0.029	0.0025	0.031	0.012

The computational approach is, that according to the chemical composition the precipitation volume and particle size is calculated using MatCalc®. Subsequently, the Zener force is calculated from the simulation results and entered into Micress®. For this purpose a Java program was written, that automatically starts the precipitation simulation with a specified chemical composition. From the results, a time precipitation profile is exported and the Zener force is calculated. As function of the time the respective Zener force is automatically and incrementally distributed to the Micress® calculations.

The chemical compositions are used to calculate the precipitation volume and size using MatCalc®. In MatCalc®, after initialization of the starting conditions, the nucleation kinetics are calculated according to the extended classical nucleation theory (CNT). The nucleation rate $\dot{m}_{nuc}$ is given by Equation (1), wherein Z is the Zeldovich factor, β^* the atomic attachment rate, N_C the possible nucleation sites, $-\Delta G^*$ the activation energy for nucleus formation, k the Boltzmann constant, T the temperature, τ the incubation time and t the time [13].

$$\dot{m}_{nuc} = Z\beta^* N_C \exp\left(\frac{-\Delta G^*}{kT}\right) \exp\left(\frac{-\tau}{t}\right) \tag{1}$$

Subsequently, the growth and the chemical composition of each formed nucleus is calculated within the software using a mean-field approach and a thermodynamic extremal principle [14].

The Zener theory is based on the assumption that the grain boundary movement is restricted due to second phase particles. Thereby a retaining force is generated that counteracts the driving force for grain coarsening, which results from an ambition to minimize the intrinsic energy and which is also related to the gain curvature. The pinning force P_Z (see Equation (2)) [7] is equal to the interfacial energy γ divided by the critical radius R_C, which depends on the volume fraction of the precipitates f, and the precipitate radius r. In Micress®, the pinning force is specified in units of the critical radius according to Equation (2).

$$P_Z = \frac{\gamma}{R_C} = \frac{3f\gamma}{2r} \tag{2}$$

Micress® is based on the phase-field approach. It is based on the description of a microstructure where a phase-field value ϕ_i is assigned to every grain. Within the grain, the value is unity, while outside the grain the value is zero. In the interface of grains i and j the equation $\phi_i + \phi_j = 1$ is satisfied. In fact, for a microstructure of N grains, the constraint $\sum_{i=0}^{N} \phi_i = 1$ is fulfilled everywhere. For grain growth the time dependent evolution of the phase-field value $\dot{\phi}_i$ is described by Equation (3) [15].

$$\dot{\phi}_i = \sum_j \mu_{ij}\sigma_{ij}\left[\left(\phi_j\nabla^2\phi_i - \phi_i\nabla^2\phi_j\right) + \frac{\pi^2}{2\eta^2}\left(\phi_i - \phi_j\right)\right] \tag{3}$$

In this equation η is the interfacial spacing and σ_{ij} and μ_{ij} are the interfacial energy and mobility respectively. The interfacial energy and mobility depend on the anisotropy of the grains in contact. Both are temperature dependent, however the temperature dependence of the mobility is more severe and mainly controls the grain growth [15]. The interfacial energy for the simulations is set to a value of $\sigma_{ij} = 0.7 \cdot 10^{-4}\,\frac{\mathrm{J}}{\mathrm{cm}^2}$ according to [16], while the interfacial mobility is adjusted as described later. The temperature dependence of the mobility is described in Equation (4) with the universal gas constant R, the interlattice distance d_{ij}, the Debye-frequency ν_D according to [16] and the activation energy Q_{ij}.

$$\mu_{ij} = \mu_0 \exp\left(\frac{-Q_{ij}}{RT}\right) \text{ with } \mu_0 = \frac{d_{ij}^4 \nu_D}{kT} \tag{4}$$

The temperature and time profile used for the coupled simulations is displayed in Figure 1.

Experimental investigations

Annealing experiments were carried out in an induction furnace. The samples were cubic with an edge length of approximately 30 mm in each direction. The annealing cycle (see Figure 1a)) consisted of a solution heat treatment, a precipitation heat treatment and an isothermal soaking treatment. The isothermal soaking at different temperatures (950 °C, 1000 °C and 1050 °C) simulates the temperature exposure during carburization, although the carbon enrichment of the material is neglected in this step. A thermocouple was used to monitor the temperature during the heat treatment and samples were quenched after reaching the soaking temperature, defining the initial grain size distribution state, after 10 000 sec and after 6 h respectively. At several positions (see Figure 1a)) test

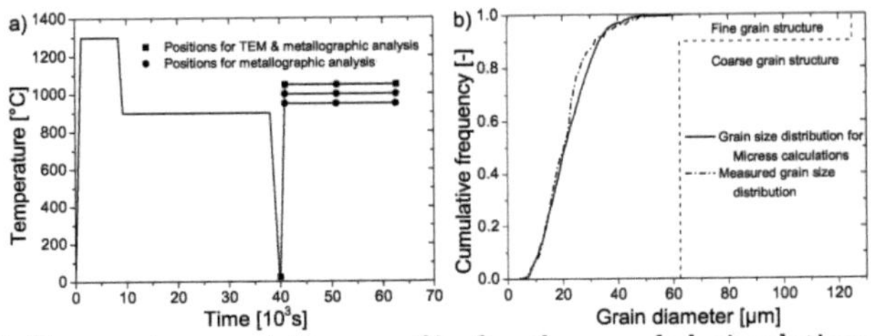

Figure 1. a) Temperature and time profile for the coupled simulations and for the experimental investigations. Positions at which samples for metallographic and TEM analysis were taken are indicated. b) Initial grain size distributions from metallographic analysis and used in the grain growth simulations.

specimens were taken for metallographic and TEM (transmission electron microscopy) investigations. The test specimens were quenched in order to achieve a martensitic microstructure making it possible to subsequently reveal the previous austenite grains and avoiding the formation of carbides. Previous austenite grains were investigated using an etchant of picric acid and the grain distributions determined. The specimens for the precipitation analysis were prepared as carbon replicas and the particle sizes determined by image contrast analysis. An example TEM image is given in Figure 2a).

The initial grain size distribution for the grain growth simulations was determined from the sample which was quenched directly after reaching a temperature of 1050 °C and an artificial microstructure was computed based on the results. The grain size distributions of the respective microstructures are displayed in Figure 1b).

Comparison of experimental and numerical results

Figure 2b) shows the comparison of the simulated and of the experimentally determined mean particle size. In the coupled simulations for variant B the mobility was adjusted, so that the simulated grain size distributions approximately match the experimentally determined. An activation energy for the mobility of $Q = 180\,\frac{\mathrm{kj}}{\mathrm{mol}}$ was determined to provide the best results. The mobility determined can be seen as an effective mobility since it also covers the deviation within the calculated and the real particle size. With the determined mobility, simulations for variant A were conducted. The results for both variants are shown in Figure 3.

Conclusions

The presented approach was successfully applied on different microalloying variants of an 18CrNiMo7-6 case carburizing steel. And industrial relevant time temperature profile was used for the experimental and numerical precipitation and grain growth simulation including a solution heat treatment, a precipitation heat treatment and a subsequent isothermal soaking.

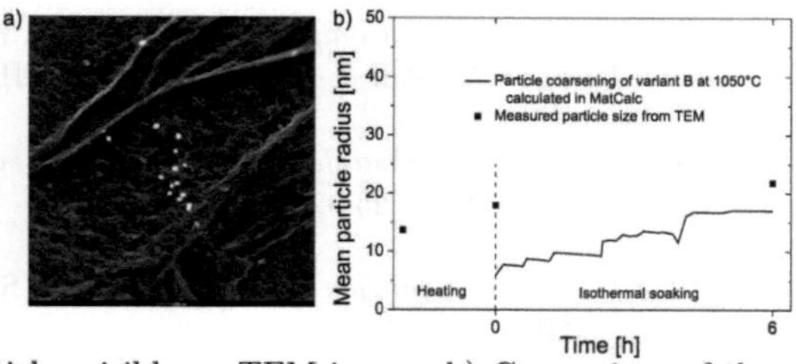

Figure 2. a) Particles visible on TEM image. b) Comparison of the measured and the simulated particle diameter as function of time for variant B

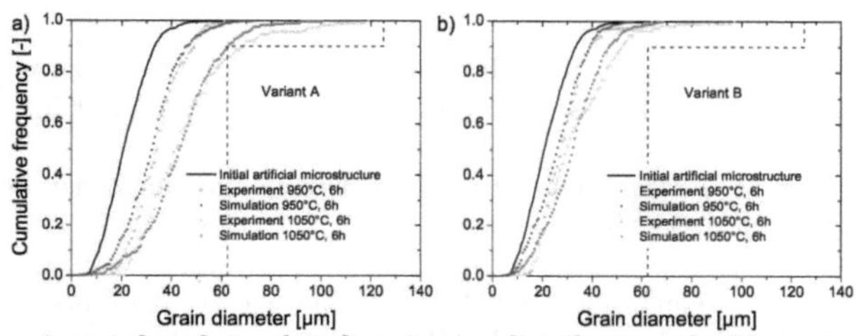

Figure 3. Experimental and simulated grain size distributions for Variant A a) and B b) for 950 °C and 1050 °C after 6 h soaking time.

An effective mobility was obtained using the modeling approach for the steel matrix without precipitations. The adjustment was only performed for steel variant B, but once the effective mobility was determined, it was possible to reproduce the grain size evolution of both microalloyed variants with quite good agreement. The evaluated mobility is however just an effective mobility since the deviation of the experimental and the simulated precipitation size is clearly noticeable. It thus also covers this deviation.

Arbitrary time temperature treatments may be analyzed and it is possible to consider various microalloying elements. With regard to fine grain stability, critical microalloying concepts or critical carburizing temperatures can be identified depending on time and precipitation state. It is thereby possible to significantly reduce the experimental effort for the alloying and processing design of microalloyed carburizing steels.

Acknowledgments

The authors would like to thank the German Research foundation DFG for the financial support of the depicted research work within the Cluster of Excellence "Integrative Production Technology for High-Wage Countries".

References

1. Christian Brecher, ed., *Integrative Production Technology for High-Wage Countries*, (Springer Publishing Company Berlin Heidelberg 2012).

2. G.J. Schmitz, U. Prahl, eds., *Integrative Computational Materials Engineering - Concepts and Applications of a Modular Simulation Platform*, Wiley-VCH Weinheim (2012).

3. Werner Theisen, Hans Berns, *Eisenwerkstoffe - Stahl und Gusseisen* (Springer Publishing Company Berlin Heidelberg, 2008), 234-240.

4. Günther Gottstein, *Physikalische Grundlagen der Materialkunde* (Springer Publishing Company Berlin Heidelberg, 2001), 145-181.

5. R. Khan, "Vacuum Gas Carburizing - Fate of Hydrocarbons", (Dr. thesis, University Karlsruhe, 2008), 1-8.

6. J. Kleff et al., "Hochtemperatur-Aufkohlen - Einflüsse auf das Verzugsverhalten schwerer Getriebebauteile", *Z. Werkst. Wärmebeh. Fertigung*, 60 (6) (2005), 311-316.

7. P.A. Manohar, M. Ferry, T.Chandra, "Five Decades of the Zener Equation", *ISIJ International*, 39 (9) (1998), 913-924.

8. B. Clausen et al., "Feinkornbeständigkeit von Bauteile aus dem mikrolegierten Werkstoff 18CrNiMo7-6 in Abhängigkeit der Prozesskette", *J. Heat Treatm. Mat.*, 65 (5) (2010), 257-268.

9. M. Maalekin et al., "In situ measurement and modelling of austenite grain in a Ti/Nb microalloyed steel", *Acta. Mat.*, 60 (2012), 1015-1026.

10. B.K. Panigrahi, "Processing of low carbon steel plate and hot strip - an overwiev", *Bull. Mater. Sci.*, 24 (4) (2001), 361-371.

11. T. Gladman, "Grain size control in steels", *The Metals Society, Metal Science Volume*, 8 (6) (1974), 167-176.

12. A.J. DeArdo, "Niobium in modern steels", *Int. Mat. Rev.*, 48 (6) (2003), 371-402.

13. J. Svoboda et al., "Modelling of kinetics in multi-component multi-phase systems with spherical precipitates I: Theory", *Mater. Sci. Eng. A*, 385 (2004), 166-174.

14. J. Svoboda et al., "Modelling of kinetics in multi-component multi-phase systems with spherical precipitates II: Numerical solution and application", *Mater. Sci. Eng. A*, 385 (2004), 157-165.

15. M. Apel et al., "Grain growth simulations including particle pinning using the multiphase-field concept", *ISIJ International*, 49 (7) (2009), 1024-1029.

16. R.G. Thiessen, "Physically based modelling of material response to welding", (Ph.D. thesis, University of Delft, 2006), 41-109.

DUCTILITY PREDICTION FOR COMPLEX MAGNESIUM ALLOY CASTINGS USING QUALITY MAPPING

Jiang Zheng [2], Mei Li [1], Joy Forsmark [1], Jacob Zindel [1],and John Allison[2]

1. Materials Research and Advanced Engineering Department, Ford Research Laboratory, Dearborn, Michigan 48121-2053, USA,
2. Department of Materials Science and Engineering, University of Michigan, Ann Arbor, Michigan, 48109-2136, USA,

Keywords: Magnesium alloy, High pressure die casting, Ductility, Quality mapping

Abstract

In this work, quality mapping (QM) approach was developed to predict the ductility for complex magnesium high pressure die castings. Carefully controlled casting experiments were conducted under 8 different casting conditions and the mechanical properties were experimentally determined at different locations in complex magnesium casting. Each processing condition was simulated using the MAGMAsoft and the simulation results were determined from each location. QM was established to predict the mean ductility and statistic variation in AM60 Magnesium casting.

Introduction

Magnesium alloys are the lightest metallic structural materials and are, hence, very attractive in such application as automotive and aerospace industries [1]. High pressure die cast (HPDC) Mg components in production automotive applications can have location and process-dependent mechanical behavior that need to be determined in order to simulate and predict the component performance accurately There are several commercial software that can predict the filling and solidification process for a given casting and processing condition [2,3,4]. However, few can predict the local microstructure and local mechanical properties accurately for a complex casting components.

Quality Mapping is a term used to describe the mapping of mechanical and other properties onto a casting simulation. In this study, quality mapping is used to predict the local ductility and yield stress in an AM60 Magnesium casting. Carefully controlled casting experiments were conducted and the mechanical properties were experimentally determined in different locations in a generic frame casting(GFC) with a nominal wall thickness of 2.5 mm, produced using a 900 T cold chamber high pressure die cast machine. Each processing condition was simulated using the program MAGMASOFT® and various outputs, or criteria functions, were manually determined from each location. A descriptive function was developed using a linear regression process and a program was written with MAGMASOFT® API to incorporate the coefficients calculated. This program was implemented as a postprocessor in MAGMASOFT® to show the map of the mechanical properties. In addition, the inherent statistical variability of the ductility, in particular, was addressed by examining the variability in the shot profile for castings produced under nominally similar conditions.

Experiment

Because many of the Mg castings used in current structural applications have a frame-like geometry, a generic frame casting (GFC) was developed to determine the effect of flow conditions and fast solidification on the local mechanical properties. A schematic diagram of the GFC is shown in Figure 1. AM60 alloy was used for this study and the composition is shown in Table I. Castings were produced under a variety of different processing conditions according to a Design of Experiment (DOE). Details of the processing condition are presented in Table II.

Table I Chemical Composition of Alloys [1]

Alloy	Al wt.%	Zn wt.%	Mn wt.%	Si wt.%	Fe wt.%	Cu wt.%
AM60	5.69	0.057	0.37	0.006	0.0053	0.0034

Flat axial tensile bars were excised from various locations in the castings to determine variability of local properties, as shown in Figure 1. The bars were 25.4 mm gauge, subsize samples, machined and tested according to ASTM standard B557 under quasi-static loading conditions, using extensometry to measure strain. Between 11 and 20 replicates were recorded for each location and processing condition to ensure adequate statistical sampling.

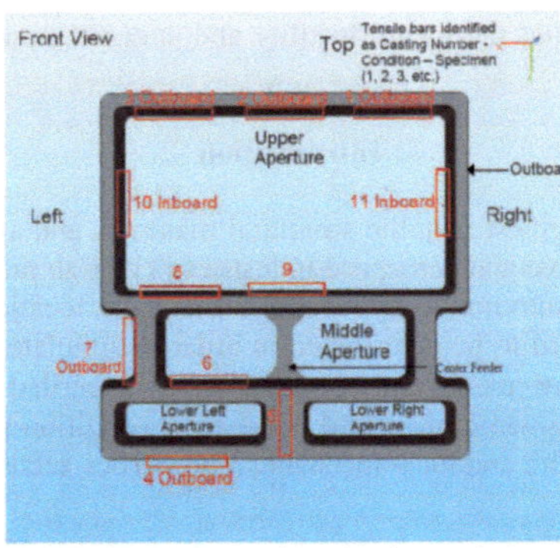

Figure 1. Geometry and locations of excised sample bars on GFC casting

Table II Actual values used in MAGMAsoft simulation [1]

Designation	Alloy	Fast Shot m/sec	Melt Temp. Degree ℃	Gating Configuration	Center Feeder
A	AM60	5.2	675	No plug	No
J	AM60	6.47	691	No plug	Yes
K	AM60	5.3	711	No plug	Yes
D	AM60	6.6	738	No plug	No
M	AM60	5.2	677	plug	Yes
G	AM60	5.08	731	plug	No
Q	AM60	6.09	727	plug	Yes
U	AM60	5.84	720	plug	Yes

Simulation

All processing conditions were recorded for use in the simulation later. The filling profile is important for MAGMAsoft simulation, which represents the melt flow from the gate to complete part fill. The filling profile was defined based on the shot-trace of each casting recorded by computer. Under each processing condition, the filling profiles of median, upper and lower bound conditions were determined using the sets of shot-traces, as shown in the Figure 2.

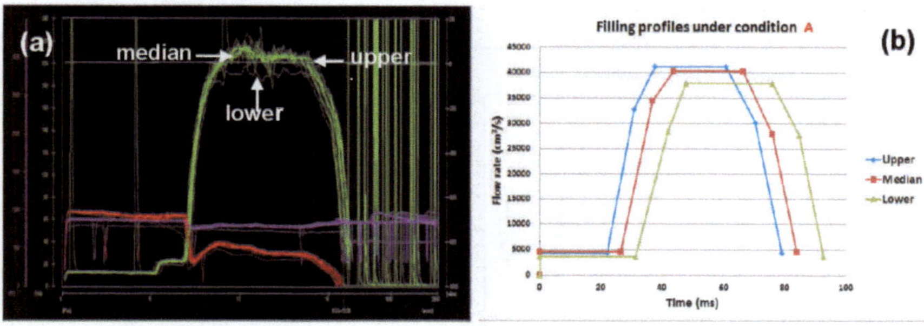

Figure 2. Shot-trace and filling profiles of condition A
(a) shot-traces for the set of castings, (b) filling profiles used in MAGMAsoft simulation under median, upper and lower bound conditions.

MAGMAsoft version 4.4 was used to investigate the filling and solidification processes for each condition in the DOE for the GFC casting. This version of MAGMASOFT generates a series of output data called criteria functions that quantify local aspects of flow and solidification conditions. The criteria functions investigated in this analysis were: flow length (FL), Air Entrapment (AE), Temperature at 100% Fill (100% T), Solidification Time (ST), Material Age (MA), and Air Contract (AC) [5]. The Flow Length calculates the distance that the melt has travelled from the inlet in each filled element. The Air Entrapment criterion function calculates the concentration of gas that has been trapped in the melt during the collapse of the air cavities. Temperature at 100% Fill is used to view the temperature distribution at the end of filling. Solidification Time displays the time from the start of the solidification (the end of filling) to the time when the temperature falls below the solidus temperature. Liquidus to Solidus shows the time it takes for regions to go from the liquidus temperature to the solidus temperature (the freezing range). Material Age shows the average age of the melt in each filled volume, which could be used to indicate the last solidified metal which usually has more defects (air entrapment or shrinkage porosity). Finally, the Air Contact Function calculates an average contact time between the melt and the air in each filled volume, which might be used to indicate the oxidation tendency for free surface.

The ductility of Mg casting would be predicted by establishing the quality mapping according to the criteria functions. For the quality mapping:

$$\text{Ductility} = C_0 + C_1{*}CF_1^{\,C2} + C_3{*}CF_2^{\,C4} + C_3{*}CF_2^{\,C4} \qquad (1)$$

Where CF is a criteria function that is predicted through the MAGMAsoft simulation and C_i is a set of constants.

Simulation and Experimental Results

Examples of some of the results of the excised sample testing for the GFC are shown in Figure 3. The results are presented in a box plot format. Box plots show the statistical variation in the data collected. The boxes consist of the first and third quartiles with the center line in the box showing the location of the median. The whiskers show the standard deviation and any outliers are represented with asterisk.

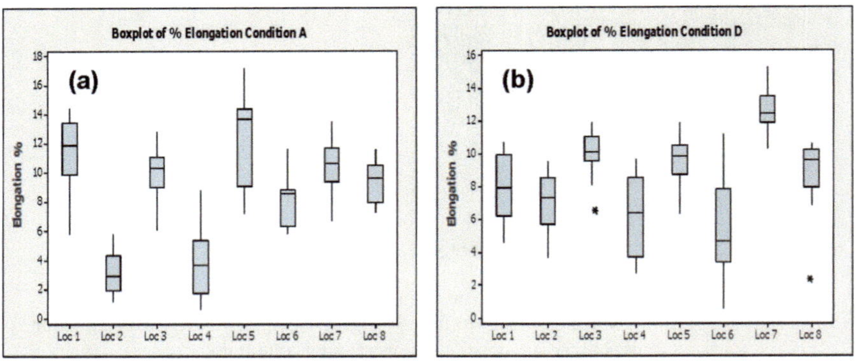

Figure 3. The strain at failure for the excised samples from the GFC casting produced under (a) condition A and (b) condition D

Criteria functions were collected from the MAGMAsoft simulations, as shown in Figure 4 for each location and processing condition. Based on the experience and some researchers' work [6,7,8,9], it was determined that values for flow length, solidification time, air contact, air entrapment, and temperature at 100% fill would be collected for each location.

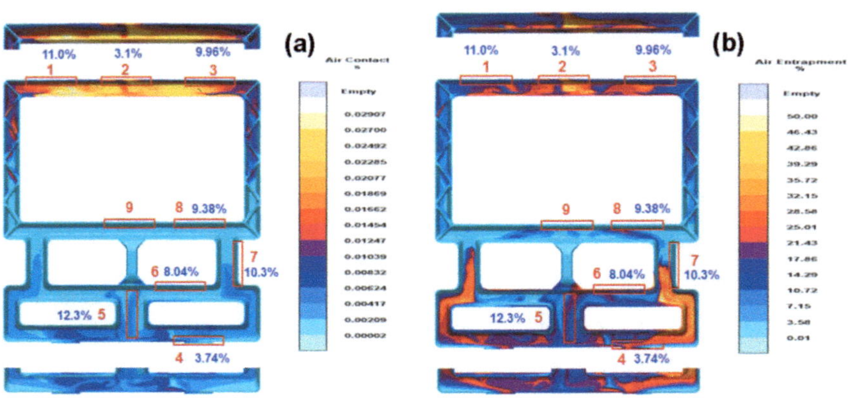

Figure 4. Simulation results (a) air contact and (b) air entrapment

Mean Ductility Prediction

All of the values for the different criteria functions were collected. However, these criteria functions had different orders of magnitude of their values and different units, which necessitated the use of normalization factors in the casting. Each criteria function was divided by a value specific to the criteria function. The maximum values were used in most cases. For example, the longest distance from the middle of the gate to the overflow on the upper window is 900mm, which was used to normalize. Additionally, typical air contact value is less than 0.02s. Liquidus to solidus was normalized to 4 sec, which is the maximum time of liquidus to solidus in the casting. Air entrapment was normalized to 35%. It was determined that these criteria functions that would be most critical were Air entrapment (AE), Air contact (AC), Flow Length (FL), Liquidus to solidus (LS), and Temperature at 100% FILL. The quality mapping was established to generate the correlation between the criteria functions and ductility by using ModelFRONTIER.

$$\text{Strain} = 26.934 - 19.594 * \text{LSnorm}^{0.19185} - 0.8745*\text{FLnorm}^{0.91371} - 5.8476*\text{AEnorm}^{4} - 7.0792*\text{ACnorm}^{0.92341} + 9.6375*\text{Tnorm}^{4} \quad (2)$$

The criteria functions were normalized as follows:
STnorm = Liq to Sol/4; AEnorm = AE/35; FLnorm = FL/900; ACnorm = AC/0.02; Tnorm = (T100%-620)/620

Figure 5 shows the measured strain versus predicted strain for the GFC castings. The fit for the strain equation was good with $R^2 = 0.57$.

In case of the condition A, the comparison of test and predicted elongation suggests a reasonable match, as presented in Figure 6. The quality mapping has been successfully tested on a production component.

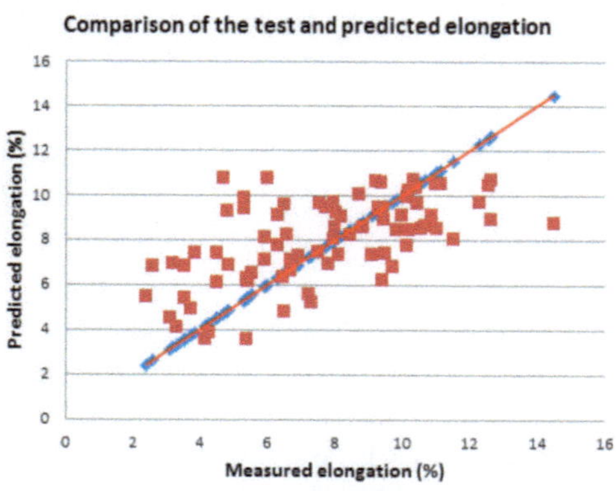

Figure 5. Measured elongation versus predicted elongation for AM60 castings

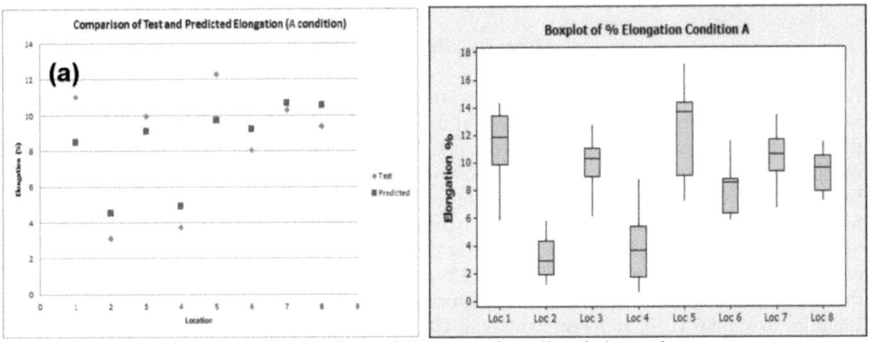

Figure 6. Comparison of the test and predicted elongation
(a) comparison of the test and predicted elongation (b) box plot of elongation

Statistic Variation Prediction

It's well established that the variation of ductility is attributed to the variation of processing parameter. For the high pressure die-casting, there are various processing parameters, such as initial temperature of the melt, initial temperature of the die, heat transfer coefficient of the boundaries and filling profile, etc., which might result in the different ductility. Among these factors, the filling profile plays the most important role in ductility control. In the case of GFC casting, under each processing condition, ductility was predicted by using formula (1) based on the filling profiles of median, upper and lower bound conditions (see Figure 2), which was used to estimate the statistic variation. Figure 7 shows the comparison of the test and predicted statistic variation for condition U. The trend appears to be good in the Figure 7 (a) and (b). Therefore, it could be believed that the quality mapping can predict the statistic variation.

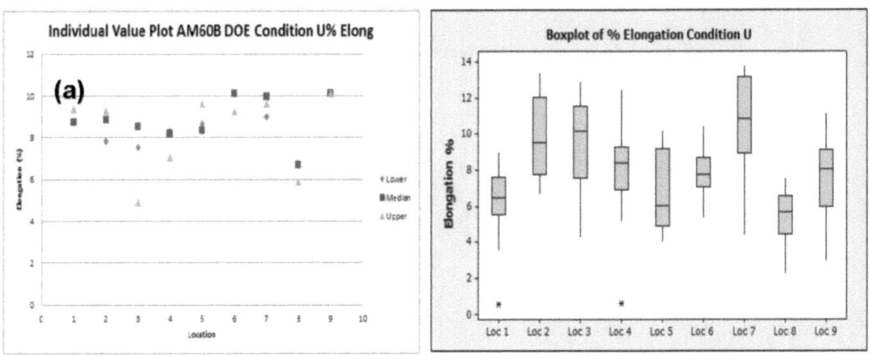

Figure 7. Comparison of the test and predicted statistic variation (a) comparison of the test and predicted statistic variation (b) box plot of elongation

Summary

In this work, a quality mapping approach was developed to predict mean ductility and statistic variation of ductility in complex magnesium castings.

1) Quality mapping for mean ductility was established using GFC castings and the predicted ductility is reasonable agreement with measured data from similar locations.
2) Initial results using Quality Mapping to predict the statistic variation for GFC casting are promising but additional work is required.

References

[1] Davis, J., SAE technical paper #910551. International Congress and Exposition. SAE International, 1991.
[2] Rudolf Seefeldt, Jörg C. Sturm, Alexander Pawlowski, Two for One – transferring proven filling characteristics from a single to a two cavity die casting die, Casting Plant & Technology 3(2007)
[3] Jiaqi Wang, Paixian Fu, Hongwei Liu, et al., Shrinkage porosity criteria and optimized design of a 100-ton $30Cr_2Ni_4MoV$ forging ingot, Materials and Design, 35(2012): 446-456
[4] Ole Köser, Badarinath Kalkunte, Dominic Brach, "Process development of Ingot casting using simulation approach", (Paper presented at the 1st International Conference on Ingot Casting, Rolling and Forging, Brussel-Saal, 2012), 1-5
[5] MAGMAsoft version 4.4 user manual
[6] Dorum, C., et al., Numerical Modelling of the Structural Behavior of Thin-Walled Cast Magnesium Components. International Journal of Solids and Structures,42(2005):2129-2144.
[7] Weiss, U., et al., Advanced Failure Modelling and Local Quality Mapping for Magnesium Castings in Crash Applications. Ford Research and Advanced Engineering Technical Reports, 2009. SRR-2009-0116.
[8] Dorum, C., et al., Numerical Modelling of Magnesium Die-Castings Using Stochastic Fracture Parameters. Engineering Fracture Mechanics, 76(2009): 2232-2248.
[9] Hepp, E., O. Lohne, and S. Sannes. Extended Casting Simulation for Improved Magnesium Die Casting in Magnesium: Proceedings of the 6th International Conference. 2004.

2nd World Congress on Integrated Computational Materials Engineering
Edited by: Mei Li, Carelyn Campbell, Katsuyo Thornton, Elizabeth Holm, and Peter Gumbsch
TMS (The Minerals, Metals & Materials Society), 2013

ADVANCED DILATOMETRY AND CALORIMETRY FOR THE VALIDATION OF MATERIALS MECHANICAL AND TRANSFORMATION MODELS

Michael Reich[1], Benjamin Milkereit[1,2], Matthias Krawutschke[1,2], Jan Kalich[3], Christoph Schick[2], Olaf Kessler[1]

[1]Chair of Materials Science, University of Rostock, Germany
[2]Polymer Physics Group, University of Rostock, Germany
[3]Institute of Manufacturing Technology, Dresden University of Technology, Germany

Keywords: dilatometry, calorimetry, steel, martensite, tempering, 22MnB5

Abstract

Dilatometry and calorimetry are in situ methods to detect changes microstructure during heating or cooling of materials. They have been used for the analysis design of a short time heat treatment of the martensitic, press hardening, automobile body steel 22MnB5. The use of these both complementary experimental methods allows a critical consideration of the established results.

Introduction

The martensitic, press hardening steel 22MnB5 is suitable for a wide range of applications in automotive bodies. The high strength-to-mass ratio opens up many attractive mass reduction opportunities. Press-hardened, high strength, deep-drawing sheet steel is state of the art in automotive engineering for reducing mass of complex components while simultaneously improving performance characteristics [1]. Components produced in this fashion are presently assembled using serial production techniques such as welding or bonding. A variety of assembly techniques are necessary because different material combinations become both more plentiful and also more specifically optimized. One possibility, to locally improve formability of press hardened 22MnB5 during mechanical joining, is rapid heating and cooling of the localised volume. At low peak temperatures below A_{c1} early stages of tempering and therefore also a softening of the initial martensitic microstructure takes place. This allows an easier plastic deformation in manufacturing processes. Several solid-state phase transformations are caused by heating of an initial martensitic microstructure. The different transformations are complex and overlap each other. The stages of tempering of iron-carbon martensite are generally categorized as five stages (e.g. [2]). Microstructural transformations lead to a change in specific volume of a specimen and can thus be characterised by dilatometry. Differential scanning calorimetry (DSC) monitors heat effects associated with phase transitions as a function of temperature. The measured heat flow difference between a sample to investigate and an inert reference sample is equivalent to enthalpy changes caused by microstructure change if normalised by scanning rate and sample mass (e.g. [3]). In literature, a wide range of experimental techniques has been applied to investigate the structural changes of martensite during tempering, in most cases for steel with a carbon content of about 1 mass% (e.g. [2, 4, 5]). In the material to investigate in this work, 22MnB5, the carbon content is about five times lower. Hence very sensitive measurement setups are needed. For the investigation of phase transformation of aluminium alloys, DSC has proven to be suitable to analyse phase changes of mass fractions of few percent during heating

[6] as well as during cooling from solution annealing (e.g. [3, 7]). Quenching and deformation dilatometry is a powerful technique to investigate phase transformation by related volume changes as well as mechanical properties e.g. of undercooled states (e.g. [8]).

Experimental

Material

The chemical composition determined by optical emission spectroscopy (OEM) of two different investigated batches of steel 22MnB5 are given in Table I. The two measurement techniques utilised (dilatometry and calorimetry) require different sample geometries. Hence, for this investigation on 22MnB5, two different sheet thicknesses and therefore also two different batches were investigated: for dilatometric measurements 2 mm sheet and for calorimetry 1 mm sheet respectively. At the moment, two methods for press hardening of steels like 22MnB5 exist: the direct and the indirect procedure. The direct procedure is the mostly used method. Thereby the forming process and the heat treatment are one single step. In this work, a direct press hardening process without deformation of the sheet was performed as follows: Placing the sheet in a furnace at 950 °C, austenitizing at 950 °C for 4 minutes, transfer from furnace to press tool, at closed press tool a contact pressure of 30 MPa for 20 seconds is present achieving a required cooling rate of min. 27 K/s down until 120 °C, further cooling to room temperature.

Table I: Chemical Composition of Steel 22MnB5, mass-fraction in %.

	C	Mn	Si	P	S	Al	N	Cr	Ti	B
batch 1 (DSC)	0.242	1.26	0.301	0.018	0.005	0.045	0.009	0.130	0.027	0.003
batch 2 (dilatometry)	0.222	1.18	0.255	0.020	0.007	0.045	0.009	0.189	0.036	0.003

Disc shaped DSC specimen with a diameter of about 6.3 mm where punched out from the press hardened sheet metal with 1 mm thickness. Samples for dilatometry were machined by fine eroding from a 2 mm sheet. These specimens were cuboids with a length of 10 mm and a cross section of 2x4 mm^2. For optimal evaluation, the calorimetric and dilatometric methods require baseline measurements to subtract effects caused by the devices. Therefore measurements of inert reference materials have proven to be suitable (e.g. [3]). For the measurement of press hardened 22MnB5, normalised samples have been produced as an inert reference material. Normalizing was performed at 920 °C for 30 min. The samples were cooled in the furnace over 24 hours. The whole process was done in vacuum. The initial hardness of the press hardened steel batches is about 540 ± 12 HV1. The initial hardness of the normalized 22MnB5 is about 146 ± 2 HV1.

Dilatometry

Press hardened and normalized states were investigated during heating in a temperature range from room temperature (RT) to 580 °C in DSC and 750 °C in dilatometry respectively. Continuous dilatometric analysis was performed using a quenching and deformation dilatometer type Bähr 805 A/D in its most sensitive mode, having a resolution for sample elongation of 0.05 μm / 0.05 K. The dilatometric specimens were induction-heated to a peak temperature at different constant heating rates. The specimens were heated from 30 to 750 °C with heating rates ranging over more than four orders of magnitude from 0.05 to 1000 K/s. The sample was mounted between quartz pushrods connected to a linear variable differential transformer (LVDT) for length measurement. Vacuum was kept for all testing times to minimize oxidation and decarburization of the outer sample surfaces. The temperature of the sample was measured with thermocouples spot welded to its outer surface. The samples expansion was recorded as a function of temperature.

At least three press hardened and one to two normalized samples were heated per rate. The changes in specific length of normalised and hardened samples during the continuous heating experiment with a heating rate of 0.5 K/s from room temperature to 750 °C are shown in Figure 1 A). For the hardened initial state a slope change is obvious. The normalized initial state shows no clear slope change in the investigated temperature range. In order to determine the range of the transformation temperatures, an interpretation of the difference between both elongations is appropriate. An example is given by the red line in Figure 1 A). To allow easy curve comparison with DSC-results, the first derivative of elongation difference with respect to temperature has been calculated (Fig. 1 B). For the evaluation of characteristic temperatures, the first significant negative difference between normalised and press hardened state as well as peak temperature are determined. Generally, the heating tests in the dilatometer are characterised by a very good repeatability.

A B

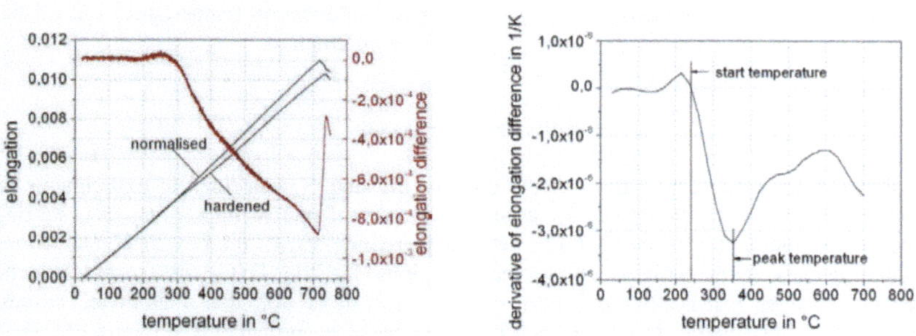

Figure 1: Comparison of dilatometric curves for normalised and hardened condition of 22MnB5 using the example of heating with 0.5 K/s.

Differential Scanning Calorimetry (DSC)

DSC experiments were mainly performed using a Perkin-Elmer Pyris 1 DSC. Samples were packed in pure Al crucibles and heated with varying heating rates to a temperature of 580 °C; limited to surely avoid melting of Al crucibles. Heating rates were varied between 0.5 and 5 K/s. Because the heat flow signal is depending on scanning rate and sample mass, calculation of specific heat is essential to be able to compare measurements at different heating rates [3]. By scanning a press hardened 22MnB5 sample versus a normalised 22MnB5 reference and subtracting a baseline measurement (scanning a normalised sample versus a normalised sample) for each heating experiment, the excess specific heat was calculated (Fig. 2). The used device in its special configuration (two stage mechanical cooler, block temperature -76 °C, star shaped guard ring inserts, N_2 purge 20 ml/min) was not able to control the heating rate exactly to the programmed value. This has been taken into account for specific heat calculation. Characteristic temperatures were evaluated as shown in Fig. 3. Peak separation was performed by fitting straight lines to the different peak flanks or shoulders. The peak temperature was evaluated respectively. Basically three different reaction peaks called "a, b and c" were observed in the heating rate range of 0.5 to 5 K/s as exemplarily shown for a scan with 0.5 K/s in Fig. 3. These three reactions are partially overlapping. Start temperature of peak "a" as well as peak temperature can be trusted as physically correct if thermal lag of the device is taken into account (thermal lag is maximal 10 K at 5 K/s for the used device). The temperature, which represents end of peak "a" as well as start of peak "b" is only an approximation.

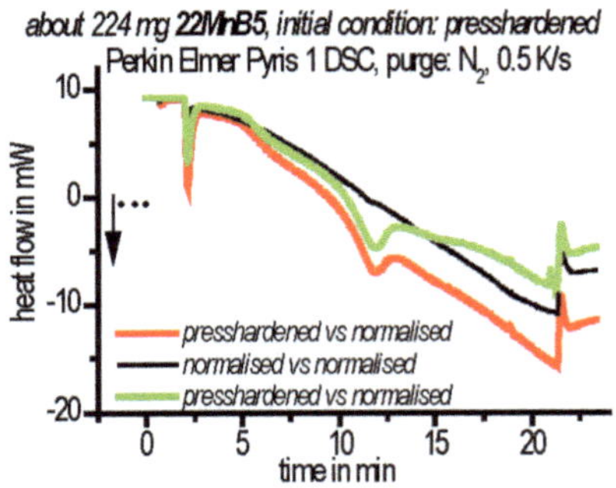

Figure 2: Exemplarily heat flow curves of sample and baseline measurements.

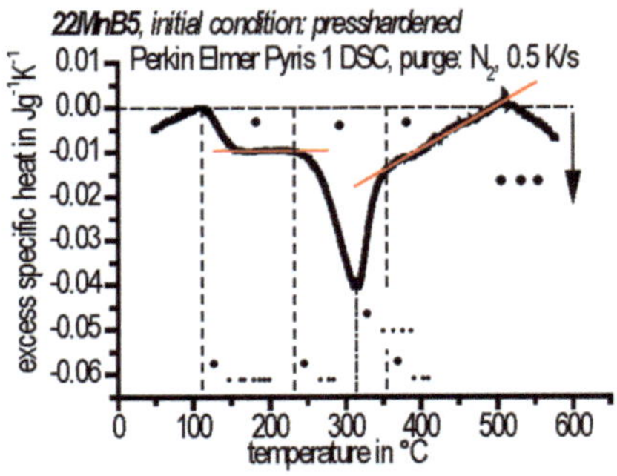

Figure 3: Example for evaluation of characteristic reaction temperatures out of the DSC measurements.

Results and Discussion

Fig. 4 shows selected heating curves of A) dilatometric and B) calorimetric experiments as functions of temperature. The curves are arranged in order of increasing cooling rate, starting with the slowest rate on top. The dotted line near each curve represents the zero level. Deviations undershooting this level indicate exothermal reactions for the DSC curves and volume differences between martensitic and normalised initial states for the dilatometer curves. Comparing the dilatometric and calorimetric results basic accordance is obvious. At slow heating, a reaction peak appears at about 350 °C. This peak is shifted to higher temperatures with increasing heating rate. Between 0.05 and 1000 K/s a shift of about 250 K is observed. It can be concluded, that microstructure modifications occurring during tempering are significantly depending on heating rate. For both methods, a clear end of the whole tempering process is not obvious. Rather it seems, that tempering is not completely finished at the highest investigated temperature. However, there are also some significant differences to observe. For a better comparison, Fig. 5 shows measurements at 0.5 K/s from dilatometry as well as from DSC. One significant difference is the number of reactions which are traceable. From the DSC curves, at least two reactions are obvious. Rather from the dilatometric curves the first reaction is not detectable due to its very small volume changes. Regarding the above mentioned tempering states, carbon clustering (state 0) has neither been detected by calorimetry nor by dilatometry. Although clustering probably will occur, its thermal and volume effects are too small for this low carbon steel. The first reaction, only visible in DSC, starting at about 150°C during heating with 0.5 K/s is attributed to the precipitation of ε/η transition carbide (state 1). The second reaction, visible by both methods, starting at about 250°C during heating with 0.5 K/s is attributed to the precipitation of Fe_3C carbide (state 3). Retained austenite (state 2) is not expected in steel 22MnB5. The results have been summarized in a continuous heating tempering diagram (Fig. 6), showing start temperatures of first and second reaction as well as peak temperature of second reaction depending on heating rate. All reactions are significantly shifted to higher

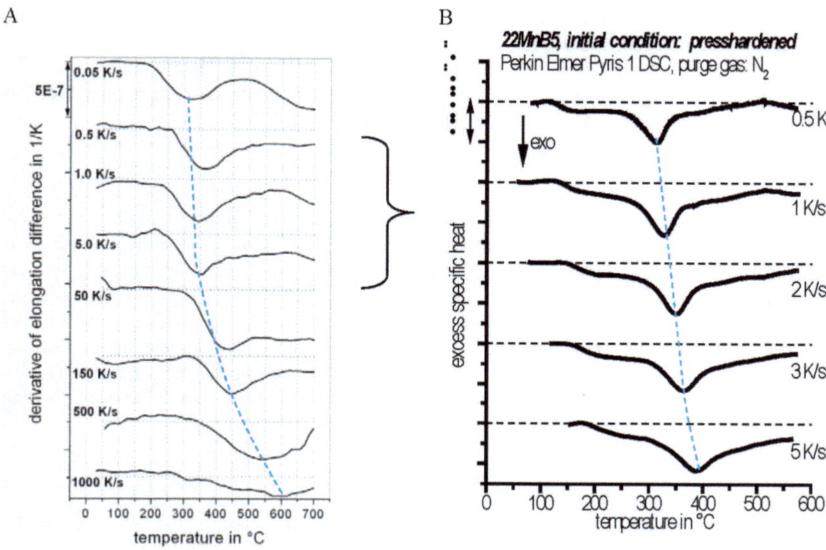

Figure 4: Selected dilatometric (A) and calorimetric (B) heating curves of 22MnB5 (initial condition press hardened) for all investigated heating rates.

temperatures with increasing heating rate. This diagram can be used to design well directed, short time heat treatment cycles of press hardened steel 22MnB5. The main reaction peak, which represents the most intensive reaction during the heating process, is detectable in both types of measurement. However, there is a difference in the peak temperatures. This difference might result from the different steel batches and will be further investigated.

It is obvious form dilatometer as well as from calorimeter results, that the whole tempering process is a complex sum of several overlapping reactions. Until now it is only roughly known

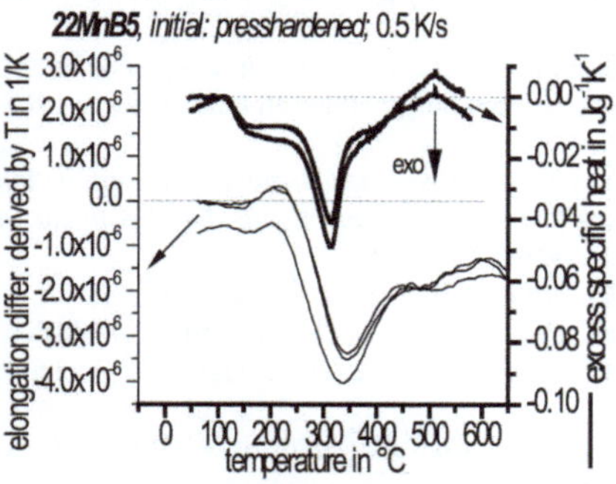

Figure 5: Comparison of dilatometric (left y-axis, fine lines) and calorimetric (right y-axis, fat lines) measurements at 0.5 K/s.

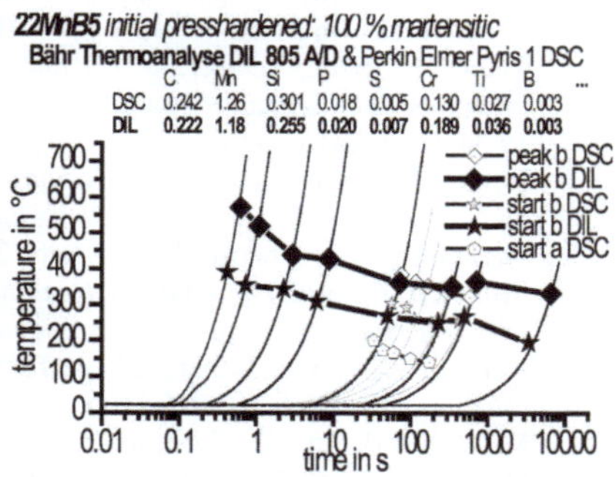

Figure 6: Continuous heating tempering diagram of 22MnB5.

which reactions are involved in steel 22MnB5. A direct comparison with literature is difficult, because most of the research done before have investigated steels with carbon concentrations of 1 % or even more. Hence, a first step to be able to perform reliable kinetic analysis will be intensive structure analysis. In a second step determination of activation energies by isoconversion methods (e.g. [14]) seem to be appropriate.

Summary

The use of complementary experimental methods (dilatometry, DSC,) allows a critical consideration of the established results and the design of a short time heat treatment for a press hardened steel 22MnB5. Calorimetry and dilatometry are essential in this investigation, because some of the phase transformations show a significant heat release but almost no length change, however others are attended with a distinctive length change. On the basis of the experimental results, the dependency of characteristic phase change temperatures on heating rate is verified. The results have been summarized in a continuous heating tempering diagram. This indicates a possible process window to improve formability of press hardened steel 22MnB5 during mechanical joining by designing appropriate short time heat treatments.

Acknowledgement

The authors gratefully acknowledge funding of this work by the Deutsche Forschungsgemeinschaft, project KE616/19-1 and FU307/9-1, in the framework of the Priority Program 1640 "Joining by Plastic Deformation".

References

1. Lenze, F.J., T. Heller, and S. Sikora, *Herstellung von Karosseriebauteilen aus warmumgeformten Stahlwerkstoffen*, in *EFB-Kolloquium*2005, EFB e.V.: Fellbach.
2. Morra, P.V., A.J. Böttger, and E.J. Mittemeijer, *Decomposition of iron-based martensite: A kinetic analysis by means of differential scanning calorimetry and dilatometry.* Journal of Thermal Analysis and Calorimetry, 2001. **64**(3): p. 905-914.
3. Milkereit, B., O. Kessler, and C. Schick, *Recording of continuous cooling precipitation diagrams of aluminium alloys.* Thermochimica Acta, 2009. **492**(Special Issue "Calorymetry on a Nano-scale"): p. 73-78.
4. Cheng, L., et al., *Tempering of iron-carbon martensite; dilatometric and calorimetric analysis.* Metallurgical transactions. A, Physical metallurgy and materials science, 1988. **19 A**(10): p. 2415-2426.
5. Primig, S. and H. Leitner, *Separation of overlapping retained austenite decomposition and cementite precipitation reactions during tempering of martensitic steel by means of thermal analysis.* Thermochimica Acta, 2011. **526**(1-2): p. 111-117.
6. Milkereit, B., et al., *Continuous Heating Dissolution Diagrams of Aluminium Alloys*, in *13th International Conference on Aluminum Alloys (ICAA13)*, H. Weiland, A.D. Rollett, and W.A. Cassada, Editors. 2012, TMS (The Minerals, Metals & Materials Society): Pittsburgh PA USA. p. 1095-1100.
7. Milkereit, B., et al., *Continuous cooling precipitation diagrams of Al-Mg-Si alloys.* Materials Science & Engineering A, 2012, **550**: p. 87-96.
8. Reich, M. and O. Kessler, *Mechanical properties of undercooled aluminium alloys and their implementation in quenching simulation.* Materials Science and Technology, 2012, Vol 28 (7), p 769-772.

2nd World Congress on Integrated Computational Materials Engineering
Edited by: Mei Li, Carelyn Campbell, Katsuyo Thornton, Elizabeth Holm, and Peter Gumbsch
TMS (The Minerals, Metals & Materials Society), 2013

THE 3D X-RAY CRYSTAL MICROSCOPE: AN UNPRECEDENTED TOOL FOR ICME

Gene E. Ice[1], John D. Budai[1], Eliot D. Specht[1], Bennett C. Larson[1], Judy W.L. Pang[1], Rozaliya Barabash[1], Jonathan Z. Tischler[2], and Wenjun Liu[2]

[1]Oak Ridge National Laboratory; 1 Bethel Valley Road; Oak Ridge, TN, 37831, USA
[2]Advanced Photon Source, Argonne National Laboratory;
9700 Cass Avenue; Argonne, IL 60439 USA

Keywords: Strain, Stress, Local Structure, Nye Tensor, Nondestructive Evaluation

Abstract

There is a long-standing debate over the length scales needed to understand the behavior of materials and the role of surfaces, defects, and inhomogeneities. Indeed the properties of most materials are ultimately determined by defects –including grain boundaries and surfaces-that are either introduced during processing or in-service, and defect density and distribution must be considered for high-fidelity integrated computational modeling and engineering. Scientists at ORNL together with partners at Argonne have developed a powerful 3D X-ray Crystal Microscope that can nondestructively characterize the local 3D crystal structure of polycrystalline materials with submicron resolution and with sensitivity to the local crystal structure, orientation, elastic strain tensor and the local Nye tensor. This emerging tool provides unprecedented tests of materials models under different processing/environmental conditions and provides new insights into the impact of unpaired dislocations, elastic strain and surfaces and interfaces. The promise of the 3D Microscope and the emergence of similar instruments at synchrotrons around the world will be discussed with respect to ICME.

Introduction

Although the emphasis in Integrated Computational Materials Engineering is on *Integration* and *Engineering*, the underlying complexity of materials modeling is a grand challenge that prevents widespread adoption of ICME [1]. To compensate, data-base methods can be used to extend our understanding of materials properties, but ultimately transformational products will depend on the ability to predict structure property relationships arising from processing and in-service. Indeed underlying ICME is a need to understand materials defects on all length scales.

The mesoscale- where the collective interactions of defects are difficult to model with atomic resolution and where the three-dimensional (3D) behavior of materials is difficult to characterize experimentally- is particularly challenging. This is a scale too large for atomic models, and often too deep for atomic characterization with electron microscopy. Here we describe the 3D X-ray Crystal Microscope [2,3] and how it can provide unprecedented information about local crystal structure and defect organization with mesoscale resolution. Because x-ray characterization is inherently nondestructive for many materials, the 3D characterization of local crystal structure can be used to test existing models and to guide the development of new models, which incorporate defect physics at a new level.

The 3D X-ray crystal microscope is based on Laue diffraction with the ability to accurately measure the energy of each Laue spot as needed [2]. Because a polychromatic beam is essential for Laue diffraction, achromatic focusing optics are essential and the 3D X-ray Crystal Microscope is a specialized version of polychromatic microdiffraction with the ability to resolve the structure of volumes both transverse to the beam and along the penetrating polychromatic x-ray beam. With Laue diffraction, 4 linearly independent reflections are sufficient to determine the unit cell shape and orientation and if the energy of at least one reflection is determined, then the unit cell volume is also measured [2]. In practice, even simple crystal structures reflect 8-20 Laue spots for x-ray bandpasses from 12-18 keV. Precision measurements of local unit cell parameters is made possible by the use of submicron focused x-ray beams, with highly accurate x-ray sensitive area detectors. The general scheme for a 3D X-ray crystal microscope is shown in Fig. 1. As illustrated, X-rays from a brilliant synchrotron source are focused to a submicron spot using wide-bandpass total-external-reflection mirror optics. Laue patterns generated by the penetrating beam, are measured by an area detector set normal to the incident beam. The overlapping patterns from each subgrain illuminated volume element along the beam are disentangled by passing an absorbing 20-80 μm diameter wire near the sample surface.[3] As the wire moves, changes in the diffraction pattern are due to rays that pass near the front or back surface of the wire [3]. By ray tracing from each pixel on the area detector, past the edge of the wire and onto the incident beam, the individual Laue patterns are determined for subgrain volumes along the beam; because the wire is much closer to the sample than to the detector, a 25-100 μm displacement on the detector has a submicron effect on the ray traced origin of the scattering.

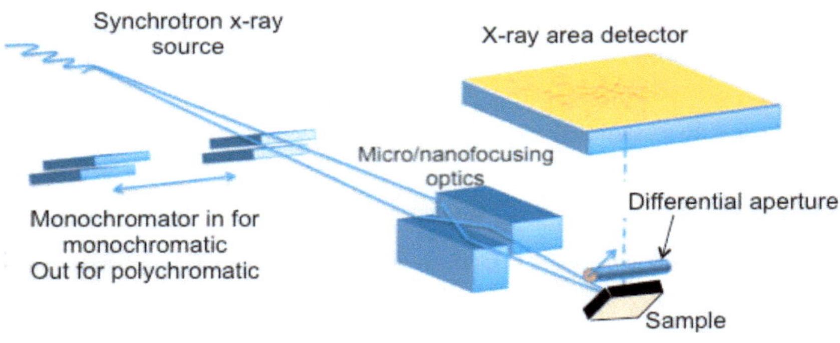

Fig. 1. Key components of a 3D X-ray crystal microscope. The ultra-brilliant x-ray source produces a polychromatic beam that is focused by nondispersive elliptical x-ray supermirrors. The beam can either be passed onto the mirrors as a monochromatic or a polychromatic beam. Diffraction from the sample is detected by an area detector and the signal from along the penetrating beam is resolved using the differential aperture, which passes close to the sample surface.

Although the method sounds simple, there is a considerable sophistication in all aspects of the instrumentation and software. Precision elliptical x-ray super mirrors and alignment hardware are needed to focus the incident beam to 200-500 nm [4,5]. Special stages are needed that can work with tens of nanometer accuracy in a "fly scan" mode for rapid data collection. Sophisticated data handling is required to process the ~1 billion bits/sec generated at a 10 frame/sec data collection rate. Over a 24 hour period, this adds up to around 80 terabits ($80x10^{12}$). In addition, a specially designed monochromator is used to precisely determine the energy of Laue spots or to obtain full reciprocal space volumes from samples [6]. This monochromator must have an absolute energy calibration and must have the ability to sweep wide energies and be inserted into the beam or withdrawn as needed. Below we describe some example applications of 3D X-ray Crystal microscopy and how they impact ICME.

Applications

Heterogeneous Deformation in Polycrystalline Materials

It is well known that polycrystalline materials deform in a heterogeneous manner. Deformation inhomogeneities arise from crystallographic orientation anisotropies, from discontinuities set by surfaces/interfaces and from inhomogeneous defect distributions or boundary conditions. Although most previous work has concentrated on ensemble average behaviors in polycrystalline materials, the development of the 3D X-ray crystal microscope now allows for a nondestructive evaluation of local crystal structure *before and after* deformation. This allows for detailed studies of the local conditions around a subgrain sample volume and how the local conditions influence deformation. Questions of interest include, how do surfaces affect deformation? How does grain boundary type influence the constitutive equations and how do surrounding grain orientations influence local deformations?

Ensemble-average strain measurements in polycrystalline materials have identified significant differences in the average strain of different phases or of grains with different crystallographic directions with respect to the macroscopic strain axis. These differences are understood in terms of compliance of the heterogeneous materials or anisotropic grains [7,8]. More recently, high-energy spatially-resolved local crystal structure measurements at the ESRF provided a first test of theories describing heterogeneous *rotations* of grains depending on their crystallographic orientation with respect to the macroscopic deformation direction [9]. These experiments identified both a strong tendency to follow the rotation direction predicted by the original grain orientation, and some anomalous behaviors. Two experiments led by Pang and co-workers now provide tantalizing insights into the fundamentally important role of surfaces and interfaces in heterogeneous polycrystalline deformation.

In one experiment [10], the local orientations of crystal volumes near grain boundaries were measured. The sample was then deformed and re-measured. The *changes* in crystal orientation across the grain boundaries were plotted against the original difference in orientation. As might be expected, there is a roughly linear relationship between the grain boundary misorientation before deformation and the *change* in misorientation after deformation. However, for some coincident site lattices, there are strong deviations from the overall trend, with a wide distribution in misorientation change including some cases with almost no change (Fig. 2 left). This points to the important role of grain boundary type in local deformation. Moreover, measurements of many deformed polycrystalline samples find a general trend for much larger curvatures in the crystallographic lattice near grain boundaries and surfaces than in grain cores. It appears that the core of individual grains tend to rotate as a unit, while the surface accommodates the rotation with respect to the surrounding grains.

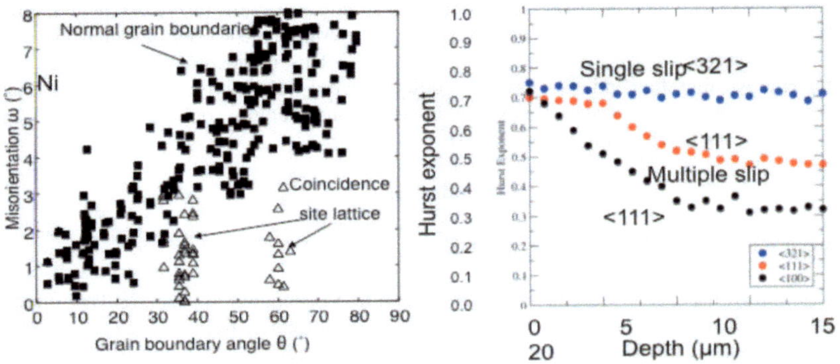

Fig 2. Left. The change in misorientation across a grain boundary follows a roughly linear relationship except near coincidence site lattices. Right. The Hurst coefficient correlating rotations near the surface of a deformed crystal has a universal behavior independent of crystal orientation at the surface, but is dependent on crystal orientation deeper into the sample.

An even more interesting set of experiments has looked at the deformation structure of a single crystal near its surface. Previous experiments with surface roughness found that roughness after deformation had a self-affine nature with a universal Hurst coefficient independent of crystallographic orientation [11]. A question of interest, was whether such a universal behavior also applied to crystallographic rotations, and if the nature of the rotations would change below the free surface. Pang and co-workers found that rotations had the same universal Hurst coefficient as roughness at the surface of crystals, independent of crystallographic orientation, but that below the surface, the Hurst coefficient changed (Fig 2 right) [12]. The depth below the surface where the Hurst coefficient changed depended on the number of slip systems activated by the deformation geometry. These results point to the importance of free surfaces, grain orientation and grain boundaries in local deformation with important implications for the constitutive equations at grain boundaries and surfaces.

Nye Tensor

The Nye Tensor is a fundamental variable in finite plasticity models, yet measurements of the full Nye tensor have not previously been possible; the Nye Tensor depends on curvature, and elastic strain in a 3D volume. Now the 3D X-ray Crystal Microscope allows for precision nondestructive measurements of local crystal curvature and the elastic strain tensor in three dimensions; this information is required for a direct comparison to finite plasticity models at a local (submicron) level. Because local curvature can arise both from unpaired dislocations, and from elastic strain gradients a simple measurement of lattice curvature is insufficient, as illustrated in a series of experiments by Larson and co-workers [13]. In these measurements single crystals of Si were bent above and below the elastic to plastic transition temperature. In measurements below the transition, maps of the Nye tensor are featureless and show structure at the uncertainty of the measurements whereas maps of the elastic strain tensor follow the anticipated patterns from bent beam mechanics (Fig 3a). For crystals bent above the elastic to plastic transition temperature the Nye Tensor maps show a strong spatial dependence whereas the elastic strain tensor m[text where caption should be here]aps are featureless (Fig 3b). This proof-of-principle experiment opens up the possibility of direct verification of plasticity models for inclusion into ICME.

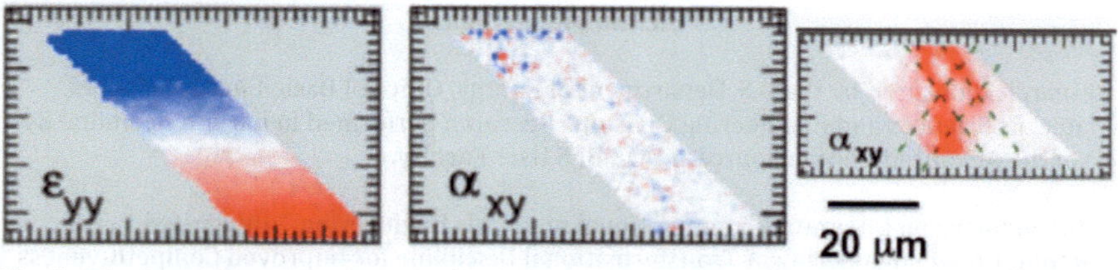

Fig. 3. Elastic strain tensor element (right) for elastically bent silicon beam, dislocation tensor middle for the same elastically bent silicon beam, and dislocation tensor from a plastically bent silicon beam [ref. 13]. The full scale color is +/- 3 x 10^{-3} for the elastic strain, and 3 mrad/µm for the middle and right images.

<u>Grain growth</u>

Thermal treatment after deformation is a classic tool for controlling the microstructure of materials and as such is an essential element that must be clearly understood for ICME modeling of materials microstructure/processing behavior. Current models of grain growth are dominated by measurements of surface grain structures, which have inherent assumptions, or by serial sectioning which provides ensemble-average information about grain sizes and distributions. However, detailed nondestructive volumetric maps before and after grain growth are just now becoming a reality. Such maps can be obtained using high-energy monochromatic three and four-dimensional (time) measurements [14] or with the 3D X-ray crystal microscope. Although the 3D X-ray crystal microscope is considerably slower in collecting data, it provides submicron resolution, and simultaneously quantifies the unpaired dislocation density. This information is unprecedented and provides a direct measurement of the grain boundary motions and grain refinement to test models. Early measurements by Budai indicate that the defect density of grains plays an important role in future grain boundary motion [15]. Future measurements will guide and test models of grain growth with unprecedented fidelity.

Summary and Resources Around the World

X-ray microscopy is an emerging field, which provides new insights into materials behavior [16]. The promise of x-ray microscopy in general, and polychromatic methods in particular is driving the development of new instrumentation around the world. The principles for the 3D X-ray Crystal Microscope were first developed on beamline 7 at the Advanced Photon Source (APS), and then transferred to a permanent station 34-ID-E at the APS. About the same time, a two-dimensional polychromatic microprobe was installed at the Advanced Light Source, and it has since been upgraded with a higher brightness super bend source and with the hardware for 3D imaging. Other polychromatic beamlines exist at the Canadian Light Source, the Pohang Light Source and the Swiss Light Source, with other development programs at the European Synchrotron Radiation Facility, Spring 8 and elsewhere. Additional beamlines are under development at the Australian Light Source and the Taiwan Light source. There are also rapid developments in high-energy monochromatic techniques, which seek to achieve 3D submicron resolution grain mapping with much faster data collection. This new instrumentation and technique development will provide unprecedented opportunities for guiding and testing mesoscale modes with direct implications for ICME.

Acknowledgement

Research sponsored by the U.S. Department of Energy, Office of Basic Energy Sciences, Materials Sciences and Engineering Division. Research performed in part on Beamline 34-ID of the Advanced Photon Source, a DOE BES User Facility.

[1] Committee on Integrated Computational materials Engineering, " Integrated Computational Engineering: A Transformational Discipline for Improved Competitiveness and National Security", *National Research Council*, 2008.
[2] J. S. Chung and G. E. Ice, "Automated Indexing for Texture and Strain Measurements with Broad-Bandpass X-ray Microbeams," *J. Appl. Phys.* 86, (1999) 5249-5256.
[3] B. C. Larson, W. Yang, G. E. Ice, J. D. Budai, and J. Z. Tischler, "Three-Dimensional X-Ray Structural Microscopy with Submicrometre Resolution," *Nature* 415, (2002) 887.
[4] G. E. Ice, J.-S. Chung, J. Z. Tischler , A. Lunt, and L. Assoufid, "Elliptical X-ray Microprobe Mirrors by Differential Deposition," *Rev. Sci. Inst.* 71, (2000) 2635-2639.
[5] W.J. Liu, G.E. Ice, L. Assoufid, C.A. Liu, B. Shi, R. Khachatryan, J. Qian, P. Zschack, J. Z. Tischler and J.Y. Choi, "Achromatic nested Kirkpatrick-Baez mirror optics for hard X-ray nanofocusing", *J. Synch. Rad* 18 (2011) 575-579.
[6] G. E. Ice, J.-S. Chung, W. Lowe, E. Williams, and J. Edelman, "Small-displacement Monochromator for Microdiffraction Experiments," *Rev. Sci. Inst.* 71, (2000) 2001-2006
[7] J. W. L. Pang, T. M. Holden, T.E. Mason and R. Panza-Giosa, "Texture and residual strain in an A17050 billet", Physica B 241 (1997) 1267-1269.
[8] J.W.L. Pang, T.M. Holden, T.E. Mason, "The development of intergranular strains in a high-strength steel", J. Strain Anal. For Eng. Design, 33 (1998) 373-383.
[9] L. Marguilies, G. Winther, H.F. Poulsen, "In situ measurement of grain rotation during deformation of polycrystals", Science 291 (2001) 2392-2394.
[10] J.W. L. Pang, W. Liu, J.D. Budai and G.E. Ice, "Inhomogeneous deformation behavior in intercrystalline regions in polycrystalline Ni", Submitted *Acta Mat.*
[11] P. Hahner, K. Bay, M. Zaiser, "Fractal dislocation patterning during plastic deformation", *Phys Rev. Lett* 81 (1998) 2470-2473.
[12] J.W.L. Pang, G.E. Ice, and W. Liu, "The role of crystal orientation and surface proximity in the self-similar behavior of deformed Cu single crystals", Mat. Sci and Eng. A 528 (2010) 28-31.
[13] B.C. Larson, J.Z. Tischler, A. El-Azab and W. Liu, "Dislocation Density Tensor Charcterization of Deformation Using 3D X-ray Microscopy", Journal of Eng. Mat. And Tech. 130 (2008) 021024-1.
[14] D. J. Jensen, E. M. Lauridsen, L. Marguilies, H. F. Poulsen, S. Schmidt, H. O. Sorensen and G.B. M. Vaughan, "X-ray microscopy in four dimensions", Mat. Today 9 (2006) 18-25.
[15] J.D. Budai, W. Yang, B.C. Larson, J.Z. Tischler, W. Liu, H. Weiland and G.E. Ice, " Three-Dimensional Micron-Resolution X-ray Laue Diffraction Measurement of Thermal Grain-Evolution in Aluminum", *Mat. Sci. Forum* 4670470 (2004) 1373-1378.
[16] G.E. Ice, J.D. Budai and J.W.L. Pang, "The Race to X-ray Microbeam and Nanobeam Science", Science 334 (2011) 1234-1239.

2nd World Congress on Integrated Computational Materials Engineering
Edited by: Mei Li, Carelyn Campbell, Katsuyo Thornton, Elizabeth Holm, and Peter Gumbsch
TMS (The Minerals, Metals & Materials Society), 2013

3D HYBRID ATOMISTIC MODELING OF β'' IN Al–Mg–Si: PUTTING THE FULL COHERENCY OF A NEEDLE SHAPED PRECIPITATE TO THE TEST

Flemming J. H. Ehlers[1], Stéphane Dumoulin[2], Randi Holmestad[1]

[1]Dept. of Physics, Norwegian University of Science and Technology (NTNU); 7491 Trondheim, Norway
[2]SINTEF, Materials and Chemistry; 7465 Trondheim, Norway

Keywords: Al–Mg–Si alloy, β'' phase, hybrid modeling, *ab initio* theory

Abstract

A key input of a truly predictive integrated computational materials engineering (ICME) scheme for an age hardenable Al alloy is the formation enthalpies – including interfacial and strain contributions – for the main hardening precipitate(s). The basic desire to compute these numbers with *ab initio* methods for essentially all relevant precipitate sizes continues to face limitations in the context of the associated requirements for the model system extensions. These obstacles manifest themselves in particular when considering a density functional theory framework based description of the full precipitate-host lattice interface – needed in order to incorporate accurately electronic interactions as well as the strain evolution along high misfit directions. Recent work within our group has made it possible to carry out this interface modeling for a fully coherent precipitate at a comparatively weak level of approximation. We describe here our first attempts to employ this scheme for 3D hybrid modeling of fully coherent needle-shaped β'', the main hardening phase in the Al–Mg–Si alloy system. Examining a physically sized precipitate, we found this structure to fully adapt to the host lattice along its main growth (needle) direction, with the cell dimensions in the precipitate cross-section falling non-negligibly below the experimental values for both compositions (Mg_5Si_6, $Mg_5Al_2Si_4$) tested. Further, the theoretical value of 107.8° for the β''–Mg_5Si_6 monoclinic angle β_P is markedly off the experimental value of 105.3°±0.5°, potentially supporting the presence of non-negligible amounts of Al in the β'' phase.

Introduction

Age hardenable Al–Mg–Si alloys are widely used by the industry due to desirable properties such as high strength-to-weight ratio, good formability, and high corrosion resistance. At peak hardness, the alloy microstructure is dominated by – and sometimes solely comprising – the fully coherent needle shaped β'' precipitate [1]. This 'single phase scenario' provides an optimal test-bed, due to its comparative simplicity, for theoretical simulations of the connection between mesoscopic (microstructure) and macroscopic (e.g. material hardness) system properties. The ultimate theoretical aim is a *parameter free* model description, where the quantities describing the drive for precipitate nucleation, growth, and coarsening are obtained from first principles. Such a scenario would remove tuning parameters, presently employed for achieving acceptable agreement between theory and experiment on the one hand, but placing obstacles on the transferability of the model scheme on the other hand.

At the heart of understanding properly precipitate evolution and influence – phase stability, interfacial energy, and the strain field generated in the host lattice – lies an accurate description of a single precipitate embedded in face-centered cubic (fcc) Al (see, e.g. [2, 3]). For the case of β" in Al–Mg–Si, the minute precipitate cross-section dimensions (typically 4×4 nm) [1] – along with the fully coherent nature of the precipitate [4] – stress the potential need for modeling the *entire precipitate cross-section* within an *ab initio* based theoretical framework, in order to achieve sufficiently accurate values for the desired parameters. In particular, the strong precipitate-host lattice misfits in the cross-section plane, responsible for limiting the precipitate size along these directions, induce significant strain levels in the two subsystems, beyond the limits of validity of linear elasticity theory (LET).

For physically sized precipitate cross-sections, a model system that ignores strain evolution along the β" needle direction (treating the precipitate as formally infinite by way of periodic boundary conditions) still easily contains beyond thousand atoms. This number is problematic from the point of view of density functional theory (DFT) studies. Recent considerations within our group [5] have established a theoretical framework for confining the DFT based computations to a narrow (≈ nm) region enclosing the precipitate-host lattice interface. In these calculations, the interface is modeled with a *sequence* of supercells, making computational demands scale linearly, with the number of cells used. As each cell in our studies contains only 44 atoms [5], the system size restriction mentioned above is essentially overcome. Further, compared to present day supercell based interface studies, the level of precision in our scheme is preserved for any given local region on the interface, *provided* that the applied cell distortion correlates with the desired position on the interface. In other words, the local interface studies are everywhere relying – through a structural boundary condition – on information obtained from the remaining parts of the β"/Al system. Such a coupling forms the basics of a hybrid scheme [6].

In this work, building upon considerations [7] for β"/Al 2D slab geometries, a proposed version of such a hybrid scheme is introduced, along with results of our first attempts to apply this scheme to a true 3D β"/Al system – with strain evolution allowed along all directions. The basic details of scheme and model system are outlined in the 'Methodology' section. We examine the two experimentally reported compositions – Mg_5Si_6 [4] and $Mg_5Al_2Si_4$ [8] – for the monoclinic β" structure (space group C2/m [4]) in the 'Results and discussion' section. Here, precipitate structural parameters and energies are compared to experimental data and a presently available, alternative model scheme [9], respectively.

Methodology

In our hybrid scheme, the full β"/Al system is investigated within LET, using the finite element commercial code LS-DYNA [10]. The subsystem (strain) interactions are described using orthotropic elasticity. Elastic constants were computed with DFT, as implemented in the plane wave (PW) based code VASP, with the Perdew-Wang generalized gradient approximation (GGA) to the exchange-correlation functional. For details on code and concepts, see [11]. With VASP used also for the interface region studies, this approach is compatible with (i) a smoothly varying stress-strain subsystem behavior across the interface region boundaries, and (ii) well defined fully relaxed bulk subsystem energies. Earlier 2D slab studies [7] suggest that structural distortions outside the interface region are well approximated within LET.

The VASP studies have employed a 225 eV PW cut-off and a (12, 22, 14) *k*-point grid ((22, 22, 22) *k*-point grid) for the bulk β" primitive unit cells (fcc Al conventional unit cell). For details on the β"/Al supercells, see [5]. Most calculated elastic constants for β"–$Mg_5Al_2Si_4$ have been outlined in [5], where also the three fcc Al elastic constants have been specified. For the 3D simulations, the constants c_{44} = 0.2635 (×10^{11} N/m²), c_{66} = 0.3234 were added for the precipitate. For β"–Mg_5Si_6, a new set of elastic constants were computed. The main change, compared to the β"–$Mg_5Al_2Si_4$ values, was a reduction in the parameter c_{55} – describing the resistance to shear strain in the cross-section plane – by ≈ 35%.

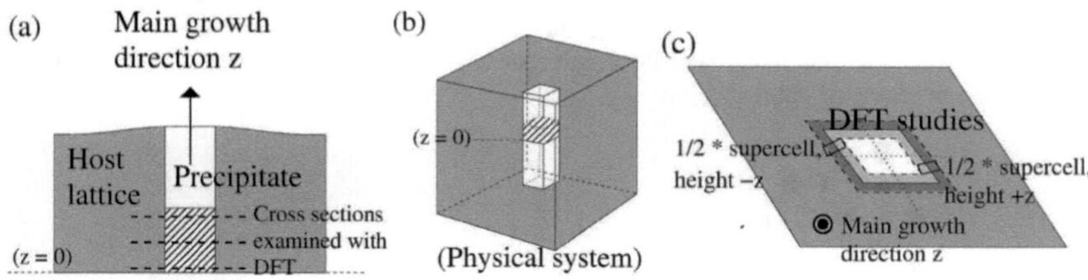

Figure 1. Schematic description of the β"/Al model system. (a) The chosen geometry subjected to LET studies. The shaded region highlights the part where results may compare well with the physical system, shown in (b). (c) Selected precipitate cross-section region from (a). DFT studies are performed in the interface region, enclosed by dashed lines. See text for details.

Figure 1 (a) displays the β"/Al model system (slab geometry) underlying our simulations. The precipitate has several unit cells width in all basis vector directions, but is only surrounded by Al in the cross-section. In simulations, the slab base (*z* = 0) is constrained, while remaining system dimensions are freely varied. Since the physical precipitate (Figure 1 (b)) is confined by the host lattice in all three dimensions, model and physical system features are expected to compare well only near the slab base (shaded region in Figure 1 (a)). The plane *z* = 0, on the other hand, represents a case of vanishing shear strain along the needle direction, which should be encountered by way of symmetry in the real system.

Figure 1 (c) presents the key elements of the hybrid scheme. In the interface region shown, LET energies are replaced by a full DFT analysis. A single interface region supercell, comprised by symmetrically equivalent local regions on either side of the precipitate (along with necessary structural modifications to ensure periodic boundary conditions) has been highlighted (full lines). In practice, the calculations revealed an additional symmetry around *z* = 0, whence the necessary information for supercell construction is obtained by changing -*z* into *z* in Figure 1 (c).

The fully coherent β" precipitate (see Figure 2 (a)) has been described in detail in [4, 8]. The experimentally reported orientation relationships with the host lattice are: $[230]_{Al}$ || $[100]_{\beta''}$; $[001]_{Al}$ || $[010]_{\beta''}$; $[-310]_{Al}$ || $[001]_{\beta''}$, while the precipitate-host lattice misfits along the precipitate basis vector directions {$\mathbf{a}_P$, $\mathbf{b}_P$, $\mathbf{c}_P$} are {3.8 (%), ≈ 0, 5.3} [4].

Results and discussion

While the precipitate cross-section comprised 4 (8) unit cells along basis vector directions $\mathbf{a}_P$ ($\mathbf{c}_P$), with the surrounding amount of Al also being well defined (11×11 times the precipitate cross-section), a series of different slab thicknesses – 10, 20, 50, and 100 unit cells – were considered in the LET simulations. The results (Figure 2 (b)) suggest that increasing the precipitate aspect ratio leads to an almost complete adaptation of β" to Al along z ($\mathbf{b}_P$) near the slab base, regardless of composition. Further, as expected, the relative importance of the needle end effects are decreasing with increasing slab thickness. An overall sensitivity to this boundary condition cannot be rigorously excluded, but it seems safe within the setup to examine system energies for the regions highlighted in Figure 2 (b). Given the presence of the symmetry plane at $z = 0$ for the system examined, the effective β" needle length reaches ≈ 50 nm for the largest model system, a typical average observed value [1].

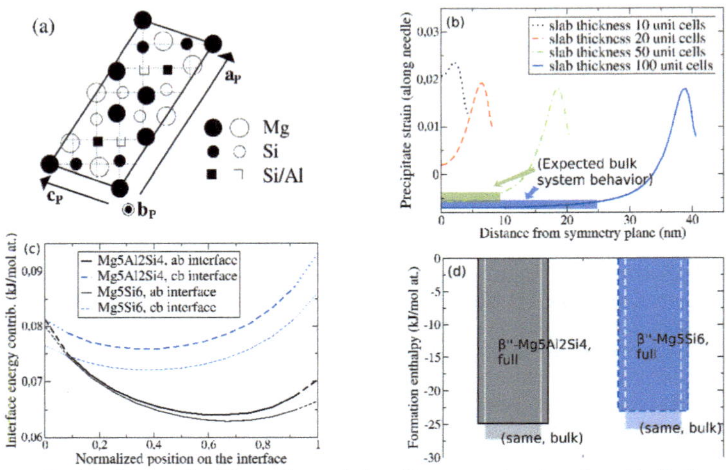

Figure 2. (a) Schematic illustration of the β" conventional unit cell. Both compositions considered in this work have been outlined (β"–Mg_5Si_6: squares Si; β"–$Mg_5Al_2Si_4$: squares Al). (b) Calculated (LET) precipitate strain along the main growth direction $\mathbf{b}_P$ for the β"–$Mg_5Al_2Si_4$/Al systems examined. Highly similar results (not shown) were obtained for β"–Mg_5Si_6/Al. (c) Calculated (DFT) interface energy contributions from the β"/Al interface regions, $z = 0$ (edges extrapolated). (d) Calculated β" formation enthalpies (bulk, full).

DFT interface region studies were carried out at $z = 0$, 10, and 20 unit cells height (up to ≈ 8 nm in Figure 2 (b)), for the second largest β"–$Mg_5Al_2Si_4$/Al slab geometry. The results obtained suggest that the misfit accumulation along the needle direction can safely be ignored within this range (energy changes within 0.3%), whence calculations at $z = 0$ (Figure 1 (c)) will suffice – within the approximation of zero needle end effects – for computing the full interface energy E_{Int}. This conclusion applies also to the strain energies for the remaining parts of the system: at $z = 20$ unit cells height, these energies have been reduced by 3% or less, compared to the $z = 0$ values.

Table I displays the calculated interface region contributions to E_{Int} for the two β"/Al interface regions considered in this work. We compare these full DFT values to those obtained in a manner mimicking a 'conventional' scheme [11], with those numbers included in square brackets in Table I. The interfacial energy σ in that approach was assumed constant [5] and the strain energy ζ was evaluated separately, solely within LET. The differences between these two sets of values is non-negligible, 16% (14%) for β"–$Mg_5Al_2Si_4$/Al (β"–Mg_5Si_6/Al), with the full DFT values being higher. Given that the interface region contributions to E_{Int} are seen to dominate, a similar relative difference applies to the full interface energy consideration.

Table I. Calculated contributions to the β"/Al interface energy

System	Method	Subsystem	E_{Int} contribution (kJ/mol at.)
β"–$Mg_5Al_2Si_4$/Al	DFT [DFT/LET]	Interface region	1.767 [1.521]
	LET	Remaining system	0.650
β"–Mg_5Si_6/Al	DFT [DFT/LET]	Interface region	1.700 [1.496]
	LET	Remaining system	0.987

With the calculated absence of any significant strain evolution along the β" needle direction, the precipitate cross section dimensions are rather well defined (though still slightly dependent on the chosen β" aspect ratio). Table II displays theoretical bulk β" conventional unit cell dimensions, along with (cross-section averaged) values obtained for the β"/Al systems of Table I. The latter values are noted to fall somewhat below the experimentally reported [4] parameters (also included), while further, the change of precipitate composition is noted to primarily affect the monoclinic angle β_P. It seems rather unlikely that the theoretical and experimental discrepancies should be resolved solely by improvements in the modeling of the chosen slab geometry. Presently, the LET modeling is missing the compressive effect of a non-zero interfacial energy, while the expansive effect of incorporating electronic interactions in the DFT interface region studies has been shown to be only a few tenths of an Å per interface [7].

Table II. Calculated and experimental β" conventional unit cell dimensions (cross-section only)

System	$\mathbf{a}_P$ (nm)	$\mathbf{c}_P$ (nm)	β_P (°)
β"–$Mg_5Al_2Si_4$ (bulk)	1.532	0.6778	105.9
β"–$Mg_5Al_2Si_4$/Al	1.496	0.6593	105.0
β"–Mg_5Si_6 (bulk)	1.511	0.6932	110.4
β"–Mg_5Si_6/Al	1.487	0.6628	107.8
β"/Al (exp.) [4]	1.516±0.002	0.674±0.002	105.3±0.5

Figure 2 (d) displays the calculated β" formation enthalpies, before and after inclusion of interface energies. While the bulk system energy is clearly dominant, the energy difference between the two compositions investigated clearly increases, by ≈ 17%, upon incorporation of subsystem interactions. The relative destabilization of β"–Mg_5Si_6 obtained can be understood primarily from the larger deviations between the monoclinic angle β_P and the experimental value of 105.3° corresponding to two fully coherent interfaces in the cross-section. An increased strain on the subsystems results, compared to β"–$Mg_5Al_2Si_4$/Al (see Table I), even though the β"–Mg_5Si_6/Al interfacial energy along the **cb** interface (see Figure 2 (c)) is arguably lower. Of course, our comparison here rests on the underlying assumption of the same (in number of unit cells) morphologies for the two compositions, which is presumably at odds with the result of a more complete system optimization. The issue addressed above for β_P however remains.

Conclusion

In this work, we have presented our first investigations of a 3D β''/Al system with an essentially *ab initio* based hybrid scheme that seeks to produce precipitate-host lattice interface energies for the full system at DFT level accuracy. The robustness of the scheme seems well supported. Examination of the two experimentally reported main candidates for the β'' composition, Mg_5Si_6 [4] and $Mg_5Al_2Si_4$ [8] provides support for the latter choice, both in the context of structure and energies. The incorporation of precipitate-host lattice interactions in the model scheme destabilize β''–Mg_5Si_6 (by 0.27 kJ/mol at.) – a feature that can presumably be attributed to the deviation in the value of the monoclinic angle β_P from the experimentally reported value (and hence from optimal coherency) for this configuration. While a series of improvements of the scheme (more physical model system, better treatment of the interface electronic effects) are arguably desired, the ability to model any local region on the interface with DFT also allows for testing issues previously ignored by necessity.

Acknowledgments

This work was financially supported by the Research Council of Norway via the MultiModAl project (project no. 205353). The DFT simulations employed access to the NOTUR facilities.

References

[1] C.D. Marioara et al., "The Influence of Alloy Composition on Precipitates of the Al–Mg–Si System,"*Metallurgical and Materials Transactions A* 36A (2005) 691-702.
[2] Y. Mishin, M. Asta and Ju Li, "Atomistic modeling of interfaces and their impact on microstructure and properties," *Acta Materialia* 58, (2010) 1117-1151.
[3] V. Vaithyanathan, C. Wolverton and L. Q. Chen, "Multiscale Modeling of Precipitate Microstructure Evolution," *Physical Review Letters* 88, (2002) 125503.
[4] S.J. Andersen et al., "The Crystal Structure of the β'' Phase in Al–Mg–Si Alloys, " *Acta Materialia*, 46 (1998), 3283–3298.
[5] F.J.H. Ehlers, and R. Holmestad, "Ab initio interface modeling for fully coherent precipitates of arbitrary size in Al alloys," *Computational Materials Science*, 72 (2013), 146–157.
[6] N. Bernstein, J.R. Kermode, and G. Csányi, "Hybrid atomistic simulation methods for materials systems," *Reports on Progress in Physics*, 72, (2009), 026501.
[7] F.J.H. Ehlers et al., manuscript in preparation.
[8] H.S. Hasting et al., "Composition of β'' precipitates in Al–Mg–Si alloys by atom probe tomography and first principles calculations," *Journal of Applied Physics*, 106 (2009), 123527.
[9] V. Vaithyanathan, C. Wolverton, and L.Q. Chen, "Multiscale modeling of θ' precipitation in Al-Cu binary alloys," *Acta Materialia* 52, (2004) 2973-2987.
[10] J.O. Hallquist, LS-DYNA Keyword User's Manual Version 971, California: Livermore Software Technology Corporation (2007).
[11] R.M. Martin, *Electronic Structure – Basic Theory and Practical Methods*, Cambridge, UK, Cambridge University Press, (2004), 119-185.

2nd World Congress on Integrated Computational Materials Engineering
Edited by: Mei Li, Carelyn Campbell, Katsuyo Thornton, Elizabeth Holm, and Peter Gumbsch
TMS (The Minerals, Metals & Materials Society), 2013

THE ROLE OF THE CALPHAD APPROACH IN ICME

F. Zhang, W. Cao, S.L. Chen, C. Zhang, and J. Zhu
CompuTherm LLC, 437 S. Yellowstone Dr., Madison WI 53719, U.S.A.

Keywords: CALPHAD, Phase Diagram, Precipitation, *Pandat*™ Software

Abstract

In this paper, the role of the calculation of phase diagram (CALPHAD) approach in integrated computational materials engineering (ICME) is discussed. Two examples are used to demonstrate how this approach can be integrated with key experiments and kinetic models in accelerating materials design and development under the framework of ICME. In the first example of developing precious metal modified nickel-based superalloys, phase diagram calculation is integrated with key experiments to quickly locate the composition region with optimum properties. In the second example, thermodynamic calculation is integrated with precipitation simulation to understand the precipitation behaviour of IN100 during a heat treatment process. The simulation clearly demonstrates the formation of primary, secondary and tertiary γ' in this alloy during the selected heat treatment process.

Introduction

Adjusting materials chemistry and processing conditions has been a common practice of materials scientists/engineers in the design of new materials and the improvement of the existing materials. Traditional approaches relying on trial-and-error are no longer viable due to limited resources. Integrated Computational Materials Engineering (ICME), which integrates materials information obtained from computational tools with engineering product performance analysis, has recently been highlighted as a methodology that can unlock great potential for significant benefits in cost-effective materials and process design.

Phase diagrams are road maps of materials scientists/engineers in understanding the effect of chemistry on the microstructure of an alloy. Although determination of phase diagrams solely by experiments is feasible and necessary for binaries and simple ternaries, it is not realistic for multi-component systems. Yet, most if not all of the commercial alloys are multicomponent in nature. The CALPHAD approach, which primarily aimed at calculating phase equilibria and thermodynamic properties of complex multi-component, multi-phase systems, emerged in the late 1950s and became an active research field since the 1970s [1-3].

The essence of the CALPHAD approach is to obtain self-consistent thermodynamic descriptions of the lower order systems: binaries and ternaries, in terms of known thermodynamic data measured experimentally and/or calculated theoretically as well as the measured phase equilibria. The advantage of this method is that the separately measured phase diagrams and thermodynamic properties can be represented by the same thermodynamic description, or refeered to as thermodynamic database, of a materials system in question. More importantly, the thermodynamic database developed according to the lower order systems enables us to calculate phase diagrams of the multi-component systems that are experimentally unavailable.

Although emerged first as an approach for multi-component phase equilibrium calculation, the CALPHAD approach has in recent years been applied to a broader field of materials science and engineering beyond phase diagrams, such as solidification, coating, joining, and phase transformation. Mobility database has been developed in a similar manner as that of thermodynamic database. Several software packages, such as *Thermo_Calc* and *Pandat*^{TM}[4], have been developed based on this approach. Such software, when integrating with suitable databases, has become a generic tool available to ICME practitioners.

In this paper, we will use two examples to demonstrate the applications of CALPHAD type of modeling tools. All the calculations in this paper are performed by using *Pandat*^{TM} software and databases developed at CompuTherm. The paper will be arranged as follows: the first part is an overview of the *Pandat*^{TM} software and its functionality, the second part gives two examples on the design of precious metal modified nickel-based superalloys and the simulation of precipitation behavior of IN100, and the last part is the conclusions.

Pandat^{TM} Software

Pandat^{TM} software is designed as a workspace, which may include multiple modules for performing variety of simulations and be easily extended. Currently, four modules are available, which are *PanPhaseDiagram* for multi-component phase diagram calculation, *PanOptimizer* for thermodynamic model parameter optimization, *PanPrecipitation* for precipitation simulation, and *PanDiffusion* for diffusion simulation. The architecture of *Pandat*^{TM} software is shown in Figure 1, which consists of three layers: the interface layer, the calculation layer and the database layer. The top layer, the interface layer, is responsible for the communication between the user and the software. It passes the user commands and calculation conditions to the calculation layer and displays the calculated results returned from the calculation layer in the format of table or graph. The middle layer, the calculation layer, processes the user commands and calculation conditions from the interface layer, loads the needed model parameters from the database layer, performs the calculation, and then returns the calculated results to the interface layer. The bottom layer, the database layer, stores all the model parameters needed by the calculation.

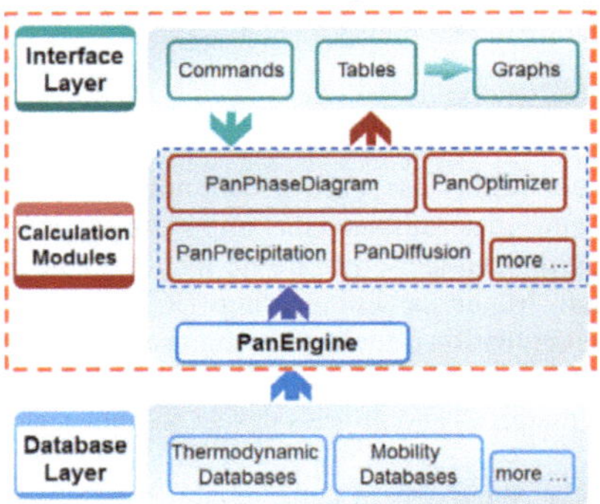

Figure 1: The architecture of *Pandat*^{TM} software

Figure 2: Applications through the integration with PanEngine API

It should mention that the calculation engine, *PanEngine*, is shared by different modules in the calculation layer as shown in Figure 1. *PanEngine*, which perfoms all the thermodynamic calculations required by different modules, is a dynamic linked library. *PanEngine*, can therefore be integrated with other models through its application programming interface

(API) and create many customized applications as shown in Figure 2. This is a good platform for the development of ICME tools.

Application of CALPHAD to Materials Design in the Framework of ICME

The CALPHAD approach is a well-developed and widely accepted tool available to ICME practitioners. In this section, we will use two examples to demonstrate its role in aiding materials design and development. In the first example, it is integrated with key experiments for accelerating alloy design of precious metal modified nickel-based superalloys, and in the second example, it is integrated with precipitation simulation for understanding the precipitation behaviour of IN100 during a heat treatment process.

Design of Precious Metal Modified Nickel-based Superalloys

Nickel-based superalloys are widely used as high temperature structrual materials, such as gas turbine engine components, due to their outstanding properties at elevated temperatures. The continual demand for improved performance of jet engines has pushed the usage of these alloys to even higher temperature. Improvement of mechanical properties at high temperature has been accomplished through the increase in refractory metal content associated with the decrease in Cr content to avoid the formation of topologically close-packed (TCP) phases. This compositional change is detrimental to the oxidation and corrosion resistance. Since platinum and other platinum-group metals (PGMs) were found to be able to enhance both oxidation and corrosion properties of nickel-based superalloys at elevated temperatures in various aggresive environments [5-7], researches have been carried out to develop PGM-modified nickel-based superalloys [8-11]. The key question is: what is the optimum amount of precious metal that should be used in the alloy? While the properties of nickel-based superalloys depend strongly on the γ/γ' coherence microstructure, it is therefore important to understand how the PGMs affect this microstructure, such as the amount of γ' and the partitioning behavior of the alloying elements. It is not hard to imagine how much work would be involved if such a understanding is established solely by experiments, especially when multi-component system is considered. The CALPHAD approach provides an ideal means to understand such effects, save experimental cost and accelerate the design of PGM-modified nickel-based superalloys.

Figure 3 shows the effect of Pt on the γ'-solvus temperature and the fraction of γ' for the base alloy: Ni-15Al-5Cr (at%). It is seen that Pt is a γ' stablizer which increases the γ'-solvus temperature. It is also found that Pt slightly decreases the solidus temperature (not shown in the figure), which leads to the reduction of heat treatment (HT) window. The calculation shows that the HT window is reduced by ~10°C when 1 at% of Pt is added. This is consistent with the experimental observation [11], and explains why Pt modification of existing Ni-based superalloys is sometimes problematic. On the other hand, Pt is found to slightly decrease the fraction of the γ' phase at low temperature, which is not favored for nickel-based superalloy design. Figure 4 shows the effect of Pt on the partitioning of Al, Cr, and Ni in the γ and γ' phases. It is clearly shown that Pt promotes the segregation of Al to the γ' phase, and Cr and Ni to the γ phase. The calculation also indicates that Pt has much less significant effect on the partitioning of Al and Ni than that of Cr. A ternary alloy is chosen in this example for the purpose of illustration. These calculations have been applied to a multi-component system based on Ni-Al-Cr-Hf-Re-Ta-W with additions of Pt, Ir, and Ru [8,11] to study the effect of PGMs on the properties of Ni-based superalloys. Figure 5 from Heidloff [11, 12] clearly demonstrates the concept of integrating CALPHAD calculation with key experiments to quickly locate the compositional region which is most promising to possess the optimum properties. In this figure, the regions with good oxidation resistance and hot corrosion property are determined by experiments for the base alloy Ni-13Al-1Re-2Ta-2W-2Ru-0.1Hf (at%) in the composition space of Cr and Pt. The region with good heat treatability is calculated by *Pandat*^{TM} software and *PanNi* [13] database, and the region with better γ′ stability is outlined by integrating calculation with experiments. The overlapped region therefore possesses the optimum properties.

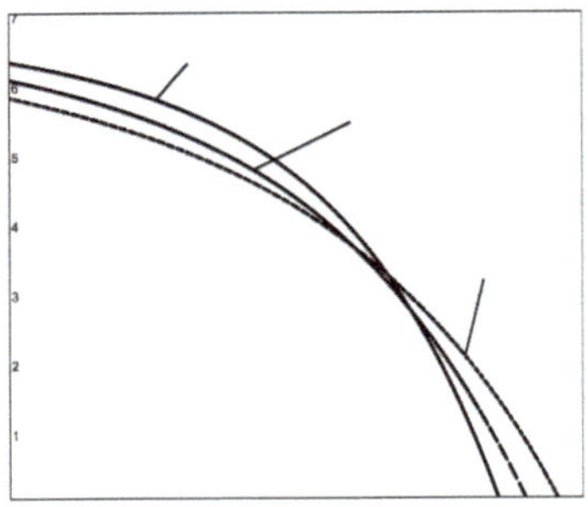

Figure 3: Effect of Pt on the γ′-solvus and the fraction of γ′ phase

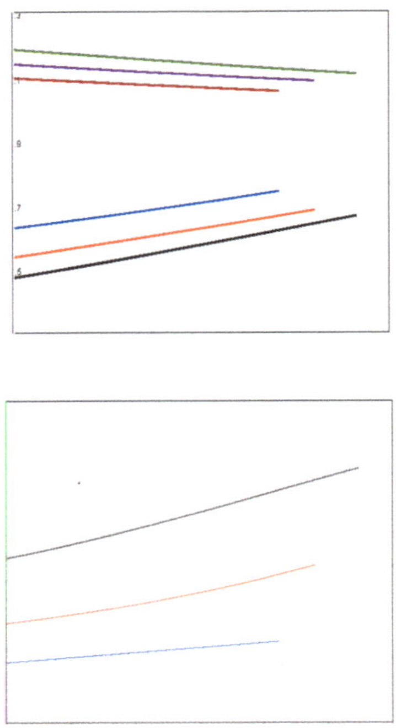

Figure 4: Effect of Pt on the partitioning coefficients (γ/γ′) of Al, Cr, and Ni

Precipitation Simulation of IN100

Precipitation hardening provides one of the most widely used mechanisms for the strengthening of many structural materials. Simulation of precipitation kinetics is therefore essential in modern materials design. However, precipitation is a highly complex process and could involve simultaneous occurrence of nucleation, growth and coarsening. Accurate modeling of the precipitation process requires a synchronous consideration of all these contributions. Moreover, the necessary phase equilibrium information and mobility data must be constantly updated during the simulation.

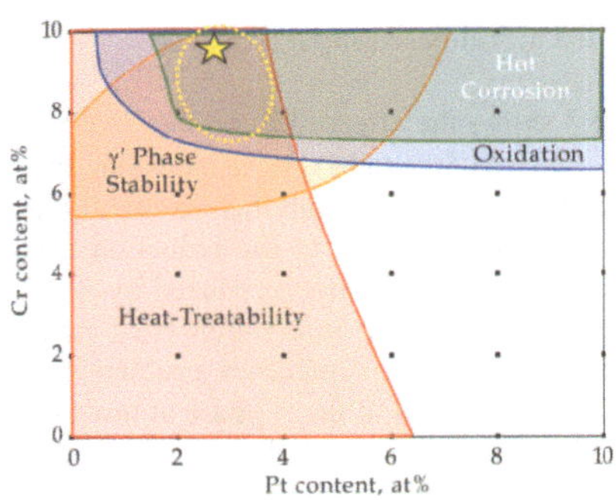

Figure 5: Courtesy from Heidloff [11, 12]: Region outlined with dotted line for a composition space of ~2-4 at% Pt and ~8-10 at% Cr is identified to satisfy all required properties for a base composition of Ni-13Al-1Re-2Ta-2W-2Ru-0.1Hf (at%)

Table I: Measured γ' sizes and fractions[14]

	Size (nm)	Volume fraction
Primary γ'	1700	20%
Secondary γ'	208	29%
Tertiary γ'	23	11%

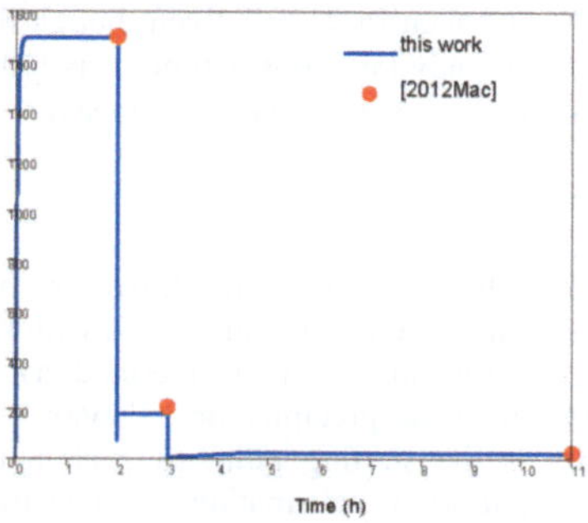

Figure 6: Comparison of simulated and experimental determined γ' size for IN100

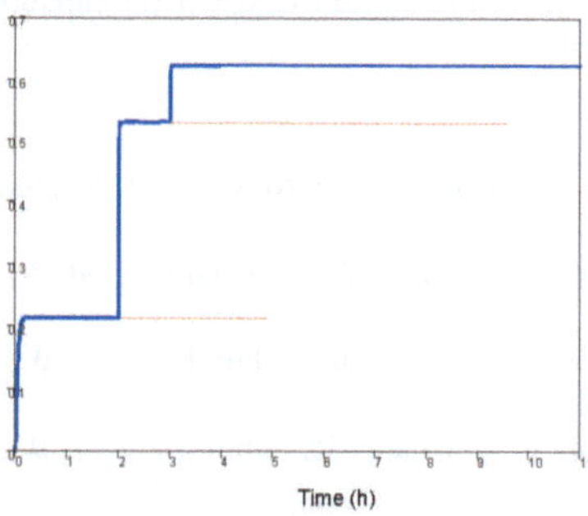

Figure 7: Simulated volume fractions of primary, secondary and tertiary γ' sizes for IN100

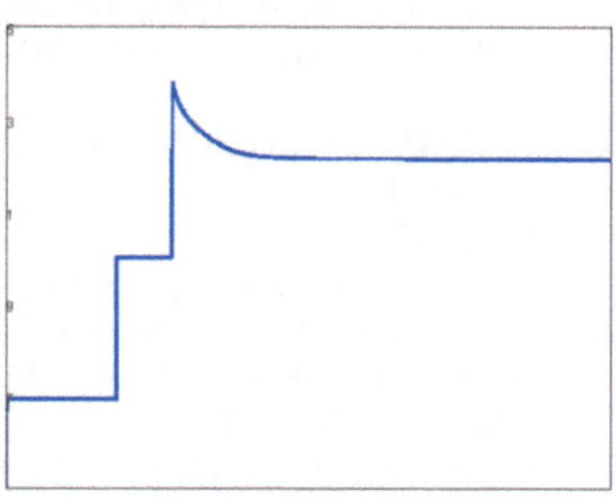

Figure 8: Simulated number densities of primary, secondary and tertiary γ' for IN100

Such a simulation, therein, requires a smooth integration of thermodynamic calculation and kinetic simulation. This is especially true for a multi-component system since multi-component phase equilibrium information is not readily accessible. The CALPHAD approach has, therefore, in recent years been extended to kinetic simulation.

In this example, precipitation behavior of the γ' phase in IN100 alloy is simulated. Experimental study was carried out by Maciejewski and Ghonem [14], and the fractions of primary γ', secondary γ', and tertiary γ' were measured due to different heat treatment processes. In this example, one of the thermal processes as used in their paper, anneal at 1149°C for 2h, oil quench to 981°C and anneal at this temperature for 1h, then fan cool to 732°C and stay for 8h, was used in the simulation. The measured sizes and volume fractions of primary γ', secondary γ', and tertiary γ' by Maciejewski and Ghonem [14] are listed in Table I. The minor components, such as C and B, and the small amounts of carbide and boride phases that may form in the microstructure are not considered in the simulation. The chemistry used in the simulation for IN100 is Ni-4.85Al-18.23Co-12.13Cr-3.22Mo-4.24Ti (wt%). It should point out that homogeneous nucleation is assumed in the simulation. The mobility database is based on the work by Campbell [15] and modified to be consistent with PanNi [13] database.

Figure 6 shows the average γ' size change with time. It clearly shows that three different size groups of γ' formed during the heat treatment process. The primary γ' is significantly larger than the secondary and tertiary γ', and the secondary γ' is about 10 times bigger than the tertiary γ'. The red points represent the average sizes of primary, secondary and tertiary γ' measured by Maciejewski [14]. The calculation agrees with the experimental data very well. Figure 7 plots the calculated fractions of primary, secondary, and tertiary γ', which are 21.2%, 31.7%, and 9.4% respectively. These calculated fractions agree with the experimental data, 20%, 29%, and 11%, very well as listed in Table I. Figure 8 shows the calculated number densities of primary, secondary, and tertiary γ' in logarithm scale. It clearly shows that the number density of secondary γ' is about 3 order of magnitude higher than that of the primary γ', while the tertiary γ' is about 2.5 order of magnitude higher than that of the secondary γ'. It should be pointed out that one minute was used in the simulation for the quench (cool down) processes from 1149°C to 981°C and then to 732°C. Even though this choice is arbitrary and the fan cool from 981°C to 732°C may take longer time, it is found that the final fraction and size of tertiary γ' reach the same results after 8 hours of anneal.

Conclusions

The CALPHAD approach plays an important role in ICME. This is illustrated by two examples in this paper. In the first example, phase diagram calculation is integrated with key experiments to accelerate materials design. In the second example, thermodynamic calculation is integrated with kinetic models to understand the precipitation behavior of a nickel alloy. Since understanding of phase stability is the starting point to explain the materials phenomena, the thermodynamic and phase equilibrium information obtained from the CALPHAD approach is essential for the development of ICME tools. The linkage of the CALPHAD modeling tools with physics-based tools, microstructural evolution prediction tools, and property prediction tools will certainly extend the current CALPHAD capability and lead to development of more powerful ICME tools.

References

1. Kaufman, L., *Computer Calculation of Phase Diagrams*. 1970: New York: Academic Press.
2. Saunders, N. and A.P. Miodownik, *CALPHAD: A Comprehensive Guide*, ed. R.W. Cahn. 1998: Pergamon Materials Series.
3. Y. A. Chang, S.-L. Chen, F. Zhang, X.-Y. Yan, F.-Y. Xie, R. Schmid-Fetzer, W. A. Oates, *Progress in Materials Science* 49 (2004) 313-345.
4. W. Cao, S.-L. Chen, F. Zhang, K. Wu, Y. Yang, Y. A. Chang, R. Schmid-Fetzer, W. A. Oates, *Calphad*, 33 (2009), 328-342.
5. D. R. Coupland, C. W. Hall, and I. R. McGill, *Platinum Metals Review*, 26 (1982), 146-157.
6. G. J. Tatlock, T. J. Hurd, and J. S. Punni, *Platinum Metals Review*, 31 (1987), pp. 26-31
7. B. Gleeson, W. Wang, S. Hayashi, and D. Sordelet, *Materials Science Forum*, 461-464 (2004), pp. 213-222.
8. F. Zhang, J. Zhu, W. Cao, C. Zhang, and Y. A. Chang, *JOM*, 62 (10) (2010), pp. 35-40.
9. A. Bolcavage and R. C. Helmink, *JOM*, 62 (10) (2010), pp. 41-44.
10. D. L. Ballard and A. L. Pilchak, *JOM*, 62 (10) (2010), pp. 45-47.
11. A. J. Heidloff, Z. Tang, F. Zhang, and B. Gleeson, *JOM*, 62 (10) (2010), pp. 48-53.
12. A.J. Heidloff, PhD Dissertation, "Development of Pt-modified γ-Ni+γ'-Ni3Al-based alloys for high-temperature applications,"Iowa State University, 2009, UMI# 3373463
13. PanNi thermodynamic database, CompuTherm LLC, Madison, WI53719
14. K. Maciejewski, H. Ghonem, *Materials Science & Engineering A*, 560 (10) (2013), 439-449.
15. C. E. Campbell, W. J. Boettinger, and U. R. Kattner, *Acta Materialia*, 50 (2002), 775-792.

2nd World Congress on Integrated Computational Materials Engineering
Edited by: Mei Li, Carelyn Campbell, Katsuyo Thornton, Elizabeth Holm, and Peter Gumbsch
TMS (The Minerals, Metals & Materials Society), 2013

SIMULATIONS OF PRECIPITATE MICROSTRUCTURE EVOLUTION DURING HEAT TREATMENT

Kaisheng Wu[1], Gustaf Sterner[2], Qing Chen[2], Herng-Jeng Jou[3], Johan Jeppsson[2], Johan Bratberg[2], Anders Engström[2], Paul Mason[1]

[1]Thermo-Calc Software Inc.;
4160 Washington Rd.; McMurray PA 15317, USA
[2]Thermo-Calc Software AB;
Norra Stationsgatan 93; SE 113 64 Stockholm, Sweden
[3]QuesTek Innovations LLC;
1820 Ridge Ave.; Evanston, IL 60201, USA

Keywords: Thermodynamics, Kinetics, Precipitation, Heat Treatment

Abstract

Precipitation, a major solid state phase transformation during heat treatment processes, has for more than one century been intensively employed to improve the strength and toughness of various high performance alloys. Recently, sophisticated precipitation reaction models, in assistance with well-developed CALPHAD databases, provide an efficient and cost-effective way to tailor precipitate microstructures that maximize the strengthening effect via the optimization of alloy chemistries and heat treatment schedules. In this presentation, we focus on simulating precipitate microstructure evolution in Nickel-base superalloys under arbitrary heat treatment conditions. The newly-developed TC-PRISMA program has been used for these simulations, with models refined especially for non-isothermal conditions. The effect of different cooling profiles on the formation of multimodal microstructures has been thoroughly examined in order to understand the underlying thermodynamics and kinetics. Meanwhile, validations against several experimental results have been carried out. Practical issues that are critical to the accuracy and applicability of the current simulations, such as modifications that overcome mean-field approximations, compatibility between CALPHAD databases, selection of key parameters (particularly interfacial energy and nucleation site densities), etc., are also addressed.

Introduction

The increasing interest in ICME techniques during recent years has boosted birth and maturity of efficient computational tools in a variety of industrial applications. TC-PRISMA[1] was developed to aid a systematic design of materials chemistry and heat treatments in order to achieve desired microstructures. It integrates seamlessly with THERMO-CALC[2] and DICTRA[2], and uses CALPHAD[3] databases for accurate thermodynamic and kinetic information in commercial alloys.
Recently, added functionality to simulate arbitrary heat treatment schedules, especially non-isothermal cooling conditions, has been included in TC-PRISMA. One of the most important applications of non-isothermal heat treatment is to produce complex multi-modal particle size distributions of γ' precipitates in Ni-base superalloys. The purpose of the current paper is to discuss this functionality and to evaluate its feasibility with several Ni-base alloys.

Model

TC-PRISMA has been developed based on Langer-Schwartz theory[4], and adopts Kampmann-Wagner numerical (KWN) method[5] for concurrent nucleation, growth, and coarsening of dispersed precipitate phases. The details of the employed model can be found elsewhere[1]. The major modification for non-isothermal conditions is the adoption of a transient nucleation model that calculates the incubation time as an integrated form of past thermal history[6]

$$\int_0^{\tau} \beta^*(t')dt' = \frac{1}{\pi Z^2(\tau)} \tag{1}$$

where τ is the incubation time, β^* is the impingement rate for solute atoms to the critical cluster, and Z is the Zeldovich factor.

TC-PRISMA uses CALPHAD databases for thermodynamic and kinetic data. In the current work, the thermodynamic database TTNI8[7] and the mobility database MOBNI1[2] have been used to carry out the simulations.

Results

<u>Effect of Alloy Compositions: Ni-Al-Cr</u>

In a recent paper, Rojhrunsakool et al. observed the composition dependence of the particle size distributions of γ' precipitates in Ni-Al-Cr alloys[8]. When subjected to the same continuous cooling condition, Ni-10Al-10Cr (at.%) alloy developed bimodal distribution of γ' precipitates, in contrast with Ni-8Al-8Cr where only single modal distribution occurred. In the current work, numerical experiments following their reported cooling treatment (continuous cooling from super-solvus temperature 1150°C to 380°C with a cooling rate of 14°C/min) have been performed. A constant interfacial energy of 0.023 J/m^2 has been used for these simulations. The results shown in Figure 1 clearly confirm their observations that Ni-10Al-10Cr generates a bimodal particle size distribution, whereas Ni-8Al-8Cr is of mono-modal type.

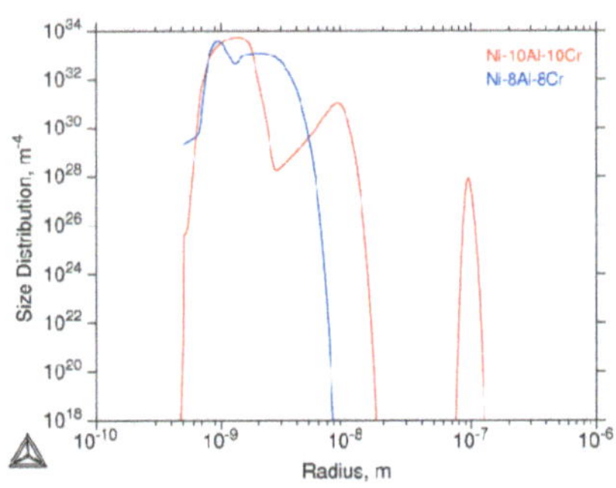

Figure 1 Calculated particle size distribution of γ' precipitates in Ni-10Al-10Cr and Ni-8Al-8Cr alloys after continuous cooling from 1150°C to 380°C with a cooling rate of 14°C/min

In-depth analyses of the simulation results have been carried out to gain more insight into the mechanisms that produce different γ' microstructures. The phase diagram in Figure 2(a) indicates that Ni-10Al-10Cr has a higher solvus temperature than Ni-8Al-8Cr, which leads to a relatively rapid growth of primary γ′ due to high atomic mobilities at high temperatures. Substantial growth of existing particles in Ni-10Al-10Cr results in severe drop of supersaturation of solutes in the matrix and brings down the driving force for nucleation to minimum values (see Figure 2(b)) so that the nucleation comes to halts (see Figure 2(c)). After each halt, further cooling increases the driving force gradually and gives rise to a new burst of nucleation when the driving force becomes large enough. In this way, multiple nucleation bursts can thus be produced, as shown clearly in Figure 2(c). For the alloy Ni-8Cr-8Ni, the nucleation of particles happens at relatively low temperatures and the growth of born particles is rather slow at low temperatures. As a result, the driving force available for nucleation is almost monotonically increasing during continuous cooling and no disruption of nucleation event occurs. Our quantitative analyses are in accord with the theoretical explanation which Rojhrunsakool et al.[8] gave to their experiments.

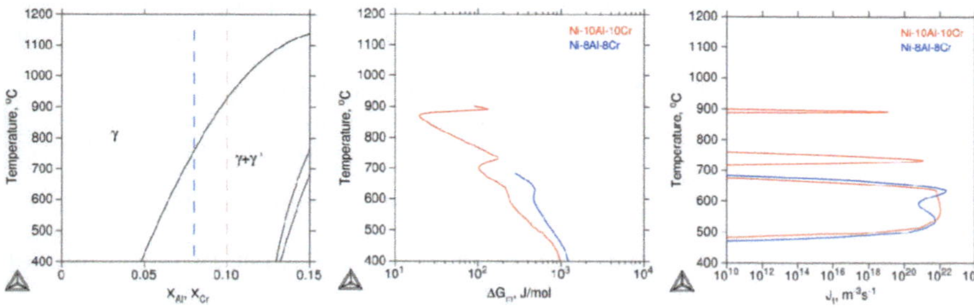

(a) (b) (c)

Figure 2 Calculated property diagrams in Ni-Al-Cr system. (a) phase diagram of Ni-*x*Al-*x*Cr (*x*=0-0.15 at.%). (b) instantaneous driving force as function of temperature. (c) nucleation rate as function of temperature

Effect of Heat Treatments: U720Li Alloy

U720Li has been chosen to evaluate the feasibility of the current tool for practical applications. The main focus of the current work is on the formation of secondary and tertiary γ′, which are defined by superalloy society as γ′ precipitates that form during continuous cooling. Meanwhile, all γ′ particles are assumed to nucleate intra-granularly. The alloy compositions employed in the current work have been listed in Table 1, and their corresponding heat treatments in Table 2, on the basis of literature data[9-12]. A constant interfacial energy of 0.025 J/m^2 has been used to perform the simulation.

It can be seen from Table 2 that there are two types of heat treatments. The first one is continuous cooling from super-solvus temperature (Alloys 1 and 2), thus with an initial microstructure of pure γ matrix. Figure 3 illustrates the calculated variation of γ′ particle size distributions with respect to the cooling rate. Within the studied cooling rate range, the results clearly indicate that decreasing cooling rate promotes the formation of multi-modal distributions, as the multiple peaks become more widely separated. Prolonged cooling time also results in larger particle sizes, as both secondary and tertiary γ′ peaks shift to the right side. The plot of average size with respect to cooling rate further confirms this effect, as seen in Figure 4(a). It should be noted that the calculation of average size in Figure 4(a) has been conducted in accordance with experimental methods which do not reflect the "actual" average size due to the complexity of the particle size distribution. Moreover, the calculations count only particles larger than 50nm, which was reported as their experimental limit[11]. One exception is those for small cooling rates (< 0.2°C/s) where only secondary γ' particles have been considered. A higher calculated value is thus to be expected since some tertiary γ' might have been included experimentally. Despite this disparity, the overall agreements between calculated and measured results are very good. The linear relationship has been obtained when both size and cooling rate are in logarithmic scale, which agrees well with experimental observations[9-11] and other numerical simulations[9, 10].

Table 1 Compositions of U720Li Alloy (wt.%)

Alloy	Al	B	C	Co	Cr	Fe	Mo	Si	Ti	W	Zr	Ni
1[9, 10]	2.53	0.014	0.014	14.43	15.92	0.09	2.96		4.96	1.26		Bal
2[11]	2.46		0.025	14.75	16.35	0.06	3.02		4.99	1.3	0.035	Bal
3[12]	2.51	0.014	0.011	14.66	16.14		2.98	0.05	5.08	1.23		Bal

Table 2 Heat Treatments of U720Li Alloy

Alloy	Heat Treatment Conditions
1[9, 10]	Cooling from 1180°C to 400°C with varied cooling rates(0.217, 0.433, 0.867, 1.625, 3.25, 6.5, 19.5, 78 °C/s)
2[11]	Cooling from 1175°C to 650°C with varied cooling rates(0.183, 0.46, 0.9417, 1.837, 2.86°C/s)
3[12]	Cooling from 1105°C to 400°C with varied cooling rates(20, 130, 280°C/min), followed by isothermal annealing at 700°C for 24hrs

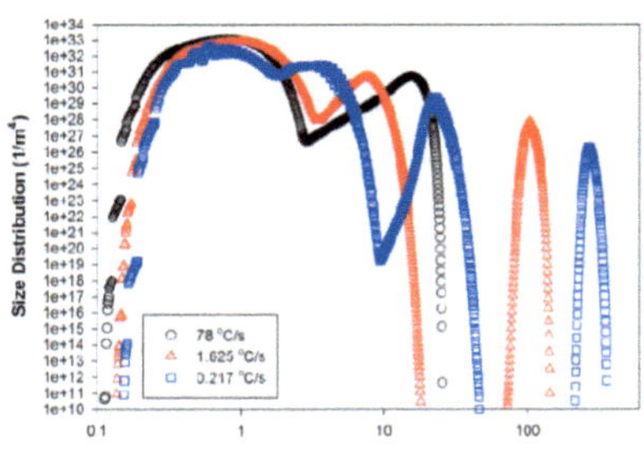

Figure 3 Simulation results for the effect of cooling rate on particle size distributions of γ' precipitates in U720Li Alloy 1

The other type of heat treatment is continuous cooling from sub-solvus temperature, followed by isothermal aging treatment (Alloy 3 in Table 2). In this case, we neglected the aging at 1105°C, prior to cooling, which produced primary γ' at grain boundaries[12], as the main focus was on the microstructure of secondary and tertiary γ'. Instead the primary γ' was assumed to reach its equilibrium fraction at 1105°C, and hence the γ matrix composition was adjusted accordingly based on equilibrium calculation. The results are compared with experiments[12] in Figure 4(b).

The same effect of cooling rate as for the previous simulation is observed. Again, satisfactory agreement has been obtained with exception for low cooling rates, where the calculation under-estimated the size. This disagreement may arise from kinetic effect of preexisted primary γ' on the subsequent precipitation, which has been neglected in the current work. The grain boundary precipitation of secondary γ' can also be a factor.

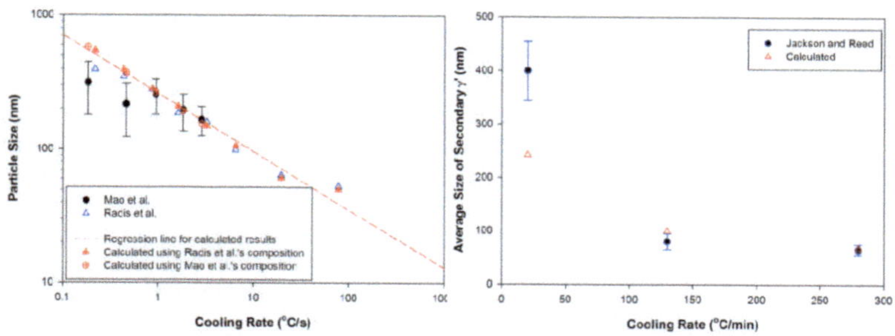

(a) (b)

Figure 4 Comparison of calculated and experimental average γ' size as function of cooling rate. (a) with Alloy 1[9, 10] and 2[11]; (b) with Alloy 3[12].

Discussion

Most theoretical explanations attribute the multi-modal distribution during continuous cooling to the multiple distinct nucleation bursts caused by alternative dominance of nucleation driving force and solute depletion[9, 10, 13, 14]. Slow cooling promotes nucleation waves as the newly formed nuclei can have sufficient time to grow, draining solute atoms substantially until further nucleation appears upon subsequent cooling. The current model keeps abreast of the same physical background. Furthermore, it enables quantitative calculations, applicable to real alloy systems with the help of multi-component thermodynamic and mobility databases. Satisfactory agreements with experimental data have been obtained.

Some input parameters are known to have significant effect on the precipitation kinetics. First of all the accuracy and compatibility of thermodynamic and mobility databases are prerequisite to provide nucleation driving force and solute diffusivities. Secondly, interfacial energy is a key parameter, and unfortunately is difficult to obtain experimentally. Interestingly in the current work, interfacial energy seems relatively insensitive to the alloy composition and temperature. This thus makes it rather feasible for this tool to tailor alloy chemistry and heat treatment. Meanwhile, homogeneous intra-granular nucleation has been proven to be a reasonable approximation in most cases, except for those at very low cooling rate, where concomitant nucleation at multiple sites, e.g. at grain boundary and within the grain, is certainly a better option. The mean field approximation adopted in the current work is another source of uncertainty. It usually assumes a near zero volume fraction of precipitate phases. However, the presented simulation results indicate that it is still valid even for alloys with more than 40% of γ' phase, partly thanks to a modified growth model in terms of volume fraction. The mean field approach may also cause faster coarsening rate in multi-modal distributions. Fortunately, for the stage of continuous cooling in absence of a dominant coarsening process, this error is less severe than during long-time isothermal aging. Multi-type nucleation and growth rate modifications, which have been under development, can be helpful to alleviate this error. Other microstructure complexities that affect the model accuracy, especially at very low cooling rates, include loss of coherency, varied interfacial energy, interface mobility, and morphology changes away from equiaxed shape, etc., which should be investigated in the future.

Summary

A non-isothermal model has been developed in TC-PRISMA and has been successfully applied to simulate multi-modal particle size distribution of γ' precipitates in Ni-base superalloys. In reasonably good agreements with experimental data, the simulations correctly predict the chemical and thermal conditions for mono- and bi-modal distributions, without requiring extensive use of adjustable parameters. The model proves to be valuable for designing alloy chemistry and heat treatment, but refinement is necessary for conditions that significantly deviate from mean field approximations.

References

1. Q. Chen et al. TC-PRISMA user's guide and examples. Thermo-Calc Software AB, Stockholm, Sweden, http://www.thermocalc.se/Library.htm; 2011.
2. Thermo-Calc Software AB. Stockholm, Sweden, http://www.thermocalc.se/.
3. N. Saunders, and A. P. Miodownik. *CALPHAD (Calculation of Phase Diagrams): A Comperhensive Guide*.(New York: Elsevier Science Ltd.; 1998).
4. J. S. Langer, and A. J. Schwartz. "Kinetics of nucleation in near-critical fluids", *Phys Rev A* 21(1980), 948-58.
5. R. Kampmann, and R. Wagner. "Kinetics of precipitation in metastable binary alloys - theory and application", *Decomposition of alloys: the early stages*, ed. P. Haasen et al.(Oxford: Pergamon Press, 1984), 91-103.
6. H. J. Jou et al. "Computer simulations for the prediction of microstructure/property variation in aeroturbine disks", *Superalloy 2004*. ed. K. A. Green et al. (Seven Springs, PA: TMS, 2004), 877-86.
7. N. Saunders et al. "The application of Calphad calculations to Ni-based superalloys", *Superalloys 2000*. ed. K. A. Green et al. (Warrendale, PA: TMS, 2000), 803-11.
8. T. Rojhirunsakool et al. "Influence of composition on monomodal versus multimodal γ' precipitation in Ni-Al-Cr alloys", *J Mater Sci* 48(2013), 825-31.
9. R. Radis et al. "Evolution of size and morphology of γ' precipitates in Udimet 720 Li during continuous cooling", *Superalloys 2008*. ed. R. C. Reed et al. (Seven Springs, PA: TMS, 2008), 829-36.
10. R. Radis et al. "Multimodal size distributions of γ' precipitates during continuous cooling of UDIMET 720 Li", *Acta Mater* 57(2009), 5739-47.
11. J. Mao et al. "Cooling precipitation and strengthening study in powder metallurgy superalloy U720LI", *Metall Mater Trans A* 32A(2001), 2441-52.
12. M. P. Jackson, and R. C. Reed. "Heat treatment of UDIMET 720Li: the effect of microstructure on properties", *Mater Sci Eng* A259(1999), 85-97.
13. P. M. Sarosi et al. "Formation of multimodal size distributions of γ' in a nickel-base superalloy during interrupted continuous cooling", *Scripta Materialia* 57(2007), 767-70.
14. Y. H. Wen et al. "Phase-field modeling of bimodal particle size distributions during continuous cooling", *Acta Mater* 51(2003), 1123-32.

2nd World Congress on Integrated Computational Materials Engineering
Edited by: Mei Li, Carelyn Campbell, Katsuyo Thornton, Elizabeth Holm, and Peter Gumbsch
TMS (The Minerals, Metals & Materials Society), 2013

DEVELOPMENT OF GRADIENT CEMENTED CARBIDES THROUGH ICME STRATEGY

Yong Du[a,*], Yingbiao Peng[a], Weibin Zhang[a], Weimin Chen[a], Peng Zhou[a], Wen Xie[b], Kaiming Cheng[a], Lijun Zhang[a], Guanghua Wen[b], and Shequan Wang[b]

[a] *State Key Lab of Powder Metallurgy, Central South University, Hunan, 410083, China*

[b] *Zhuzhou cemented carbide cutting tools limited company, Zhuzhou, Hunan 412007, China*

*E-mail: yongducalphad@gmail.com

Abstract

An integrated computational materials engineering (ICME) including CALPHAD method is a powerful tool for materials process optimization and alloy design. The quality of CALPHAD-type calculations is strongly dependent on the quality of the thermodynamic and diffusivity databases. The development of a thermodynamic database, CSUTDCC1, and a diffusivity database, CSUDDCC1, for cemented carbides is described. Several gradient cemented carbides sintered under vacuum and various partial pressures of N_2 have been studied via experiment and simulation. The microstructure and concentration profile of the gradient zones have been investigated via SEM and EPMA. Examples of ICME applications in design and manufacture for different kinds of cemented carbides are shown using the databases and comparing where possible against experimental data, thereby validating its accuracy.

Keywords: ICME, CALPHAD, Cemented carbides, gradient sintering, microstructure

1. Introduction

WC-Co-based cemented carbides have long been used in applications such as cutting, grinding, and drilling [1]. Cubic carbides or carbonitrides based on Ta, Ti, and Nb are often added in cemented carbides to increase the resistance to plastic deformation or as gradient formers [1]. Some grain growth inhibitors such as Cr and V may also be added in small amounts [2]. In order to prevent crack propagation from the coating into the substrate, a gradient layer, which is free of cubic phases and enriched in binder phase, is introduced between coating and substrate [3]. In the past decades, cemented carbides were mainly developed through a large degree of mechanical testing. However, there are numerous factors influencing the microstructure and properties of cemented carbides, such as alloy composition, sintering temperature and time, partial pressure and so on. These factors can only be varied in an infinite number of ways through experimental method. The need to describe the interaction of the various process conditions has led to the ICME approach, which presents the opportunity to limit the experiments to an economical level.

Computational thermodynamics, using, e.g. the Thermo-Calc [4] and DICTRA [5] packages, has shown to be a powerful tool for processing advanced materials in cemented carbides [6-8], which is more efficient on composition and process parameters optimization compared with expensive and time consuming experimental methods. With the development of thermodynamic (CSUTDCC1 [9-12]) and diffusivity (CSUDDCC1 [10-12]) databases, it is possible to make technical calculations on commercial products which are multicomponent alloys. On the basis of thermodynamic database, thermodynamic calculations can give an easy access to what phases form at different temperatures and alloy concentrations during the manufacture process. By combining CSUTDCC1 and CSUDDCC1 databases, DICTRA [5] permits simulations of the gradient process, which is a major advance in the understanding of the gradient zone formation in the cemented carbides. This paper is intended to 1) describe the development of the CSUTDCC1 and CSUDDCC1 databases, 2) design experiments to investigate the gradient zone formation under different alloy compositions and sintering environments, 3) validate the accuracy and reliability of the presently established databases in cemented carbides alloy design and optimization by making several case studies, e.g. the simulation of the gradient zone formation, the influence of alloy composition, sintering temperature, and partial pressure of N_2 on the stability, amount and solubility of phases, and 4) present the successful implementation of ICME in industrial cemented carbide design and manufacture.

2. Experimental procedure

The alloys were prepared from a powder mixture of WC, (Ti,W)C, Ti(C,N), (Ta,Nb)C and metallic Co powder provided by Zhuzhou cemented carbide cutting tools limited company. The composition of the sintered material is given in Table 1. After milling and drying, the powders were pressed into cutting tool inserts. Samples were dewaxed and sintered under different nitrogen partial pressures (0, 20 and 40 mbar) at 1723 K for 1 h. After sintering the samples were cut, embedded in resin and polished. SEM (Nova NanoSEM 230, USA) was employed to investigate the microstructure of the gradient zone, and EPMA (JXA-8230, JEPL, Japan) was used to determine the concentration profiles of the elements.

Table 1 Chemical composition of the investigated cemented carbides (wt%)

No.	Alloy	Ti	Co	Nb	Ta	C	N	W
1#	WC-Ti(C,N)-Co	5	7.5	-	-	6.25	0.1	Bal.
2#	WC-Ti(C,N)-NbC-Co	2.85	6.75	4.93	-	6.19	0.29	Bal.
3#	WC-Ti(C,N)-TaC-Co	2.72	6.4	-	9.3	6.02	0.22	Bal.
4#	WC-Ti(C,N)-(Nb,Ta)C-Co	3	9	0.8	6	6.06	0.08	Bal.

3. Modeling of Gibbs Energy and Atomic Mobility

In the CALPHAD approach, the Gibbs energy or atomic mobility of each phase in a multicomponent system is modeled hierarchically from lower-order systems, and the model parameters can be evaluated by considering both experimental and ab-initio calculation data.

3.1 Thermodynamic modeling

The thermodynamic models describe the thermodynamic properties of various types of phases depending on the crystallography, order-disorder transitions, and magnetic properties of the phases. With parameters stored in database, many different models [13], including the substitution solution model, sublattice model, order-disorder model, have been adopted for the phases in cemented carbide systems. The thermodynamic models for Gibbs energy of a phase can be represented by a general equation:

$$G_{\mathrm{m}}^{\theta}={}^{ref}G_{\mathrm{m}}^{\theta}+{}^{id}G_{\mathrm{m}}^{\theta}+{}^{E}G_{\mathrm{m}}^{\theta}+{}^{magn}G_{\mathrm{m}}^{\theta} \tag{1}$$

Here ${}^{ref}G_{\mathrm{m}}^{\theta}$ represents the Gibbs energy of the pure elements of the phase and ${}^{id}G_{\mathrm{m}}^{\theta}$ represents the contribution due to the ideal mixing. The term ${}^{E}G_{\mathrm{m}}^{\theta}$ represents the excess energy and ${}^{magn}G_{\mathrm{m}}^{\theta}$ the magnetic contribution. The thermodynamic model and Gibbs energy expression for individual phases are presented in one publication of our group [9] and will not be repeated here.

3.2 Kinetic modeling

In a multicomponent system, a large number of diffusivities need to be evaluated, making a database very complex. A superior alternative is to model atomic mobility instead. In this way, the number of the stored parameters in the database is substantially reduced and the parameters are independent. A detailed description for the atomic mobility is given by Andersson and Ågren [14]. The atomic mobility for an element B, M_{B}, can be expressed as

$$M_B = M_B^0 \exp\left(-Q_{\mathrm{B}}/RT\right)\left(1/RT\right) \tag{2}$$

where R is the gas constant, T the temperature, M_{B}^0 a frequency factor and Q_{B} the activation enthalpy. Both M_{B}^0 and Q_{B} are in general dependent on composition and temperature.

4. Database Development of CSUTDCC1 and CSUDDCC1

The development of CSUTDCC1 and CSUDDCC1 databases has started from the major elements in gradient cemented carbides C-Co-Cr-W-Ta-Ti-Nb-N. More elements, such as Fe, Ni, V, Zr, Mo and so on, are being added or will be introduced in the future. The path toward ICME implementation in industry will necessarily require that CSUTDCC1 and CSUDDCC1 databases provide data that is not only of high quality, but relevant to industrially complex materials.

4.1 Thermodynamic database CSUTDCC1

Developed using the CALPHAD approach, CSUTDCC1 is based on critical evaluations of binary, ternary and even higher order systems which enable making predictions for multicomponent systems and alloys of industrial importance. In total, 28 binaries, 23 ternaries and 10 quaternaries in the C-Co-Cr-W-Ta-Ti-Nb-N system have been assessed. The thermodynamic descriptions of all of the binary systems and most of the ternary systems are carefully selected from the literature together with the consideration of the validation for multicomponent cemented carbides, especially the commercial alloys, and others are assessed or reassessed in the present work. Seven ternary systems C-Cr-Ta [15], C-Cr-Ti [16], C-Cr-Nb, C-Ta-Ti [17], C-Ta-Nb, C-Ti-Nb, and Co-Cr-Ti [18] have been assessed in our group. Due to the refinement of binary systems, the compatibility among high-order systems, and/or new experimental information of technical importance, six ternary systems C-Co-Cr, C-Co-Ta, C-Co-Ti, C-Co-Nb, C-W-Ta, and Co-Cr-W have been reassessed in the present work. The details of these reassessed systems can be referred to Ref. [9].

4.2 Diffusivity database CSUDDCC1

CSUDDCC1 is a diffusivity database containing mobility data for the liquid and fcc phases as well as some technologically important phases. For fcc phase, the database contains assessed atomic mobility parameters for 5 binaries, 10 ternaries, 6 quaternaries, 4 quinaries, and 1 hexahydric in the C-Co-Cr-W-Ta-Ti-Nb-N system. Due to the lack of detailed information on liquid diffusion, the Sutherland equation, which is modified to predict the temperature-dependence diffusivity in liquid metals [19], was utilized to develop the atomic mobility database.

5. Thermodynamic Calculations

Multicomponent phase diagrams demonstrate what phases can form at different temperatures and alloy concentrations during various stages of the materials processing route, and are therefore very useful in alloy development and process design. The present thermodynamic database is validated by applying thermodynamic calculations to a few industrial applications, some examples of which are presented in the following text.

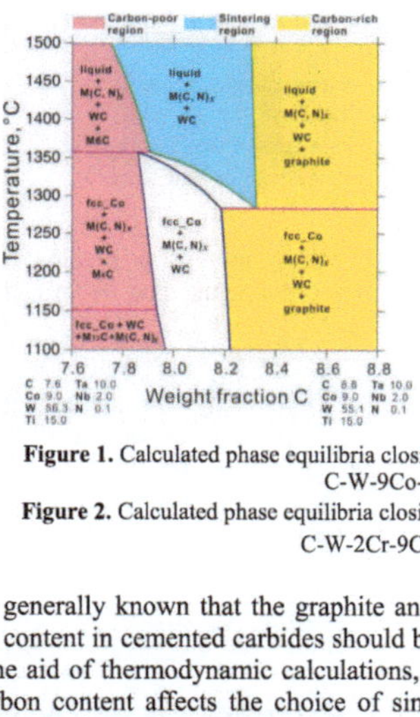

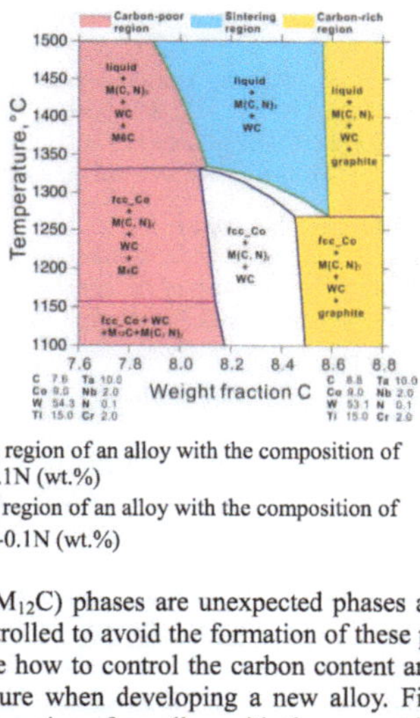

Figure 1. Calculated phase equilibria closing to the sintering region of an alloy with the composition of C-W-9Co-15Ti-10Ta-2Nb-0.1N (wt.%)

Figure 2. Calculated phase equilibria closing to the sintering region of an alloy with the composition of C-W-2Cr-9Co-15Ti-10Ta-2Nb-0.1N (wt.%)

It is generally known that the graphite and eta (M_6C or $M_{12}C$) phases are unexpected phases and the carbon content in cemented carbides should be carefully controlled to avoid the formation of these phases. With the aid of thermodynamic calculations, it is easy to see how to control the carbon content and how the carbon content affects the choice of sintering temperature when developing a new alloy. Figure 1 shows a calculated phase equilibria closing to the sintering region of an alloy with the composition of C-W-9Co-15Ti-10Ta-2Nb-0.1N (wt.%). As can be seen, the carbon content have to be carefully located in a narrow range about 0.2 wt.% in order to avoid the appearance of unwanted phases. Figure 2 presents a similar calculation by adding 2 wt.%, of Cr. From Fig. 2, it can be seen that the melting point of binder phase is decreased substantially by Cr addition and the existence of the preferable fcc_Co + $M(C, N)_x$ + WC equilibrium is broadened. On the basis of CSUTDCC1, a similar calculation can be performed on alloys with any composition, which will be a useful guidance for developing new alloys.

Knowledge of solubility of alloy elements in binder or cubic phase is the basis of studying some properties of cemented carbides, such as grain size, hardness, strength and so on. For instance, because WC is difficult to nucleate, dissolution of WC during sintering will thus result in a coarse microstructure. Based on the thermodynamic database, it is convenient to calculate the influences on the solubility of binder or cubic phase, such as alloy composition, sintering temperature, sintering partial pressure of N_2 and so on. Figure 3 shows the calculated composition of the cubic phase at various temperatures for 8C-9Co-2Cr-10Ta-15Ti-2Nb-0.1N-W (wt.%) alloy. As can be seen, the solubility of W in cubic phase increases substantially while those of Ta and Ti decrease with increasing temperature. Miscibility gaps may form in carbonitrides and has been claimed to be of technical importance. For example, the miscibility gaps in cubic phases may influence the microstructure of the "core-rim" structure and, therefore, change the mechanical properties of cemented carbides. Figure 4 presents the calculated solubility of WC in cubic carbides and the miscibility gap of fcc carbide in the C-W-Ti-N system with unit carbon activity at 1427 °C, along with the experimental data [20]. As can be seen in Fig. 4, with increasing nitrogen activity (or N_2 partial pressure), the solubility of WC in cubic phase decreases while the size of the miscibility gap in cubic phase increases.

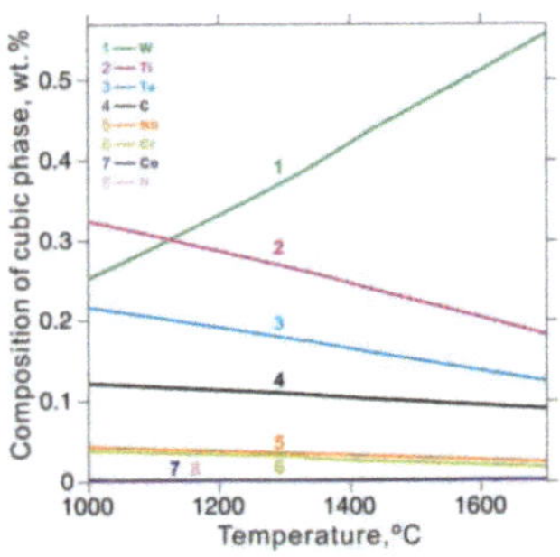

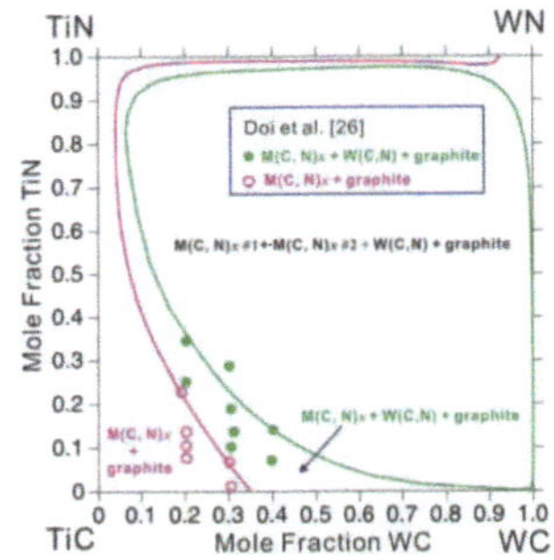

Figure 3. Calculated composition of the cubic phase at various temperatures for 8C-9Co-2Cr-10Ta-15Ti- 2Nb-0.1N-W (wt.%) alloy.

Figure 4. Calculated solubility of WC in cubic carbide and the miscibility gap of the fcc carbide in the C-W-Ti-N system with unit carbon activity at 1427 °C along with the experimental data [20].

6. Results and Discussion

Gradient cemented carbide tools coated with hard phases are widely used in processing industry. Studying the microstructure of the gradient zone experimentally and further simulating its formation will be extremely useful for materials design and processing.

6.1 Experimental results

Based on thermodynamic calculations, four WC-Ti(C,N)-Co-based gradient cemented carbides have been prepared, as seen in Table 1. Figure 5 shows SEM micrograph of the cross section of alloys 1# sintered under different nitrogen partial pressures (0, 20 and 40 mbar) at 1450 °C for 1 h. It is obvious that the near-surface of the alloy has formed the gradient zone which is enriched in binder phase and depleted in cubic carbides. It is also shown that the thickness of the gradient layer decreases with the increase of nitrogen partial pressure. According to the color contrast, some of the cubic phase has formed "core-rim" structure consisting of a (Ti, W)$(C,N)_x$ phase.

6.2 Kinetic simulations

By combining the presently established thermodynamic and diffusivity databases, DICTRA software has been used to simulate the formation of the gradient zone. Figure 6 shows the simulated phase fractions in alloy 4# after sintered for 1 h at 1450 °C under vacuum, compared with the experimental results. The present simulations indicate that the experimental data can be well reproduced based on the present thermodynamic and diffusivity databases. Figures 7(a)-(b) illustrate the simulated elemental concentration profiles for Co and Ti in alloys 1# after sintering for 1 h at 1450 °C under different nitrogen partial pressures (0, 20 and 40 mbar), compared with the measured data. This result indicates that the content of Ti is free in the near-surface zone and enrich inside the surface zone. At the near-surface zone, the content of Co increases sharply and reached a maximum value. Beneath the near-surface zone, a decrease of Co is observed, which leads to the minimum value. Above this minimum value, the content of Co increases slowly to its bulk value. The calculated thickness of the gradient layer decreases with the increasing of nitrogen partial pressure, which shows the similar diffusion behavior as the experimental results. As can be seen in Figs. 7(a)-(b), the presently obtained thermodynamic and diffusion databases can reasonably reproduce most of the experimental concentration profiles.

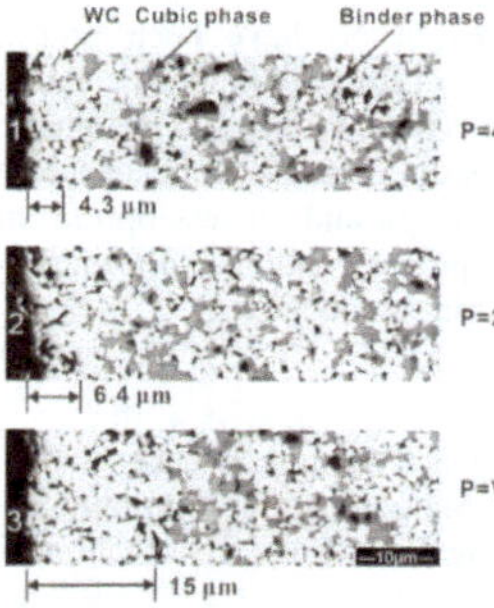

Figure 5. SEM micrograph of the cross section of alloys 1# sintered under different nitrogen partial pressures (0, 20 and 40 mbar) at 1450 °C for 1 h.

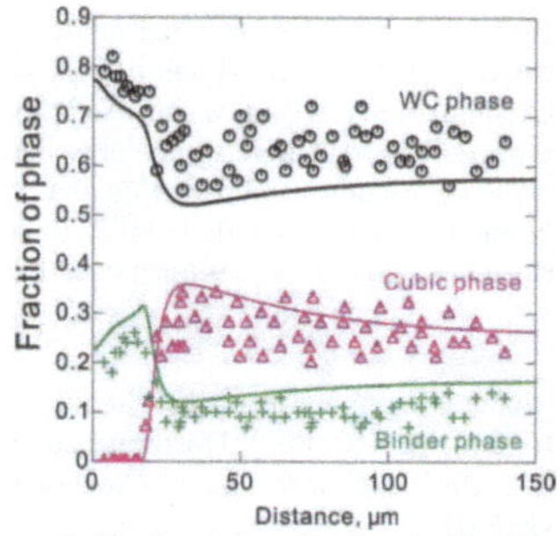

Figure 6. Simulated phase fractions in alloy 4# after sintered for 1 h at 1450 °C under vacuum, compared with the experimental results.

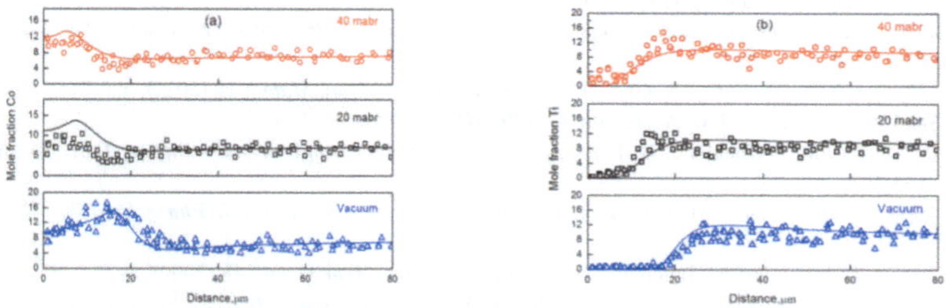

Figure 7. Concentration profile for (a) Co and (b) Ti in alloys 1# : measurement (symbols) and calculation (curve).

Figure 8. Designed and manufactured industrial cemented carbides with the integration of ICME.

6.3 ICME applications

The urgent need to develop cemented carbides with excellent performance has pull industry toward ICME. With the integration of ICME into design, several brands of cemented carbides have been developed in the present work, as shown in Figure 8. The path toward ICME implementation in cemented carbides design and manufacture can be expressed as the following steps. Firstly, alloy composition and process parameter are designed via CALPHAD-type calculations/simulations. Secondly, a series of corresponding cemented carbides are prepared under the guidance of the previous simulations. After that, the microstructure and mechanical properties of the cemented carbides are investigated experimentally, thereby validating the accuracy of the calculations/simulations. Process routes are optimized and finally chosen for industrial production of cemented carbides with excellent or special performances.

7. Summary

Thermodynamic database, CSUTDCC1, and diffusivity database, CSUDDCC1, for cemented carbides have been developed through a combination of experimental, theoretical and assessment work. Gradient cemented carbides WC-Ti(C,N)-Co, WC-Ti(C,N)-TaC-Co, WC-Ti(C,N)-NbC-Co, and

WC-Ti(C,N)-TaC-NbC-Co sintered under various partial pressures of N_2 have been prepared and investigated by means of SEM and EPMA techniques. Some examples of thermodynamic calculations and DICTRA simulations as part of the validation of these databases for different kinds of cemented carbides have been performed. Good agreements between simulated and measured results indicate the powerful ability of the presently established databases in cemented carbides design and process optimization. An ICME framework with validated ability to composition designation, process optimization, microstructure and performance prediction has been described and utilized for industrial production.

Acknowledgments

The financial support from the Creative Research Group of National Natural Science Foundation of China (Grant No.51021063), Doctoral Scientific Fund Project of the State Education Committee of China (Grant No. 20120162110051), and Zhuzhou Cemented Carbide Cutting Tools Co., LTD of China is greatly acknowledged.

References

[1] H.E. Exner, ***Int. Met. Rev.***, 24 (1979), 149-173.

[2] G.S. Upadhyaya, Editor, Cemented Tungsten Carbides: Production, Properties, and Testing, Westwood, USA: Noyes, Westwood, N. J., 1998.

[3] T.E. Yang, J. Xiong, L. Sun, Z.X. Guo, D. Cao, ***Int. J. Miner., Metal. Mater.***, 18 (2011), 709-716.

[4] B. Sundman, B. Jansson, J. O. Andersson, ***CALPHAD***, 9 (1985), 153-190.

[5] J. O. Andersson, L. Hoeglund, B. Jansson, J. Ågren, ***Fundam. Appl. Ternary Diffus., Proc. Int. Symp.***, (1990), 153-163.

[6] K. Frisk, L. Durnitrescu, M. Ekroth, B. Jansson, O. Kruse, B. Sundman, ***J. Phase Equilib.***, 22 (2001), 645-655.

[7] M. Ekroth, R. Frykholm, M. Lindholm, H.O. Andren, J. Agren, ***Acta Mater.***, 48 (2000), 2177-2185.

[8] M. Ekroth, K. Frisk, B. Jansson, L.F.S. Dumitrescu, ***Metall. Mater. Trans. B***, 31B (2000), 615-619.

[9] Y. Peng, Y. Du, P. Zhou, S. Wang, G. Wen, W. Xie, ***Int J Refract Met Hard Mater***, under review (2013).

[10] W. Zhang, Y. Peng, P. Zhou, W. Chen, Y. Du, W. Xie, G. Wen, S. Wang, ***J. Phase Equilib. Diffus.***, in press, doi: 10.1007/s11669-013-0225-2 (2013).

[11] W.M. Chen, T. Xu, W.B. Zhang, H. Zhang, Y. Du, C.Y. Tan, L. Chen, Z.J. Zhang, X.Z. Liu, ***The Chinese Journal of Nonferrous Metals***, 22 (2012), 1440-1447.

[12] W. Zhang, C. Sha, Y. Du, G. Wen, W. Xie, S. Wang, ***Acta Metallurgica Sinica***, 47 (2011), 1307-1314.

[13] H.E. Lucas, S.G. Fries, B. Sundman, Computational thermodynamics, the Calphad method, Cambridge University press, 2007.

[14] J.-O. Andersson, J. Ågren, ***J. Appl. Phys.***, 72 (1992), 1350-1355.

[15] C. Sha, M. Bu, H. Xu, Y. Du, S. Wang, G. Wen, ***J Alloys Compd***, 509 (2011), 5996-6003.

[16] J.C. Schuster, Y. Du, ***CALPHAD***, 23 (2000), 393-408.

[17] Y. Peng, P. Zhou, Y. Du, K. Chang, ***Int J Refract Met Hard Mater***, accepted (2013).

[18] P. Zhou, Y. Peng, B. Hu, S. Liu, Y. Du, S. Wang, G. Wen, W. Xie, ***CALPHAD***, 41 (2013), 42-49.

[19] W. Chen, L. Zhang, D. Liu, Y. Du, C. Tan, ***J. Electron. Mater.*** , in press, doi: 10.1007/s11664-013-2549-4 (2012).

[20] A. Doi, T. Nomura, M.-A. Tobioka, K. Takahashi, A. Hara, ***High Temp. High Press.***, 18 (1986), 443-452.

2nd World Congress on Integrated Computational Materials Engineering
Edited by: Mei Li, Carelyn Campbell, Katsuyo Thornton, Elizabeth Holm, and Peter Gumbsch
TMS (The Minerals, Metals & Materials Society), 2013

FULL-FIELD MULTI-SCALE MODELLING OF SHEET METAL FORMING TAKING THE EVOLUTION OF TEXTURE AND PLASTIC ANISOTROPY INTO ACCOUNT

Paul Van Houtte[1], Jerzy Gawad[2], Philip Eyckens[1], Albert Van Bael[1], Giovanni Samaey[2], Dirk Roose[2]

[1]Department MTM, KULeuven, Kasteelpark Arenberg 44, 3001 Leuven, Belgium
[2]Department of Computer Science, KULeuven, Celestijnenlaan 200 A, 3001 Leuven, Belgium

Keywords: sheet metal forming, FE analysis, texture and anisotropy, multi-scale modeling

Abstract

Focus is on the implementation of texture-induced plastic anisotropy in FE simulations of metal forming. The crystallographic texture can be introduced as a state variable in every integration point. A multi-scale model is then called to calculate the stress-strain response and the local texture evolution in every integration point and for every strain increment. Less calculation-intensive is to use anisotropic analytical constitutive models, identified in advance from mechanical tests. These can also be done in a "virtual" way, i. e. using measured texture data and a multi-scale model. However, texture evolution is then not taken into account. An adaptive scheme for updating the texture and the anisotropy has been developed recently. Texture and anisotropy were updated by the ALAMEL-model. Results for some sheet metal forming processes are shown. The calculation times had been reduced from months to days. Predicted fields of plastic anisotropy and textures are discussed including experimental validation.

Introduction

The crystallographic texture in the material is not constant during sheet metal forming. Its evolution depends on the local deformation history, which may differ from one location to the next. The constitutive law which describes the plastic response of the material at macro-scale depends on the texture and will also evolve in a different way in every location. Taking this into account in a finite element (FE) model for metal forming processes may increase the accuracy of the results. In the present work, a rather conventional explicit elastic-plastic FE code with a user-written routine (VUMAT) is used. The integration points represent RVEs at macro-scale consisting of 5000 crystallites. The plastic anisotropy is derived from the texture by means of the FACET model, in fact an analytical constitutive model for the anisotropic response of the material; its parameters are derived from the current texture whenever needed. This is an example of a so-called "hierarchical multi-scale model". Even so this method requires in principle an enormous computational effort as the texture and the constitutive model need to be updated after every deformation step for every integration point. It will be explained how by optimizing the code this calculation effort has been reduced without noticeable loss of accuracy. Results obtained for cup drawing of aluminium and steel sheets will be discussed.

Hierarchical Multi-Scale Modelling

The local texture of the material can be treated as a state variable in a FE model and associated to each integration point as a discrete set of crystal orientations. A polycrystal model (such as the well known Taylor model [1-2]) can then be used to (i) produce the local stress-strain relationship whenever required by the FE shell, and (ii) simulate the texture evolution for each strain increment. This is the "embedded approach", see Dawson et al. [3] for a nice overview. Texture evolution is automatically obtained. However, in practice this method is not suited for industrial problems because of the computational effort. Reducing this by working with a small discrete set with for example only 100 grains instead of 5000 is not recommended, as the resulting ODF [4] might become insufficiently accurate to be used for predictions of plastic anisotropy [5]. The present authors have chosen another method to reduce the computational effort of the embedded multi-scale method. Texture updating is still done by a polycrystal deformation model (see below). However, the latter is not used to give the stress-strain relationship of the polycrystal each time the FE shell asks for it. That task is left to the 'Facet' model [6,7], essentially an analytical model which reproduces the plastic anisotropy of the polycrystal in an approximate way. After each texture update, the coefficients of the Facet expression are updated by identification of the model using virtual mechanical tests: a series of mechanical tests is simulated by the polycrystal model. This method is an example of a "hierarchical multi-scale model". It already drastically reduces the number of times that the polycrystal code needs to be called by the FE shell. The polycrystal model chosen by the present authors was ALAMEL [2], as a good compromise between accuracy and calculation speed. The well-known Taylor model [1] is occasionally used for comparison. For a more complete discussion of the available options we refer to some more elaborate papers [6,8]. Note that the ALAMEL model is a statistical model belonging to the class of "grain interaction models", the GIA [9] model being another one

Further Code Optimization

The first tests have been performed on a cluster computer with 256 Opteron dual-core processors. It was decided not to spread the execution of FE code itself (Abaqus Explicit) over several nodes. It was however decided to use extra nodes for tasks that can be run separately, such as texture updates and identification of the Facet model in particular integration points. This already led to a significant reduction of the wall clock time needed for the calculation. 5939 updates needed for example to be done in a particular case. This would have taken a wall clock calculation time of 43 days on a single node. This was reduced to 1.57 days by the method mentioned above. But adaptive scheduling of the updates gave the largest benefit. The idea was, only to perform an update in a given integration point when certain criteria were satisfied instead of doing it after every strain increment. A good example for such requirement is, that the total strain P_{cr} undergone since the last update should be equal or larger than a given critical value (about 3 % was used in the present work). It is indeed so that the change of the deformation texture and the associated plastic anisotropy is very small indeed after a single strain increment of the kind used in explicit simulations. Consider the following 'real' application: the simulation of a cup drawing process with ~3000 integration points and 11 x 10^6 grains in all discrete textures together. Without code optimisation, this would require 230 days of calculation on a single node. With P_{cr} approximately equal to 0.03, the total number of updates was 33445, and the total wall-clock 5 days on the cluster computer.

Some Results for Cup Drawing

Cup drawing may not be the most exciting industrial drawing process, but it nevertheless often used to study the effect of work hardening and texture on metal forming. Texture-induced plastic anisotropy usually causes the development of "ears" at the rim of a drawn cup. Quantitative prediction of these is quite a challenge for any model. Cup height profiles have been experimentally obtained and simulated for two materials. The results given below have been borrowed from [6,10].

Grade	Thickness (mm)	Additional Elements (mass %)	Blank (mm)	Punch	Die	BHF (kN)	μ
Steel: DC06	0.70	0.02 % C, 0.25 % Mn, 0.30 % Ti	100	50R7	52R6	10	0.10
Aluminium alloy: AA31303-O	0.85	0.2 % Si, 1.0 % Mn, 0.7 % Fe	100	50R5	52.5R10	5	0.05

Table 1. Cup drawing tests: material, diameter of the circular blanks, punch diameter (R means: fillet radius), blank holder force (BHF). μ: friction coefficient used during the simulation.

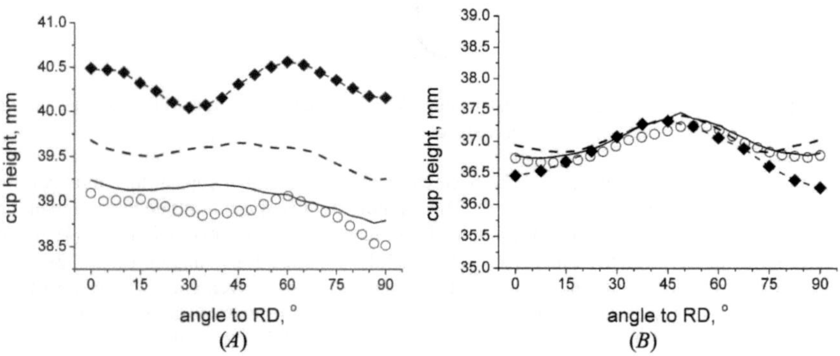

Fig. 1 After [10]. Comparison between experimental and simulated cup height profiles. (*A*) DC06, (*B*) AA3103-O.
Dashes with filled diamonds: experimental cup profiles, average of four measurements in (*A*). *Full line*: simulation with ALAMEL without texture updating. *Circles*: also ALAMEL, with texture updating. *Dashes*: Taylor, without texture updating.

Table 1 describes the experimental cup drawing tests which have been carried out. Typical industrial materials which do not have a particular strong earing tendency were used. That makes it more difficult for models to make useable predictions of the earing profile. The finite element mesh used represented one quarter of the blank. There were three layers of brick elements with reduced integration (C3D8R). The total number of elements was 3096. Each integration point had its own texture, described by 5000 discrete orientations. Three simulations have been done for each material: one which used ALAMEL as polycrystal deformation model for the virtual tests needed for the identification of the anisotropic constitutive model as well as for texture updating; one which used ALAMEL only for the identification, and one which used the Taylor model only for the identification. Table 1 gives the values of the friction coefficients used during the simulations. The experimental and simulated earing profiles (cup wall heights) are shown by Fig. 1. All simulations used the experimentally measured textures of the blanks as starting textures. Let us first look at the results for DC06 (Fig. 1a). The experimental curves have local maxima at 0° and 60°. The one at 0° would correspond to 2 ears in a full 360° representation and the one at 60° to 4 ears. This makes 6 ears in total. The simulations without texture updating both lead to a flatter shape with valleys at 90° (i.e. TD). However, a 6-ear behaviour only appears in the simulation with texture updating using ALAMEL, with ears at RD and at 60°. The shape of this curve is qualitatively similar with the shape of the experimental findings. The main difference is an underestimation of the global cup height by about 1.5 mm. No effort has been done to estimate (for example by reverse engineering) the value of the friction coefficient, which had been arbitrarily set to 0.1 For the AA3103-O material, there is almost no difference between the simulated curves; they reproduce the experimental maximum at 45° (Fig. 1b).

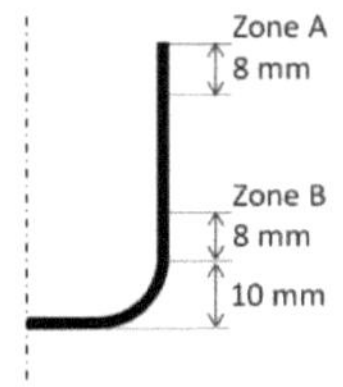

Fig. 2 After [10]. Cross section of a drawn cup. A and B are the locations of the samples cut out for texture measurements.

Samples for texture measurements have been cut out from the drawn cups at locations A (close to the rim) and B (bottom of the rim) (Fig. 2). Four pole figures of each sample were measured at half-thickness by means of X-ray diffraction. These have then been converted into an orientation distribution function (ODF) called $f(g)$ [4]. g stands for three Euler angles φ_1, Φ, φ_2 which together describe the angular orientation of the crystal lattice of a given grain with respect to the external axes of the sample. For cubic materials (and after an arbitrary forming operation), their range is 0°-360°, 0°-90° and 0°-90°, respectively [4]. Steps of 5° are used in the present work; that means that the experimentally measured ODF and the predicted ones need to be equal to each other on 26353 locations in orientation space in order to be able to make the statement "measured and predicted ODF are equal to each other". The following error measure is introduced in order to quantify the difference between the measured and predicted ODFs:

$$D_{\mathrm{ODF}} = \frac{\int \left(f_{\mathrm{sim}} - f_{\mathrm{exp}}\right)^2 dg}{\int f_{\mathrm{exp}}^2(g)dg} \quad (1)$$

This measure is called "normalized ODF difference". The quadratic difference integral in this expression is divided by the integral of the square of the experimental ODF in order to achieve normalization with respect to the intensity of the measured texture. D_{ODF} would be zero in case the two textures are identical. Table 2 gives its values for the simulations. The values obtained for DC06 and AA3103-O in zones A and B and along 3 directions with respect to the rolling direction are given in Table 2. The ALAMEL model was used to predict the textures. Fig. 3

gives the φ_2=45° sections of the experimental and simulated ODFs at Zone A for 90°, so that the reader can have a visual impression of what a D_{ODF} value of 0.23 might mean. More details about the texture results can be found elsewhere [6].

Material	Zone	0°	45°	90°
DC06	A	0.16	0.09	0.23
	B	0.16	0.16	0.26
AA3103–O	A	0.25	0.15	0.16
	B	0.16	0.18	0.15

Table 2. D_{ODF} values for the final simulated textures and the final experimental textures for DC06 and AA3103-O along 3 directions with respect to the rolling direction. Texture evolution: calculated using ALAMEL.

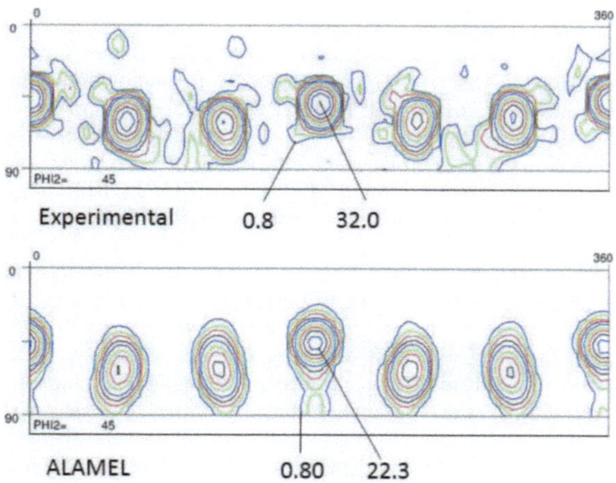

Fig. 3 After [10]. DC06, measured and simulated textures, φ_2=45° sections of the ODF at location A, 90° from RD (fig. 2).

Conclusion

It has been demonstrated that fully coupled multi-scale simulations of sheet metal forming operations are now feasible. They lead to texture and anisotropy prediction in every integration point of the mesh. The computer platforms required for this are not excessively powerful. This goal has been achieved by (i) adaptive updating of the local texture and anisotropy and (ii) using a polycrystal model which represents a good compromise between accuracy and calculation speed. The results presented suggest that certain fine features of the deformation pattern in metal forming, such as 6-ear earing profiles in cup drawing, can actually only be predicted by a fully coupled multi-scale models as the one presented here.

Acknowledgements

The authors gratefully acknowledge the financial support from the project IDO/08/09, funded by K.U.Leuven, and from the Belgian Federal Science Policy agency, contracts IAP7/19 and IAP7/21. They also thank the companies TataSteel, ArcelorMittal and Aleris for the sheet materials and cup drawing experiments. GS is Postdoctoral Fellow of the Research Foundation – Flanders (FWO).

References

1. G.I. Taylor, "Plastic strain in metals," *J. Inst. Metals*, 62 (1938), 307-324.
2. P. Van Houtte, S. Li, M. Seefeldt, L. Delannay, "Deformation texture prediction: from the Taylor model to the advanced Lamel model," *Int. J. Plasticity*, 21 (2005), 589-624.
3. P.R. Dawson, S.R. MacEwen, P.D. Wu, "Advances in sheet metal forming analyses: dealing with mechanical anisotropy from crystallographic texture," *International Materials Reviews*, 48 (2003), 86-122.
4.H.J. Bunge, *Texture analysis in material science* (London, UK: Butterworths, 1982).
5. L.S. Toth, P. Van Houtte, "Discretization techniques for orientation distribution functions.,"*Textures and Microstructures*, 19 (1992), 229-244.
6. P. Van Houtte , J. Gawad, P. Eyckens, B. Van Bael, G. Samaey, D. Roose, "Multi-scale modelling of the development of heterogeneous distributions of stress, strain, deformation texture and anisotropy in sheet metal forming," *Procedia IUTAM 3* (2012), 67–75. (Available on-line at http://www.sciencedirect.com/science/journal/22109838/3, accessed on January 26, 2013)
7. P. Van Houtte, S.K. Yerra, A. Van Bael, "The Facet method: A hierarchical multilevel modelling scheme for anisotropic convex plastic potentials," *Int. J. Plasticity*, 25 (2009), 332-360.
8. P. Van Houtte, J. Gawad, P. Eyckens, B. Van Bael, G. Samaey, D. Roose, "A full-field strategy to take texture induced anisotropy into account during FE simulations of metal forming processes," *JOM: the Journal of the Minerals, Metals and Materials Society,* 63/11 (2011), 37-43.
9. O. Engler, M. Crumbach, S. Li, "Alloy-dependent rolling texture simulation of aluminium alloys with a grain-interaction model," *Acta Mater*, 53 (2005), 2241-2257.
10. P. Van Houtte, J. Gawad, P. Eyckens, B. Van Bael, G. Samaey, D. Roose, "An efficient strategy to take texture-induced anisotropy point-by-point into account during FE simulations of metal forming processes," *Materials Science Forum*, 702-703 (2011), 26-33. (Textures of Materials, Proc. ICOTOM 16, Part 1. Tewari, A., Suwas S., Srivastava, D., Samajdar, I, Haldar, A., Editors. IIT, Mumbai, India.).

2nd World Congress on Integrated Computational Materials Engineering
Edited by: Mei Li, Carelyn Campbell, Katsuyo Thornton, Elizabeth Holm, and Peter Gumbsch
TMS (The Minerals, Metals & Materials Society), 2013

Integrating Quench Modeling into the ICME Workflow

Andrew Banka[1], Jeffrey Franklin[1], William Newsome[1]

[1]Airflow Sciences Corporation 12190 Hubbard St.; Livonia, MI 48150, USA

Keywords: quenching, CFD, boiling

Abstract

ICME offers the promise of accelerating design cycles while improving the performance of manufactured components. In order for ICME to be effective, all portions of the design materials specification, and manufacturing process need to be included in the analysis.

Heat treating, including a liquid or gas quenching step, is an integral component of the production of most metal components. The rapid cooling that occurs during quenching largely defines the solid phase distribution, microstructure, residual stress, and distortion in the as-quenched part. While Computational Fluid Dynamics (CFD) can predict the heat fluxes during gas quenching operations, the complexity of phase change boiling phenomena makes the prediction of liquid quenching operations difficult. For this reason, the quenching process is often omitted from the ICME workflow.

Development of suitable CFD boiling models for practical heat treating operations requires the collection of relevant data and development of physics-based relationships that reflect the observed behavior. While significant work has been done in this area for nucleate boiling of water at saturated conditions, additional data is needed to cover the full range of boiling regimes, other quench liquids (e.g., oil and polymer), and subcooled conditions.

In order to collect the data needed to develop improved CFD boiling models, an experimental facility has been constructed. A description of the test facility is provided along with data collected for a range of boiling conditions.

Introduction

Many precision mechanical and structural components are made from material that is forged, heat treated, and machined in several steps. Plastic strains that occur during forging and heat treating lead to residual stresses and distortions in the part. As the part is machined and material is removed, the residual stresses in the remaining material cause the part to deform. Successive machining passes ensure that the final part geometry reflects the designers intent, though it may still contain residual stresses that can affect the service life of the part. The difference in the amount of material in the forged and finished part, along with the removal of that excess material, represent significant factors in the final cost of those components.

The goal of ICME is to apply computational methods to the materials and manufacturing processes in order to optimize and streamline those processes for lower production cost and better part quality. Computational tools provide significant cost and timing benefits over physical trials to achieve those same goals. While well validated tools exist for many steps of the manufacturing process, there is a distinct lack of ability to predict the surface heat fluxes that occur during the quenching step of heat treating operations. Due to this gap in the simulation technology, those heat flux rates are typically determined through experimental trials using parts instrumented with thermocouples. In addition to the cost associated with those trials, the accuracy of the resultant heat flux rates is limited by both the number of thermocouples and the inversing process used to convert the transient temperature data to surface heat fluxes.

As an alternative to this experimental approach, computational fluid dynamics offers a way to predict the surface heat transfer rates, which can then be incorporated into the ICME workflow. While CFD is an established method of predicting surface heat transfer rates for single phase situations, most quenching operations feature a vaporizable quenchant, and phase change heat transfer is an important contributor to heat flux rates.

Many CFD studies of boiling behavior have been focused on detailed simulations of individual bubble formation [1] or have used generalized approximations based on limited data for special cases (e.g., pool boiling). The former approach is too detailed to be applied to industrial processes, while the latter approaches are insensitive to issues such as surface orientation and fluid velocity. A new approach is needed that better reflects the physics controlling boiling heat transfer while avoiding the fine detail that would make the simulation impractical for industrial use.

One key to the development of accurate and efficient boiling simulation methods is access to boiling process data spanning the range of conditions in industrial quenching operations. Since quenching operations start with a hot part inserted into subcooled liquid, the data must include film boiling, film breakdown, transition boiling, critical point, and nucleate boiling. Accurate representation of the early stages of the quenching operation are critical to the accuracy of the balance of the transient quenching simulation.

There is extensive literature covering nucleate boiling through both experimental and theoretical approaches. The bulk of these studies, however, are for water at or near saturated conditions. Since quenching operations are typically performed with significant subcooling, this information is not directly applicable to quenching operations. Published literature for film and transition boiling are less common than for nucleate boiling, and again focus on the near saturated conditions of water. The authors have found no published information on boiling heat transfer in oils.

In order to cover these gaps in the data, a test facility has been designed and constructed to collect the breadth of data needed to formulate a robust boiling model that covers the range of boiling phenomena found in quenching operations but can also be efficiently incorporated into a CFD modeling framework.

CFD Simulation Approach

One of the challenges inherent in a simulation of quenching operations is the disparity of length scales. While the quench tank may have dimensions in the tens of feet, and the parts may be on the order of single feet, the bubbles that emanate from the surface are on the order of millimeters or less. Inclusion of the entire quench tank in the modeling process is critical to proper prediction of the flow fields in the vicinity of the parts. With a model of that physical size, fully resolving the small scale behavior near the wall becomes problematic.

When CFD methods were first being developed, the problem of small-scale, near-wall decay of turbulence was addressed through the creation of "wall models" [3] that apply the effect of the near-wall behavior to the first computational cell. Leveraging that approach, the authors have proposed the development of boiling wall models that encompass the effect of phase change boiling by modifying the surface heat flux calculations included in the CFD simulation. Inherent in this choice is the assumption that the quench media is sufficiently subcooled, such that any vapor released from the surface will condense within a short distance. Where the fluid reaches saturation in stagnant zones near the hot surfaces, a suitable modification of the method can be made.

For best accuracy, the forms of the heat transfer relationships should reflect the physics of the boiling phenomena. While those physics are quite complex, ultimate control over the resultant heat flux rate can only depend on those variables available through the CFD model, namely, surface temperature, fluid temperature, fluid velocity, and turbulence level.

A preliminary implementation of this boiling wall model approach has provided encouragement for its continued development. Figure 1(a) below shows a comparison between measured temperatures for a simple quench probe and simulation results in which unknown modeling parameters were adjusted to provide a good match. Using those same modeling parameters for a flat disk as shown in Figure 1(b), good correlation was achieved for the side of the disk, which is very similar in orientation to the quench probe. The correlations for the top and bottom of the disk, however, were not as good, suggesting that the physics controlling the phase change heat transfer were substantially different in those locations. Nevertheless, these results suggest that accurate models can be constructed if there is sufficient data to guide their development. The lack of published data for the range of conditions expected in quenching operations is the motivation for the development of our experimental facility.

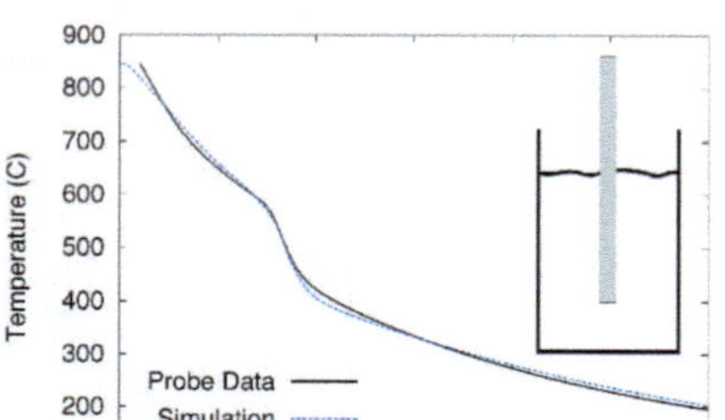

(a) Quench probe data and simulation

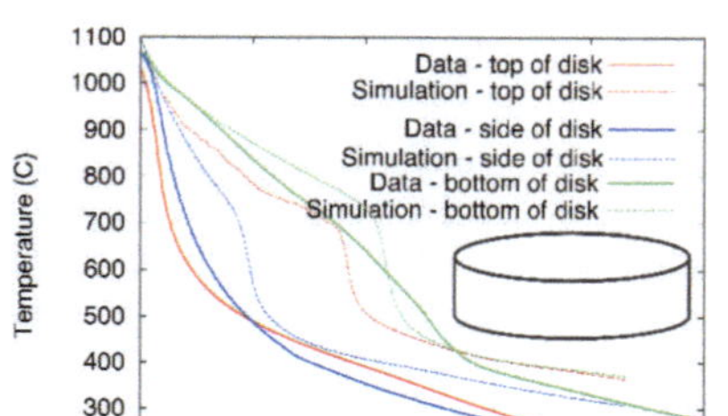

(b) Disk data and simulation

Figure 1: Comparison of Preliminary Simulation Implementation and Test Data

Quench Test Facility

The quench test facility consists of a closed-loop circuit that passes flow over a heated surface, as shown schematically in Figure 2(a). Open loop control of the flow rate is via a VFD-controlled pump, with the flow rate measured by a Krohne 3-beam ultrasonic flow meter. An auxiliary pump cycles the quench liquid through a liquid-to-liquid heat exchanger to control the temperature of the quenchant in the reservoir. Flow passes through a settling chamber on the inlet side of the 76 mm square test section to ensure smooth, low turbulence flow.

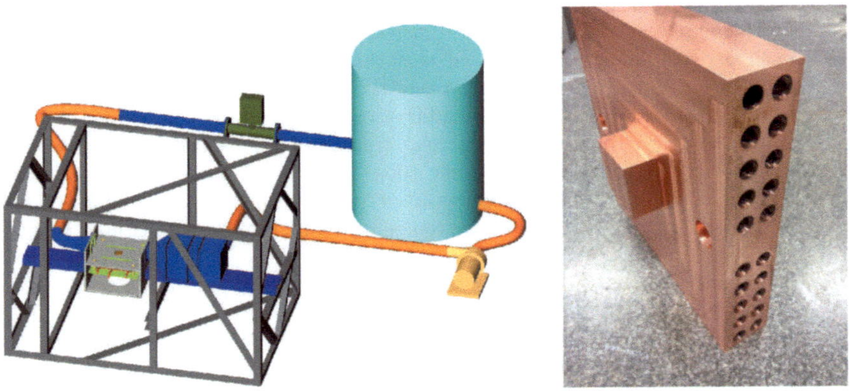

(a) Schematic of flow loop design

(b) Copper heater block

Figure 2: Quench Test Facility

The heated test surface is a 51 mm square located midway along the test section on one side. This surface is the exposed portion of the large copper block shown in Figure 2(b). Twenty Dalton 1350W cartridge heaters are inserted into the block for a total heat input of 27 kW, or a maximum theoretical surface heat flux of 10.5 MW/m^2. Staying at least 100 °C below its melting point, the maximum internal temperature of the copper block is around 950 °C, which leads to a maximum projected surface temperature during film boiling of 760 °C.

Temperature and heat flux at the wetted surface are measured through an array of fine (0.5mm) thermocouples passing into the copper block from the back side. The thermocouples are clustered at five locations beneath the wetted surface, forming a cross. At three of these locations, the cluster consists of thermocouples at three depths below the surface (2.5 mm, 18.5 mm, and 34.5 mm), allowing for a second order projection of the values at the wetted surface. The other two locations have just two thermocouples (2.5 mm and 18.5 mm), limiting projections to first order.

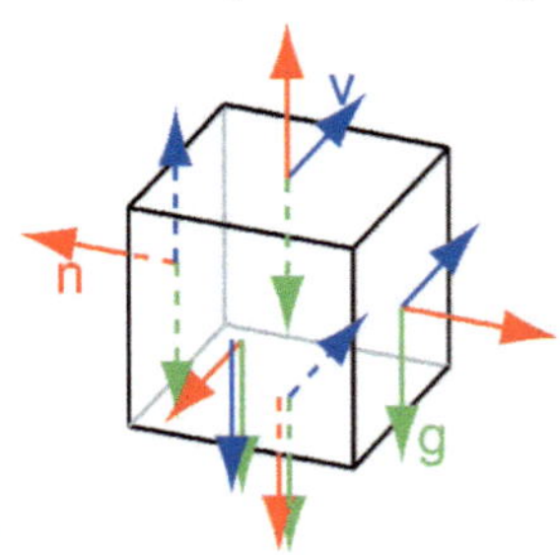

Figure 3: Possible orientations of heated surface normal (n), gravity (g), and velocity (v) vectors.

Boiling phenomena, particularly in the nucleate range, are expected to be influenced by both gravity and velocity through their effect on bubble departure. In order to capture these effects, a total of five different orientations are included in the test matrix, as shown in Figure 3, that encompass all orthogonal combinations of these vectors relative to the surface normal. These orientations will be achieved by rotating the test section cage.

The primary outputs from these tests are the surface heat flux rates and visual observations of the boiling behavior that will aid in the development of mechanistic boiling models (e.g., bubble growth rates, bubble density, bubble departure size, film thickness).

Test Data

Figure 4(a) presents a sampling of data collected in the nucleate boiling regime for a range of velocities. In this graph, the color of the symbols represents the flow velocity, while the symbol shape represents the location of the data on the test surface. Solid lines connect data points collected at the same operational conditions. All data are for water at a temperature of 20 °C. The test surface was on the bottom of the test section and flow was horizontal.

These data suggest that, at this level of subcooling, the fluid velocity has relatively little effect on the boiling heat flux at a given surface temperature. The clustering of the symbols also suggests that the boiling behavior is consistent across the entire 51 mm square test surface.

Comparison of these data with the data of Maurus et al [5], shows that the measured heat fluxes are significantly higher. The main differences between the two tests is the degree of subcooling. While Maurus's data is for 20 °C subcooling, the present data has 80 °C of subcooling.

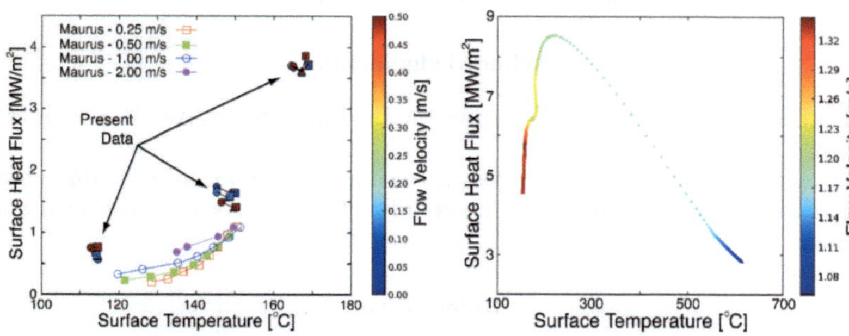

(a) Data collected in the nucleate boiling range compared to the data of Maurus

(b) Data collected in the transition through nucleate ranges

Figure 4: Preliminary experimental data

Figure 4(b) presents a set of data collected while the heated surface was moving through transition boiling and the critical heat flux into the nucleate boiling range. These data were collected by establishing film boiling and then reducing the heat input to the copper block, allowing it to cool down and transition through the different boiling modes. The water velocity was not kept constant during this test. Data points were collected at 10 Hz.

The data of Jeshar et al [4] for quenching of a 30mm nickel sphere suggest that the critical heat flux from the data of 4(b) at the given fluid velocity is in line with expectations. The data of Groeneveld [2] from the 2006 CHF look-up table suggest a critical heat flux at the test conditions of figure 4(b) of ~7.2 MW/m^2, which is comparable to the ~8.5 MW/m^2 values obtained from the current testing. Note that the Groeneveld value was taken from the "use with caution" portion of the look-up table, and pertains to a vertical 8 mm diameter tube.

Summary

The full potential of ICME can only be realized if all steps in the manufacturing process are included in the analysis workflow. One notable exception in the established workflow is the numerical simulation of the quenching operation that is key to many heat treating operations. Due to the complexity of the phase change heat transfer that occurs in liquid quenching operations and the lack of suitable models of that process, current analyses have relied on experimental trials.

A novel boiling model approach has been proposed by the authors that aims to accurately capture the effect of boiling on surface heat transfer rates without physically resolving the associated fine scale detail. Key to the development of those models is access to boiling heat flux data that covers the range of conditions expected during quenching operations.

An experimental test facility has been constructed to collect that fundamental data. Data collected to date are in line with published values for similar conditions, but also show the strong effect of the degree of subcooling on the resultant heat fluxes.

Acknowledgements

This material is based upon work supported by the United States Air Force under Contract No. FA8650-12-C-5110.

Any opinions, findings and conclusions or recommendations expressed in this material are those of the author(s) and do not necessarily reflect the views of the United States Air Force.

References

[1] A. Esmaeeli and G. Tryggvason. Computations of film boiling. part i: numerical method. *International journal of Heat and Mass Transfer*, 47:5451–5461, 2004.

[2] D.C. Groeneveld, J. Q. Shan, A.Z. Vasic, L. K. H. Leung, A. Durmayaz, J. Yang, S. C. Cheng, and A. Tanase. The 2006 chf look-up table. *Nuclear Engineering and Design*, 237:1909–1922, 2007.

[3] W. M. Kays and M. E. Crawford. *Convective Heat and Mass Transfer*. McGraw-Hill, United States of America, 1987.

[4] Božidar Liščič, Hans M. Tensi, Lauralice C. F. Canale, and George E. Totten, editors. *Quenching Theory and Technology*. Elsevier Science Publisher B.V., 2nd edition, 2010.

[5] Reinhold Maurus and Thomas Sattelmayer. Bubble and boundary layer behavior in subcooled flow boiling. *International Journal of Thermal Sciences*, 45:257–268, 2006.

2nd World Congress on Integrated Computational Materials Engineering
Edited by: Mei Li, Carelyn Campbell, Katsuyo Thornton, Elizabeth Holm, and Peter Gumbsch
TMS (The Minerals, Metals & Materials Society), 2013

Modeling Crack Propagation in Polycrystalline Alloys using a Variational Multiscale Cohesive Method

Veera Sundararagahavan[1], Shang Sun[2]

[1]Department of Aerospace Engineering, University of Michigan, 3025 FXB building, 1320 Beal Ave, Ann Arbor, MI 48109, USA
[2]Department of Naval Architecture Engineering, University of Michigan

Keywords: Crack initiation, crack propagation, grain boundaries, transgranular transition, intergranular crack.

Abstract

Crack propagation in polycrystalline grains is analyzed using a novel multiscale polycrystalline model. The approach combines reduced order descriptors of microstructures with explicit representation of polycrystals at critical areas (eg. crack tips). For the critical areas, refined meshes are employed to discretize each crystal. The crack propagation in the microstructure is calculated using the variational multiscale method which allows for arbitrary transgranular and intergranular crack paths. The computational load is reduced substantially by combining probabilistic representation of the macroscale problem with exact resolution of the crystals at the crack tips. One example of grain boundary failure is demonstrated in this paper, showing exceptional mesh convergence and efficiency of the numerical approach.

1. Introduction

In recent years, cohesive interface models have emerged as attractive methods to numerically simulate fracture initiation and growth by the finite element method (Ortiz and Pandolfi, 1999 [1]; Zavattieri and Espinosa, 2001 [2]). Typically, a cohesive interface is introduced in a finite element discretization of the problem by the use of special interface elements, which obey a nonlinear interface traction separation constitutive relation. However, the crack paths are not known in advance and thus, these interface elements are practically needed at every element–to–element interface. This makes the problem both computationally expensive and also results in unsatisfactory mesh convergence properties. In this work, we pursue the use of variational multiscale cohesive model (VMCM) to embed the cohesive elements selectively in critical regions. VMCM, which is an enhanced finite element method containing elements that handle discontinuities, originates from the variational multiscale formulation presented by Garikipati 1998, 2002 [7, 8]) and extended towards modeling failure by Rudraraju et al [5, 6]. In this approach, the displacement discontinuities are represented over unstructured meshes using specially constructed element shape functions that are discontinuous. The approach can be used to embed the cohesive zone models within an element allowing us to simulate arbitrary crack paths that include both intergranular and transgranular failure in polycrystals. have recently demonstrated the approach for predicting failure in laminated fiber reinforced composites and the predictions showed good correlations with experimentally measured crack paths. In this work, VMCM is extended to modeling of failure in polycrystalline microstructures. In Section 2, the VMCM scheme is briefly introduced. The multiscale polycrystalline model used in this investigation is described and crack propagation simulation examples are shown in Section 3. Conclusions are given in Section 4.

2. Introduction of VMCM

A mathematical model of the variational multiscale cohesive method (VMCM) is briefly described here. The crack opening configuration is shown at Fig. 1. The displacement discontinuity at the crack is represented by $[|\boldsymbol{u}|]$. The domain is represented as Ω with boundary Γ and the crack surface is represented by Γ^c.

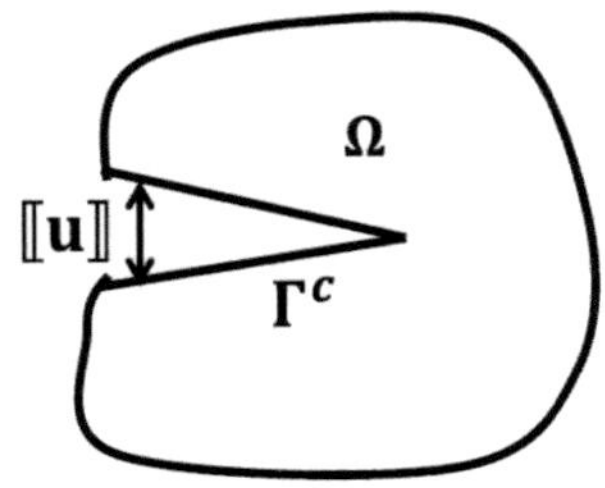

Figure 1: *Representation of crack opening* $[|\boldsymbol{u}|]$*, domain* Ω *and crack surface* Γ^c.

The traction acting on the surface of an emerging crack is represented as a vector:

$$\boldsymbol{T}^c = \boldsymbol{\sigma n} = T_n \boldsymbol{n} + T_m \boldsymbol{m} \tag{1}$$

where (T_n, T_m) are the normal and tangential tractions, $\boldsymbol{\sigma}$ is the stress tensor and $(\boldsymbol{n}, \boldsymbol{s})$ are unit vectors normal and tangential to the crack surface, respectively. The traction components are assumed to be linearly related to the displacement jump $[|\boldsymbol{u}|]$ using softening moduli $\mathcal{H}$:

$$T_n = T^c_{n0} - \mathcal{H}_n[|\boldsymbol{u}|] \cdot \boldsymbol{n}, T_m = T^c_{m0} - \mathcal{H}_m[|\boldsymbol{u}|] \cdot \boldsymbol{m} \tag{2}$$

where T^c_{n0}, T^c_{m0} are the critical values of normal and tangential tractions that leads to the formation of a new crack and the crack is fully formed when the surface eventually becomes traction free $T_n = T_m = 0$.

VMCM approach involves the augmentation of an additional degree of freedom $[|\boldsymbol{u}|]$ in elements that exceed the critical values of normal or tangential traction during loading. This additional degree of freedom is represented within the cracked element using a special discontinuous shape function that ensures that the displacement jump is localized to that particular element. The analysis procedure involves the solution of two coupled equations for the overall displacements (d) and the displacement jump $[|\boldsymbol{u}|]$:

$$\int_\Omega \boldsymbol{B}^T \mathbb{C} : (\boldsymbol{Bd} - \boldsymbol{G}[|\boldsymbol{u}|]) dV = \int_\Gamma \boldsymbol{N}^T \cdot \boldsymbol{T} dS \tag{3}$$

$$\boldsymbol{H}^T \mathbb{C} : (\boldsymbol{Bd} - \boldsymbol{G}[|\boldsymbol{u}|]) = \boldsymbol{T}^c \tag{4}$$

The first equation is the usual finite element weak form with augmented degrees of freedom in the cracked elements and the second equation is the traction–displacement relation (given by Eq. (1)). Here, $\boldsymbol{N}$ and $\boldsymbol{B}$ are the usual interpolation functions for a three noded triangle. The special interpolation functions $\boldsymbol{H}$ and $\boldsymbol{G}$ describe the jump discontinuity and are described in more detail in Rudraraju et al [5, 6]). The strain in the cracked element is described by $\epsilon = (\boldsymbol{Bd} - \boldsymbol{G}[|\boldsymbol{u}|])$. The stress is written as $\boldsymbol{\sigma} = \mathbb{C} : \epsilon$ where $\mathbb{C}$ is the stiffness tensor.

In the overall model, the microstructure is explicitly modeled only at critical regions such as notch tips or other stress concentrators (eg. nanoindenter tips). In other regions, the microstructure is represented using an orientation distribution function (ODF) $\mathcal{A}(\boldsymbol{g})$ that gives the probability density of crystals with orientation $\boldsymbol{g}$. Details of the reduced order representation based on ODFs can be found in our recent publication, Sun and Sundararaghavan (2012) [4]. All the simulations were carried out using an in-house finite element code. A standard NewtonRaphson scheme was used for solving the system of non-linear equations given above.

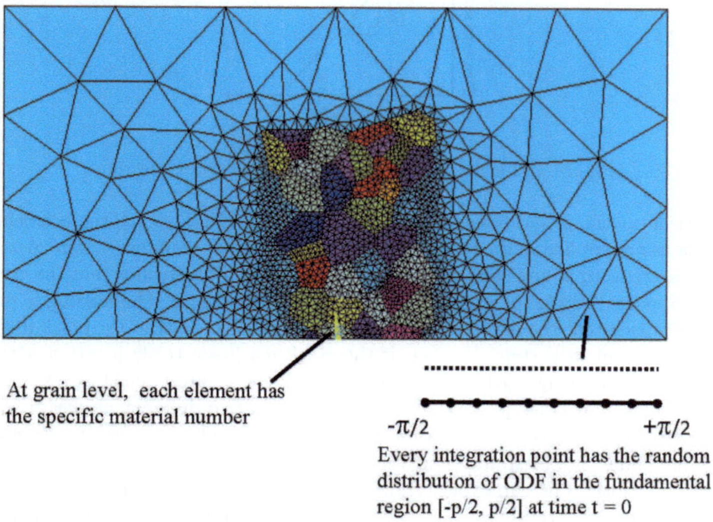

Figure 2: *The configuration of multi-scale model. The model is divided into two scales. Microscale is critical areas contains explicit meshes of crystals. Macroscale elements have an underlying ODF that statistically represents a large aggregate of crystals.*

3. Multiscale crack propagation model

We employed the novel multiscale approach for modeling crack propagation in polycrystalline microstructures. The model configuration is shown in Fig. 2. Here, the entire computational region is divided into two levels: macro-scale and micro-scale. At the micro-scale level, aggregates of grains are explicitly modeled to track propagation of cracks. During deformation, the microstructure evolution at the macroscale is statistically captured by evolving the ODF. There is a gradual transition of mesh density from microscale to the macroscale wherein the far field elements are much bigger than the microscale elements. This section presents numerical simulations to demonstrate crack path formation under tension under elastic regime. A small notch is used to initiate a crack in the polycrystal region. The tractionseparation laws are implemented to allow mixed mode failure (normal and shear separation, ie. Mode–I and Mode–II). The crack is always assumed to grow along a direction perpendicular to direction of maximum tensile stress. A FCC material is simulated with $c_{11} = 2\ GPa$ and $c_{12} = c_{44} = 1\ GPa$ with loading under plane stress conditions. The linear cohesive law employed in the crack simulation is shown in Fig. 3.

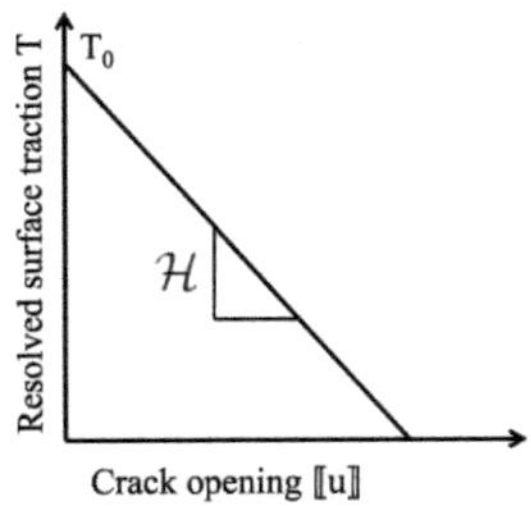

Figure 3: *Linear traction law for normal crack opening.*

The tensile specimen and boundary conditions are shown in Fig. 4. Displacement boundary conditions are applied in the model. In this case, the grain boundaries were given a much lower critical stress to failure compared to the grains itself. This ensures that the cracks preferentially propagate along the grain boundaries as seen in Fig. 4. The loading is performed incrementally such that the next incremental load is applied only after the crack propagation stops in the current load step. Further, if the crack reaches a triple junction, the crack is assumed to advance along a grain boundary that has the greatest tensile stress. The simulation of intergranular crack propagation was done using two different meshes to show that the crack path is converged. The converged load displacement curves can be seen in Fig. 5.

By giving higher T_{n0} to grain boundaries such that the failure stress for grains and grain boundaries are closer, one can expect transgranular failure modes. Simulation of transition from intergranular and transgranular crack is illustrated in Fig. 6. As soon as the load is applied the crack begins to propagate along the grain boundaries. However, as a triple junction is reached, the crack stops and an intergranular crack is formed. Within the grain, normal stresses along six directions were sampled to identify the direction of the advancing crack. The angle with maximum normal traction was chosen as the new propagation direction. As soon as the intergranular crack reaches another grain boundary with an easier propensity to form a crack, the crack transitions back to intergranular mode. The red lines in Fig. 6(left) denotes the transgranular crack path, while the orange ones shows intergranular crack path.

4. Conclusion

Modeling failure at microstructural scales is valuable for predictive modeling of component life, and ensuring structural integrity in aerospace structures. Since failure is usually localized at regions of stress concentrators such as notches, cracks and contact surfaces, a multiscale model is developed in this work which is based on a multiresolution mesh. The model includes an explicit microstructure representation at critical regions, while at regions farther away from the stress concentration, a reduced order model that statistically captures the effect of microstructure is employed. A novel method for modeling crack propagation based on a VMCM framework was developed. In this approach, discontinuous crack elements

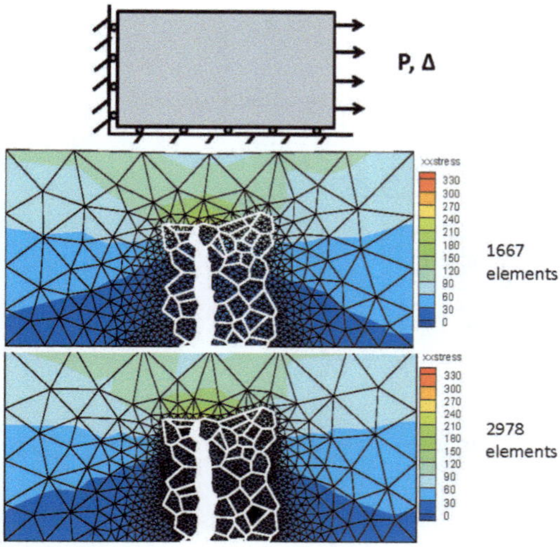

Figure 4: *Tensile convergence test for polycrystalline crack propagation. boundary condition of loaded specimen (top) and stress (σ_{xx}) contours for two different mesh densities (bottom).*

are progressively embedded during loading in regions with critical stresses. To demonstrate applicability of the finite element based methodology, seamless modeling of both intergranular and transgranular crack propagation modes in a complex grain structure are illustrated. The approach is currently being extended towards modeling fracture paths during fatigue loading (experiments in Dunne et al (2007) [3])

References

[1] Ortiz, M. and Pandolfi, A., Int. J. Numer. Methods Eng., 44(1999), p. 1267–1282.

[2] Zavattieri, P.D. and Espinosa, H.D., Acta Mater., 49(2001), p. 4291-4311.

[3] Dunne, F. P. E., A. J. Wilkinson, and R. Allen., Int. J. Plast. 23(2007), p. 273-295.

[4] Sun, S. and Sundararaghavan, V., Acta Mater., 60(2012), p. 5233–5244.

[5] Rudraraju, S., Salvi, A., Garikipati, K., and Waas, A. M., Int. J. Solids Struc., 47(2010), p. 901-911.

[6] Rudraraju, S., Salvi, A., Garikipati, K., and Waas, A. M., Compos Struct, 94(2012), p. 3336-3346

[7] Garikipati, K., and Hughes, T. J., Comput. Methods. App. Mech. Eng., 159 (1998), p. 193-222.

[8] Garikipati, K., Comput. Model. Eng. Sci. 3(2002), p.175-184.

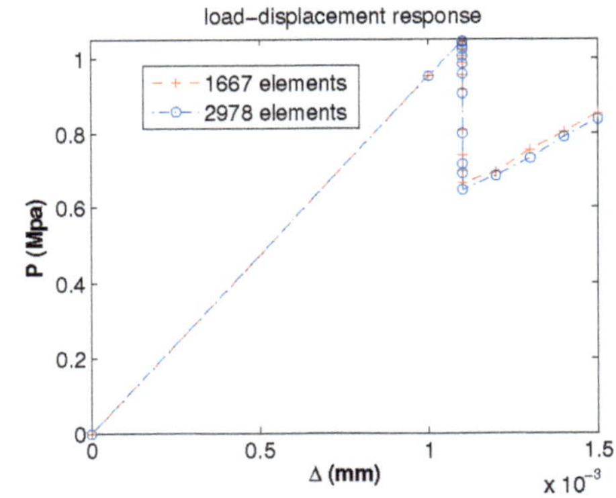

Figure 5: *Convergence of load displacement response for the case shown in Fig. 4*

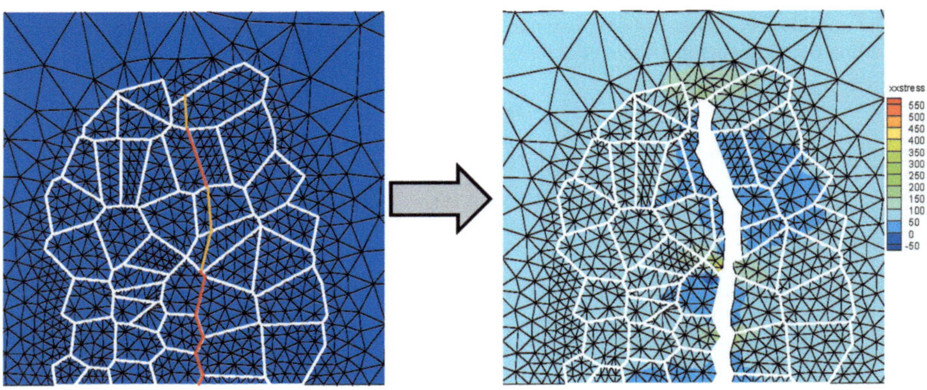

Figure 6: *Example showing transition between transgranular and intergranular crack propagation. Red and orange lines at the left figure shows the transgranular and intergranular cracks respectively.*

2nd World Congress on Integrated Computational Materials Engineering
Edited by: Mei Li, Carelyn Campbell, Katsuyo Thornton, Elizabeth Holm, and Peter Gumbsch
TMS (The Minerals, Metals & Materials Society), 2013

A COUPLED APPROACH TO WELD POOL, PHASE AND RESIDUAL STRESS MODELLING OF LASER DIRECT METAL DEPOSITION (LDMD) PROCESSES

Michael Vogel[1], Mushtaq Khan[1], Juansethi Ibara-Medina[1]
Andrew Pinkerton[2]
Narcisse N'Dri[3], Mustafa Megahed[3]

[1]Manufacturing and Management Group, School of Mechanical, Aerospace and Civil Engineering, The University of Manchester, Sackville Street, Manchester M13 9PL, United Kingdom
[2]Lancaster University, Lancaster LA1 4YR, United Kingdom
[3]ESI GmbH, Kruppstr. 82-100/ETEC V5-105, 45145 Essen, Germany

Keywords: Laser Cladding, melt pool modeling, CFD/FEM coupling

Abstract

This paper describes a complete CFD model of the laser metal deposition process. The model covers the complete process, starting from the simulation of powder particles in the deposition head and finishing with the final part. Individual phenomena that are considered in the gas-phase stage of the model include the ricocheting of particles within the head, the flow of powder particles, their interaction with the laser and powder catchment/bouncing. Phenomena considered in the liquid phase (melt pool) stage of the model include particle enthalpy effects, buoyancy, temperature-dependent material properties and Marangoni forces. The CFD model is coupled with a metallurgical database to predict the phase and material properties of the solidified deposit and heat affected area and residual stresses in the part. Modelled and experimental characteristics of multi-track deposits of M2 steel show good agreement.

Introduction

In the past years Laser Cladding has been used processing various materials including nickel alloys, copper alloys, Titanium and stainless steel [1-5]. The applied processing parameters such as laser power, focus diameter, scanning speed, powder mass flow rate, shielding and shroud gas flow rate and preheating are known to interfere with the final clad characteristics. The parameter can change the clad geometry, the wetting angle to the substrate, porosity, surface roughness and residual stresses and metallurgical properties as well [6-9]. However, the involved physical process shows many complex and nonlinear effects like thermo-capillary flow (Marangoni), metallurgical and physical phase changes, surface tension, absorption and radiation that increases the effort to accurately predict the clad characteristics.

A widely used material in automotive and tool application is high-speed steel (HSS). The steel AISI M2 high-speed steel is known for it's mechanical properties depending strongly on the metallurgy and microstructure of the material. This microstructure can be controlled by the cooling rate that is applied to the material during processing steps towards the final part [10-12].

Depending on the processing conditions a complex microstructure containing different types of carbides (e.g. MC, M_2C, M_6C, $M_{23}C_6$, M_7C_3 and $M_{23}C_6$) has been reported. Related to very high cooling rates in laser Cladding process the resulting microstructure is prone to metastable phases such as Austenite and Martensite [13]. Further phase transformations may occur during reheating of clad structure initiated by the successive application of new clads.

When modeling Laser Cladding process was first considered, a basic understanding of the process has been derived from experimentally achieved process maps by Weerasinge and Steen [14]. In the beginning only sub-processes were considered for modeling, e.g., multiple analytical models of the powder stream which then developed into models of powder flow and heating of single particles [15, 16, 17]. Recently, developed models using computational fluid dynamics (CFD) methods show potential to implement effects such as ricocheting of particles within the nozzle and at the substrate and heating of particles passing the laser beam [18, 19].

This work shows a model of Laser Cladding implementing an integrated approach based on process data available from user parameters and equipment settings. Some previously established data such as surface absorption coefficients of laser beam has been taken from literature sources.

Procedure

Material

In this study AISI M2 High Speed Steel (HSS) powder supplied by LPW Technologies was used as material for deposition, with particle sizes in the range 44-88μm. The chemical composition of the powder is summarized in table I. Annealed M2 HSS blocks (50 mm x 50 mm x10 mm) have been used as the substrate material. First the upper surface of the substrate has been grit blasted and degreased to increase absorptivity and to remove contaminants before cladding was applied.

Table I. Chemical composition of M2 HSS powder (%wt).

Fe	C	V	Mo	Cr	W	Mn	Si	S	O	N
Bal	0.9	2.0	5.0	4.4	6.1	0.4	0.3	0.01	0.02	0.13

Numerical Modeling

The numerical model presented in this work is based on computational fluid dynamics methods (CFD) for the powder flow rate model and the melt pool dynamics model and on finite element method (FEM) for the solidification and metallurgical models. The different methods are coupled into an integrated model to predict the melt pool temperature and dimensions, cooling rates and resulting metallurgical phases after solidification of the liquid metal from initial process parameters. A schematic diagram of the modeling domains is shown in Figure 1.

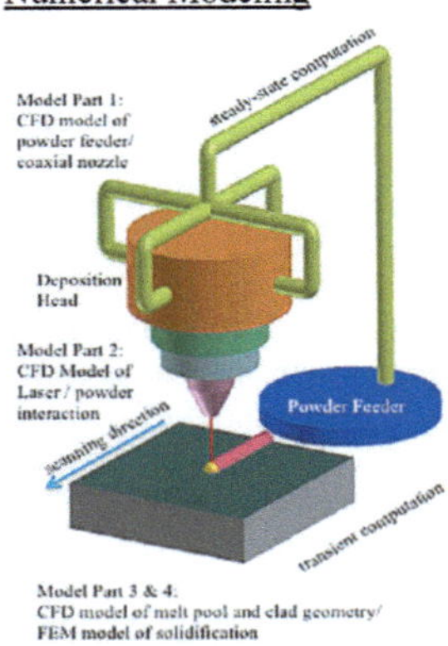

Figure 1: Schematic diagram of a laser cladding setup and model-part domains

The modeling details for powder tracking, the interaction of the powder particle concentration with laser irradiance and the weld pool physics was summarized in [18, 23]. The coupling of CFD and FEM models was described in [22, 24]. Due to space limitations, this paper focusses on the prediction of the metallurgical properties and the resultant residual stresses. The metallurgical model inside the FEM system is based on the model of Leblond [20]:

$$\frac{dP(T)}{dt} = f(\dot{T})\frac{P_{eq}-P(T)}{\tau(T)} \tag{1}$$

with P phase proportion, T temperature, t time, $\dot{T}$ heating or cooling rate, P_{eq} phase proportion at equilibrium and τ a delay time as a function of temperature. Martensitic transformation was computed using the equation of Koistinen-Marburger [20]:

$$P(T) = 1 - e^{-b(M_S-T)} \tag{2}$$

with P phase proportion of Martensite, b Koistinen- Marburger coefficient, M_S Martensite start temperature, T temperature.

Experimental Verification

A Laserline LDL160-1500 diode laser (Laserline GmbH, Germany), with maximum power of 730W and wavelength range of 808-940nm is used for the cladding process. The laser beam has a "top hat" intensity profile. A powder feeder unit, SIMATIC OP3, with rotating disc feeder is used to control the flow of powder to the coaxial deposition head, consisting of collimating and focusing optics and a coaxial nozzle. As shielding and carrier gas Argon gas is used.
The process parameters are listed in table II.

Table II.1: Process parameters

Carrier gas flow rate	6L/min
Shielding gas flow rate	4L/min
Nozzle offset distance	8 mm

Table II.2: Process parameters

Sample No	Specific Energy (J/mm^2)	Mass flow rate (g/sec)
Sample 1	45.625	0.27
Sample 2	45.625	0.16
Sample 3	91.25	0.27
Sample 4	91.25	0.16

Results and Discussion

The powder tracking, the amounts of lost powder during the cladding process, the powder heating were validated in our previous work [18]. The predicted weld pool shape for multiple tracks was also investigated intensively, validation was presented in [22].

Figure 2 shows a typical comparison of a specimen with the cladding cross sections obtained using CFD. The clad shape is predicted accurately providing the basis for further analysis of the results using FEM to obtain the grain structure as well as the residual stresses.

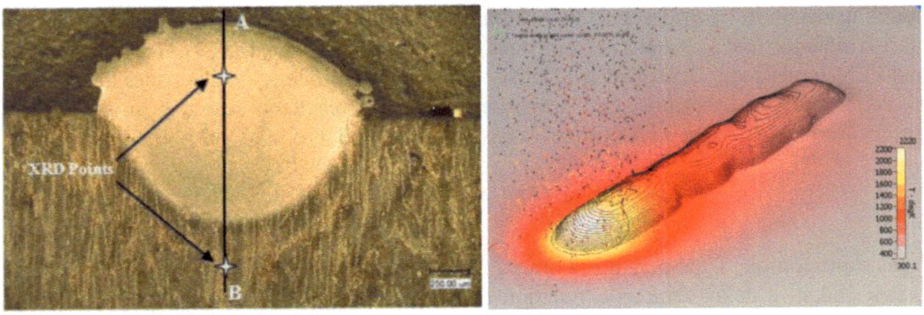

Figure 2: Cross-section image of clad (left: experiment, right: CFD).

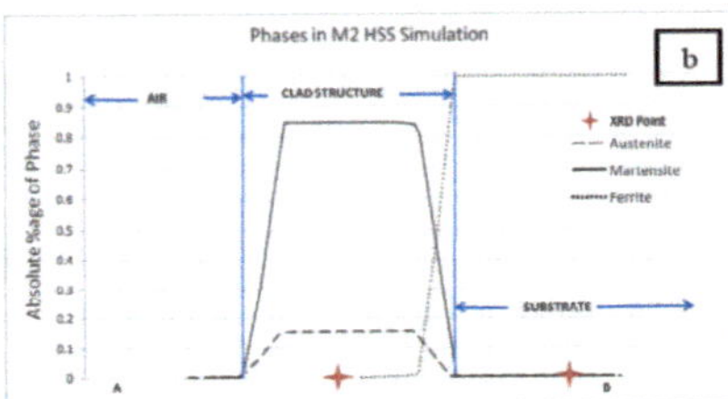

Figure 3: Simulation results of metallurgical

Figure 2 also indicates the position of 2 points: A in the clad and B in the substrate, where metallurgical simulations were performed to predict the phases at these points as shown in figure 3. The metallurgical phase evolution along the middle of the clad, starting from the substrate towards the clad surface, shows clearly the base material proportion of Ferrite until the edge of the heat affected zone (HAZ) where the metallurgical phases change to the martensitic proportion of the clad. This is perfectly in agreement with the CCT diagram of M2 steel. Due to the high cooling rate inside the molten clad the martensitic phase reaches almost full dominance inside the clad. Only little austenite remains next to the Martensite.

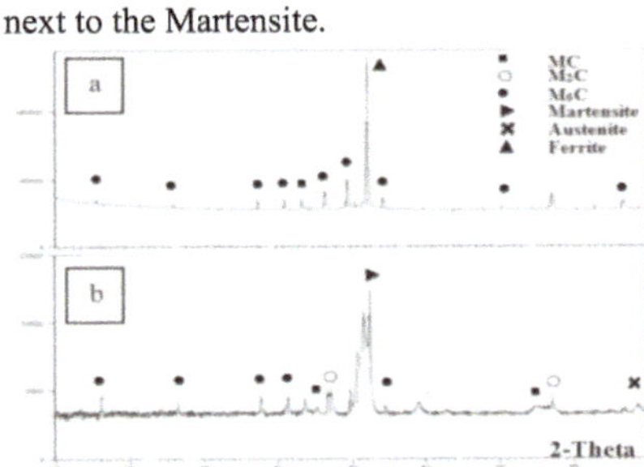

Figure 4: XRD scan results (a: substrate, b: clad)

Figure 4 shows the corresponding comparison using XRD-analysis at points A and B of the clad and the substrate. Both results show good agreement with the simulation results at the same positions in the model. X-ray Diffraction (XRD) spectra provide phase information: Peak height is related to the amount of phase found in the evaluated location. In the substrate results one can see the main peak of Ferrite along with other small peaks of different carbides. In the clad the main peak is much wider and also some carbide positions have changed. This means that carbon from previously located at grain boarders were dissolved during the melting phase of the clad and contributing to building Martensite.

Figure 5 shows the numerically predicted shape of a cross shaped cladding track and the corresponding image of the experimental profile. The clad shape and the thermal history are transferred to an FEM code to determine the von Mises stresses – also shown in Fig. 5. The stress distribution is very similar for both tracks, each showing a distinct valley in the center, where the tracks cross. The reheating and changed cooling rate at that position leading to a relaxation of the local stressed.

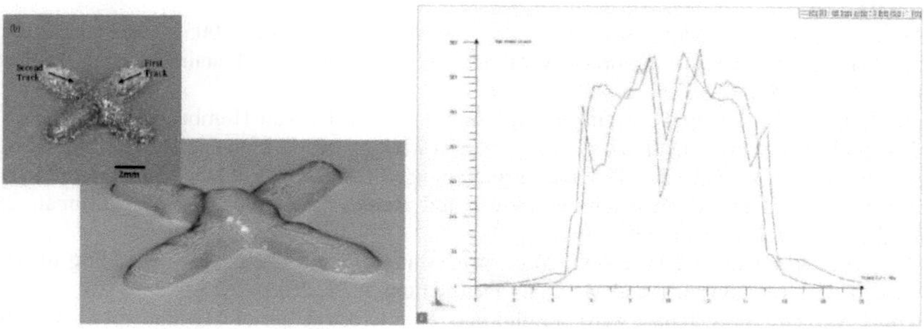

Figure 5: Image of cross clad tracks as compared with numerical prediction and the corresponding von Misses stresses along the center-line of each track

Conclusion

The presented model of an integrated CFD – FEM method to simulate the Laser Cladding process shows good agreement with experimental results. This work summarizes the validation of the metallurgical component building on the results obtained from the CFD model. The coupled models aim for an accurate prediction of the melt pool, clad geometry and metallurgical properties of the clad and the substrate, as well as for the stress and strain field in the part. This is achieved starting from only commonly available initial conditions such as Laser power, velocity, powder feed rate and shielding gas flow rate. The model predicts the clad geometry and the metallurgical phase proportions inside the clad. Based on these results the stress and strain evolution is computed in a mechanical computation step.

Acknowledgements

The authors gratefully acknowledge the support of the European Commission via EU FP7 IAPP Agreement 230756.

References

1. Pinkerton, A.J. and Li, L. (2004), Multiple-layer cladding of stainless steel using a high-powered diode laser: an experimental investigation of the process characteristics and material properties, Thin Solid Films, 453-454, 471-476.

2. Dutta Majumdar, J., Pinkerton, A.J., Liu, Z., Manna, I. & Li, L. (2005) Mechanical and electrochemical properties of multiple-layer diode laser cladding of 316L stainless steel, Applied Surface Science 247, 373–377.

3. Shah, K. (2011), Laser Direct Metal Deposition of Dissimilar and Functionally Graded Alloys, PhD thesis, university of Manchester, UK

4. Syed W.U.H., Pinkerton, A., Liu, Z. & Li, L. (2007) Single-step laser deposition of functionally graded coating by dual wire-powder or powder-powder feeding--A comparative study, Applied Surface Science 253, 19, 7926-7931

5. Pleterski, M., sek, J. T., Muhic, T. & Kosec, L. (2011), Laser Cladding of Cold-Work Tool Steel by Pulse Shaping, Journal of Materials Science Technology, 27, 8, 707-713.

6. Alimardani, M., Fallah, V., Hajepour, A. & Toyserkani, E. (2010) The effect of localized dynamic surface preheating in laser cladding of Stellite 1, 204 , 3911–3919.

7. Zhou, S., Dai, X. & Zeng, X. (2009) Effects of processing parameters on structure of Ni-based WC composite coatings during laser induction hybrid rapid cladding, Applied Surface Science 255, 8494-8500.
8. Majumdar, D., Pinkerton, A.J., Liu, Z., Manna, I. & LI, L. (2005) Mechanical and electrochemical properties of multiple-layer diode laser cladding of 316L stainless steel, Applied Surface Science 247, 373-377.
9. Liu, Z., Chua, C., Leong, K., Kempen, K., Thijs, L., Yasa, E., Van-Humbeeck, J. & Kruth, J. 2011. A preliminary investigation on Selective Laser Melting of M2 high speed steel, Innovative Developments in Virtual and Physical Prototyping, Proceedings of the 5th International Conference on Advanced Research in Virtual and Rapid Prototyping, Leiria, Portugal, 28 September - 1 October, 2011, 339–346.
10. Niu, H.J. & Chang, I.T.H. (2000) Microstructural evolution during laser cladding of M2 high speed steel, Metallurgical and Materials Transactions A 31, 2615- 2625.
11. Zhou, X., Feng, F. & Jian, J. (2011) Solidification microstructure of M2 high speed steel by different casting technologies, China Foundry 8, 290-294.
12. Zhou, X., Fang, F. & Jiang, J. (2011), A study on the microstructure of AISI M2 high speed steel manufactured by continuous casting, Advance Materials Research 146-147, 1211-1215.
13. Colaço, R. & Vilar, R. (1998) Effect of the processing parameters on the proportion of retained austenite in laser surface melted tool steels, Journal of Materials Science Letters 17, 563-567.
14. Weerasinghe V. M. & Steen W. M. (1987) Laser Cladding with Blown Powder, Metal Construction (October), 581-585.
15. Lin J. (1999) Concentration mode of the powder stream in coaxial laser cladding, Optics and Laser Technology 31 (3), 251-257.
16. Liu C.Y. & Lin J. (2003) Thermal processes of a powder particle in coaxial laser cladding, Optics & Laser Technology 35(2), 81-86.
17. Pinkerton A. J. & Li L. (2005) Multiple-layer laser deposition of steel components using gas- and wateratomised powders: The differences and the mechanisms leading to them, Applied Surface Science 247 (1-4), 175-181.
18. Ibarra-Medina, J. & Pinkerton, A. J. (2011) Numerical investigation of powder heating in coaxial laser metal deposition, Surface Engineering 27(10), 754-761.
19. Kovalev, O.B., Zaitsev, A.V., Novichenk, D. & Smurov, I. (2011) Theoretical and experimental investigation of gas flows, powder transport and heating in coaxial laser direct metal deposition (DMD) process, Journal of Thermal Spray Technology 20(3), 465-478.
20. Leblond, J.B., & Devaux, J. (1984) A new kinetic model for anisothermal metallurgical transformations in steel including effect of austenite grain size, Acta Metallurgica 32, 137 – 146
21. Benyounis, K.Y., Fakron, O.M. & Abboud, J.H. (2009) Rapid solidification of M2 high-speed steel by laser melting, Materials and Design 30, 674-678.
22. Vogel, M; Khan, M.; Ibarra-Medina, J.; Pinkerton, A. J.; N'Dri, N.; Megahed, M. (2012) A verified model of transient and residual stresses in laser direct metal deposition.
In: Proceedings of the 31st International Congress on Applications of Lasers and Electro-optics (ICALEO). Laser Institute of America, 2012. p. Paper 304.
23. Ibarra-Medina, J., Vogel, M. & Pinkerton, A. J. (2011) A CFD model of laser cladding: From deposition head to melt pool dynamics, In: Proceedings of the 30th International Congress on Applications of Lasers and Electro-optics (ICALEO). Laser Institute of America, p. 378-386.
24. Vogel, M. , Ibarra-Medina, J. , N'Dri, N. & Pinkerton, A. J. (2011) An integrated model of laser clad formation using CFD and FE methods In: Surface modification technologies XXV : Proceedings of the 25th International Conference on Surface Modification Technologies. Sudarshan, T. S. & Nylen, P. (eds.). India: Valardocs

2nd World Congress on Integrated Computational Materials Engineering
Edited by: Mei Li, Carelyn Campbell, Katsuyo Thornton, Elizabeth Holm, and Peter Gumbsch
TMS (The Minerals, Metals & Materials Society), 2013

PREDICTION OF THE UNCERTAINTY IN THE RESPONSE OF LIGHTWEIGHT STRUCTURES CONSISTING OF SOLID FOAMS

Jörg Hohe[1], Carla Beckmann[1]

[1] Fraunhofer-Institut für Werkstoffmechanik IWM
Wöhlerstr. 11, 79108 Freiburg, Germany

Keywords: Solid Foams, Disordered Microstructures, Uncertainties, Probabilistic Models, Sandwich Construction

Abstract

The present study is concerned with an integrated computational scheme for prediction of the uncertainties in the structural response of sandwich structures with metal foam core, resulting from the disorder in the foam microstructure. A stochastic homogenization scheme is utilized for determination of the effective elastic properties of the foam core. The individual cells of the microstructure are considered as testing volume elements to which the homogenization equations are applied. The advantage of this scheme is that both, the scatter in the effective material properties as well as the spatial correlation of the varying effective properties are addressed. Based on the results, a random field formulation for the effective material of the core layer is employed to assess the uncertainty in the macroscopic structural response of a sandwich beam. It is observed that significant effects of the disordered microstructure are to be expected especially in the strength and structural integrity assessment.

Introduction

Sandwich structures are important components in many types of lightweight structures. The typical sandwich structure consists of three principal layers where two thin, high-density face sheets are adhesively bonded to a thick core made from a low density material. Within the principle of sandwich construction, the face sheets carry all in-plane and bending loads whereas the core keeps the face sheets at the desired distance and transmits all transverse normal and shear loads (e.g. Vinson [5]). Typical face sheet materials are composite laminae or lightweight metals such as aluminum sheets. The core consists of a low density material such as a polymeric or metallic foam, honeycomb or balsa wood. In this context, solid foams are advantageous materials since they may easily be processed to any desired shape. Further benefits derive from their inherent good thermal and acoustic damping characteristics. Their main disadvantage is their random microstructure leading to a distinct scatter in their effective properties and thus uncertainties in the response of the entire structure (e.g. Ramamurty and Paul [4]).
The present study is concerned with a numerical scheme for the prediction of the uncertainties in the mechanical response of sandwich or other lightweight structures made at least partially of solid foams using an integrated two-step procedure. In a first step, the effective material properties of the solid foams and their uncertainty are determined. For this purpose, a probabilistic homogenization procedure is utilized (Beckmann and Hohe [1]). This scheme is based on the analysis of the mechanical response of subsets of a large scale, statistically representative volume element for the foam microstructure. As a result, the probability distributions for the effective properties and the autocorrelation properties for a random field

description of the effective material response of the solid foam are obtained. In a second step, the random field model is employed in a numerical analysis of the entire structure on the macroscopic level. In a case study, a single edge clamped sandwich beam with a foam core is considered. Whereas the material uncertainty is found to cause only minor scatter in the deformation of the beam, significant uncertainties are observed in the strength of the structure.

Probabilistic Homogenization

General Procedure

For the determination of the effective material properties of the solid foam used as a core material, a numerical homogenization scheme based on the concept of the representative volume element is employed. Hence, a representative extract Ω^{RVE} of the given microstructure is considered together with a similar volume element Ω^{RVE^*}, consisting of the effective material with yet unknown properties. The strain and stress states of both volume elements are defined to be equivalent on the intermediate (mesoscopic) level of structural hierarchy, if their volume averages

$$\bar{\varepsilon}_{ij} = \frac{1}{V^{\mathrm{RVE}}} \int_{\Omega^{\mathrm{RVE}}} \varepsilon_{ij} \, \mathrm{d}V = \frac{1}{V^{\mathrm{RVE}}} \int_{\Omega^{\mathrm{RVE}^*}} \varepsilon_{ij}^* \, \mathrm{d}V \tag{1}$$

$$\bar{\sigma}_{ij} = \frac{1}{V^{\mathrm{RVE}}} \int_{\Omega^{\mathrm{RVE}}} \sigma_{ij} \, \mathrm{d}V = \frac{1}{V^{\mathrm{RVE}}} \int_{\Omega^{\mathrm{RVE}^*}} \sigma_{ij}^* \, \mathrm{d}V \tag{2}$$

are equal. The parameters of the effective material replacing the real microstructure in the subsequent macroscopic analysis of the considered structural problem are determined such that Eqns. (1) and (2) are satisfied for a number of independent reference stress or strain states. In the context of the present study, the representative volume element Ω^{RVE} is deformed by a number of independent reference strain states. For each strain state, the stress response is computed using a finite element analysis. Subsequently, the effective material parameters are computed from the relations between the (applied) effective strains and the computed effective stress components.

Computational Foam Model

The determination of effective material properties based on the concept of the representative volume element requires the generation of an appropriate computational foam model. To ensure that the model is statistically representative, the foam is characterized by stochastic methods in terms of the probability distributions of the main stochastic variables characterizing the microstructure. These stochastic variables are the parameters defining the cell size distribution, the material distribution over the cell surfaces and – most important – the distribution of the local relative density. Subsequently, artificial foam models are computed, using the stochastic information about the microstructure in conjunction with random number generation.

The microstructure of the representative volume element Ω^{RVE} is generated by means of a Voronoï process in Laguerre geometry (e.g. Fan et al. [3]). For this purpose, a number of n nuclei is randomly positioned within a rectangular space Ω^{RVE}, each of which is supplied with a non-intersecting sphere with radius r_i. Subsequently, the volume of the i-th cell is defined as the set

$$\Omega^{(p)} = \left\{ x_i \mid x_i \in R^3, r^{\mathrm{L}}(x_i, x_i^{\mathrm{nuc}(p)}) \le r^{\mathrm{L}}(x_i, x_i^{\mathrm{nuc}(q)}), q \ne p \right\}, \; p, q = 1, \ldots, n \tag{3}$$

of all spatial points with a distance

$$r^{\mathrm{L}}(x_i, x_i^{\mathrm{nuc}(p)}) = \left((r^{\mathrm{E}}(x_i, x_i^{\mathrm{nuc}(p)}))^2 - r_i\right)^{1/2} \tag{4}$$

to the nucleus i in Laguerre geometry, which is smaller than the distance to all other nuclei. In Eq. (4), $r^{\mathrm{E}}(x_i, x_i^{\mathrm{nuc}(p)})$ is the Euclidean distance of the spatial points x_i and $x_i^{\mathrm{nuc}(p)}$. By an appropriate choice of the distribution of the sphere radii r_i, the cell size distribution can be controlled.

The numerical analysis of the microstructure is performed by means of the Finite Element Method using standard three-node displacement-based tetrahedral shell elements for the cell walls. Typical representative volume elements consist of about 256 cells in order to account properly for all possible interaction effects of neighboring cells in the microstructure.

Probabilistic Enhancement

The application of the concept of the representative volume element requires the characteristic size l of the volume elements Ω^{RVE} and $\Omega^{\mathrm{RVE*}}$ to be much larger than the characteristic length l^{micro} of the microstructure

$$l >> l^{\mathrm{micro}} \tag{5}$$

in order to account for all relevant microstructural details, i.e. to ensure that Ω^{RVE} is statistically representative. On the other hand, the representative volume element has to be much smaller than the characteristic length L of the entire structure under consideration

$$l << L \tag{6}$$

since it represents the material response of a single material point in the macroscopic structure, assumed to consist of the substitute effective medium. If both inequalities, (5) and (6), cannot be satisfied simultaneously, no well-defined representative volume element and thus no well-defined effective material properties exist. This problem might occur e.g. for the case of a sandwich structure with a foam core, where the smallest relevant characteristic length of the structure is the panel thickness, which is typically in the range of $L \approx 20$ mm, ..., 100 mm whereas the largest characteristic microstructural length is the cell diameter, which – for metallic foams – might be in the range of $l^{\mathrm{micro}} \approx 1$ mm, ..., 4 mm.

In order to deal with this problem, the authors have proposed a probabilistic homogenization procedure (Beckmann and Hohe [1]). The basic idea of this procedure is to employ a large-scale, statistically representative volume element Ω^{RVE} satisfying inequality (5), although this might involve a characteristic length l, which does not satisfy inequality (6). Hereby, it is ensured that the volume element accounts for all relevant microstructural mechanisms and effects. In a second

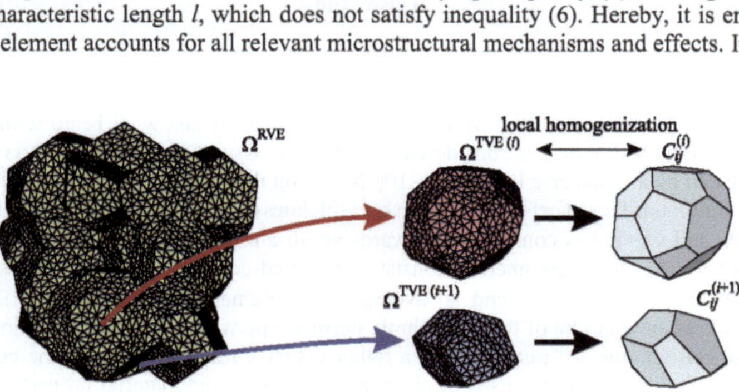

Figure 1: Testing volume elements.

step, the volume element Ω^{RVE} is subdivided into a number of n small-scale *testing volume elements* Ω^{TVE}, satisfying inequality (6). The individual testing volume elements $\Omega^{\mathrm{TVE}(k)}$ are not statistically representative, however, the entire set of n testing volume elements forming the entire representative volume element Ω^{RVE} is statistically representative. The smallest feasible testing volume elements are the individual cells (see Figure 1). Applying the homogenization equations (1) and (2) to the testing volume elements $\Omega^{\mathrm{TVE}(k)}$ instead of the entire representative volume element yields n sets of local effective material properties. By assessing the results statistically, the probability distributions of the effective properties are obtained.

Macroscopic Structural Analysis

For reasons of numerical efficiency, the structural analysis on the macroscopic level is performed in terms of the effective material rather than by a direct model of the microstructure. Within the concept of integrated computational materials engineering, the macroscopic analysis is based on the effective material properties determined in the previous mesoscopic analysis. Since the material response of the micro-heterogeneous material is uncertain, the material is described in terms of a random field using the stochastic properties determined in the homogenization analysis.

The crucial point is the computation of the random field such that the stochastic properties of the random field and the results of the stochastic homogenization analysis are identical. In the present study, a three-step procedure is employed. In a first step, material parameters are assigned to finite element integration points on a grid with a width slightly larger than the decay length of the spatial correlation of the effective material properties. Since the material properties on these grid points are entirely uncorrelated, material properties can be assigned to these points without consideration of the neighboring spatial positions. Nevertheless, the material parameters assigned to the grid points must possess the same probability distribution as computed in the stochastic homogenization analysis.

In the second step, the material properties for the integration points at the intermediate spatial positions are interpolated from the material properties assigned to the nearest neighboring grid points in the first step. In the third step, the material parameters at the intermediate spatial positions are varied randomly within a range given by the distance dependent semi-variance.

Case Study

Structural Example

As an example for the application of the proposed methods, a sandwich beam with a length of 500 mm according to Figure 2 is considered. The beam is assumed to be clamped on its left hand side and loaded by a transverse load of $F = 100$ N/mm on the free edge. A symmetric lay-up with two equal aluminum face sheets with a face sheet thickness of $t^{\mathrm{f}} = 1$ mm and the elastic constants $E^{\mathrm{f}} = 70$ GPa and $\nu^{\mathrm{f}} = 0.3$ is considered. Towards an all-aluminum design, the core layer with a thickness of $t^{\mathrm{c}} = 25$ mm is assumed to consist of a closed-cell aluminum foam with an average relative density of $\rho / \rho_0 = 0.08$ and an average cell volume of $V^{\mathrm{cell}} = 3.9\ \mathrm{mm}^3$. The cell size distribution is assumed to be of the logarithmic normal type with a standard deviation of 40% of the average cell volume V^{cell}, constituting a rather disperse microstructure of the core material. The cell walls are assumed to consist of aluminum with the same material properties as the face sheets. For simplicity, a constant, non-varying cell wall thickness is assumed.

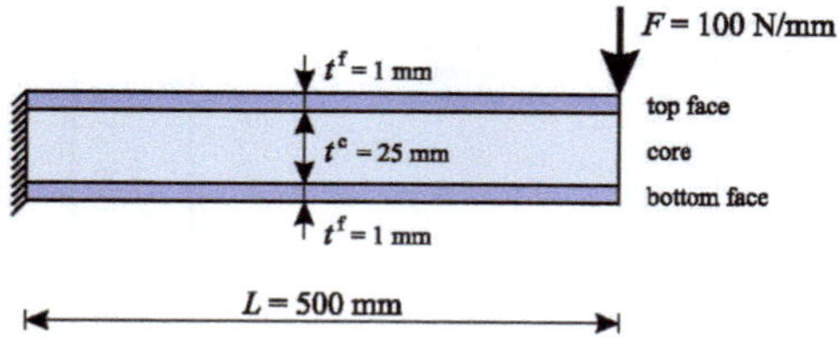

Figure 2: Single edge clamped sandwich beam.

Foam Material Properties

In a first analysis, the components C_{ij} of the effective elastic stiffness matrix for the core material are determined. The results for the cumulated probability distributions of the components C_{11}, C_{44} and C_{23} as examples for the normal, shear and normal coupling stiffness, respectively, are presented in Figure 3. For assessment of the generation of the random field representation, their properties are added as dashed lines. Similar results are obtained for all other normal, shear and normal coupling stiffness components. For all components, distinct uncertainties with scatter band widths in the order of the respective median stiffness are obtained. The random field representation is found to be in good agreement with the results of the probabilistic homogenization.

Macroscopic Structural Response

The random field representation of the effective material response for the foam core is employed in the structural analysis of the sandwich beam according to Figure 2. To assess the scatter in the structural response, a Monte-Carlo analysis based on 20 repeated generations of the random field is performed. For comparison, the classical deterministic response is computed, based on the average material properties determined in the stochastic homogenization analysis.
The results are presented in Figure 4. As a measure for the scatter in the stiffness of the structure, the vertical load point deflection u_3 is employed. It can be observed that only insignificant scatter of this integral property is obtained. The classical deterministic approach based on the average material properties provides a reasonable estimate. More distinct effects are observed for the maximum values of the equivalent stress σ_e and the hydrostatic stress σ_h of the core as the relevant parameters in a strength and structural integrity assessment. For the core, the random

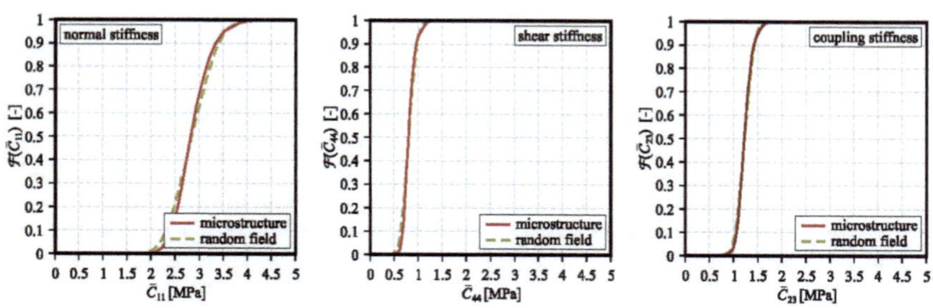

Figure 3: Probability distributions of normal, shear and coupling stiffness components.

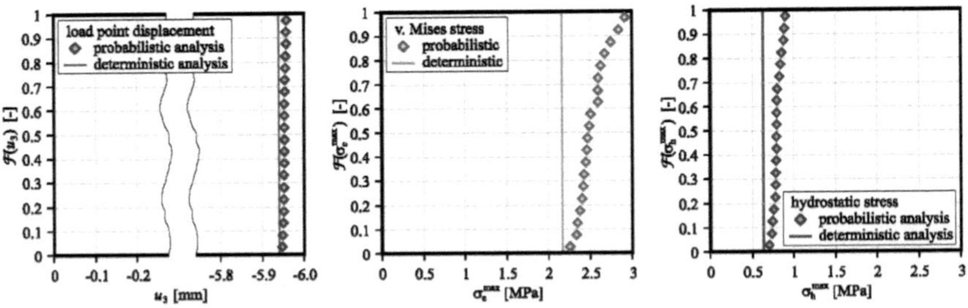

Figure 4: Probability distributions of structural response.

field formulation results in an increase of the median equivalent stress by about 13% with a scatter band width of 18% of the corresponding median value. Similar results are obtained for the hydrostatic stress, revealing that the microstructural disorder might have distinct effects on the integrity of structures consisting partially of foamed materials.

Conclusions

The results obtained in the present study reveal that care has to be taken in classical deterministic approaches for the integrity assessment of structures consisting partially or in total of micro-heterogeneous materials with disordered microstructure. Whereas the deterministic approach might still provide reasonable estimates for the deformation, inadequate results might be obtained in a strength assessment. The deterministic approach is unable to predict the uncertainty and might even underestimate the median value of the relevant strength parameters. The integrated probabilistic approach of the present study provides a tool for a direct prediction of the uncertainty effects on the structural level caused by the microstructural disorder of its constituents.

Acknowledgement

The present work has been funded by the Deutsche Forschungsgemeinschaft (DFG) under grant no. Ho1852/6-2. The financial support is gratefully acknowledged.

References

[1] Beckmann, C., Hohe, J.: *Assessment of material uncertainties in solid foams based on local homogenization procedures*, Int. J. Solids Struct. **49** (2012) 2807-2822.
[2] Beckmann, C., Hohe, J.: *Effects of material uncertainty in the structural response of metal foam core sandwich beams*, J. Sandwich Struct. Mat. (submitted for publication).
[3] Fan, Z., Wu, Y., Zhao, X., Lu, Y.: *Simulation of polycrystalline structure with Voronoi diagram in Laguerre geometry based on random closed packing of spheres*, Comput. Mat. Sci. **29** (2004) 301–308.
[4] Ramamurty, U., Paul, A.: *Variability in mechanical properties of a metal foam*, Acta Mat. **52** (2004), 869–876.
[5] Vinson, J.R.: *Sandwich structures,* Appl. Mech. Revs. **54** (2001) 201–214.

2nd World Congress on Integrated Computational Materials Engineering
Edited by: Mei Li, Carelyn Campbell, Katsuyo Thornton, Elizabeth Holm, and Peter Gumbsch
TMS (The Minerals, Metals & Materials Society), 2013

BUILDING 3D MICROSTRUCTURE DATABASE USING AN ADVANCED METALLOGRAPHIC SERIAL SECTIONING TECHNIQUE AND ROBUST 3D SEGMENTATION TOOLS

Umesh Adiga, Murali Gorantla, James Scott, Daniel Banks, Yoon-Suk Choi

Abstract

Building a 3D microstructure database is gaining rapid attention as it provides material specific key knowledgebase toward ICME approaches. Recent advances in automated metallographic serial sectioning technology such as RoboMet.3D, has significantly improved metallographic procedures and control systems for the acquisition of high quality metallographs in the form of serial sections. However, the image registration for 3D reconstruction and the segmentation of actual microstructural attributes from the raw data is still challenging, particularly for polycrystalline materials. The present study demonstrates systematic efforts to improve the automated metallographic serial sectioning technique for the 3D microstructure data. In particular, comprehensive image-processing tools were developed for the segmentation of complex 3D microstructures. Its capability of handling various complex 3D microstructure data (from polycrystals to composites) is also demonstrated. Geometric and gray-scale features are calculated using voxel values and their scale-space relational analysis. We propose to build a volumetric image database that allows qualitative and quantitative comparison of different materials and their structural properties measured using image processing algorithms and micrographs.

1. Introduction

Microstructure is the structure of a processed surface of material as revealed by the high resolution microscopy. It is generally established that the microstructure of a material such as metallic, polymeric, ceramic, composite, etc., influence physical properties such as strength, toughness, ductility, hardness, corrosion resistance, high/low temperature behavior, wear resistance, etc. These properties in turn govern the application of these materials in industrial practice. The microstructures appear as tightly bound grains where each individual grain has same molecular structure with respect to crystallographic orientation of the atoms. When the RoboMet.3D technology, shown in Figure 1, is used to collect microstructural data in the form of a sequence of high resolution images, the process of chemical etching and polishing generally result in clusters of precipitates near the grain boundary. Several precipitate structures of varying size and shape can also be observed well within the grains. Figure 2 shows a Nickel alloy micrograph where on can visually observe several grains at different gray levels that re separated from neighboring grains either by the gray level intensity or by the cluster of precipitates in the form of dot like structures. Precipitates of varying size and shape can also be observed within the relatively homogenous regions of the grains.

Figure 1: RoboMet.3D technology for collecting 3D data in the form of high resolution sequence of optical micrographs (www.ues.com/content/robomet3d).

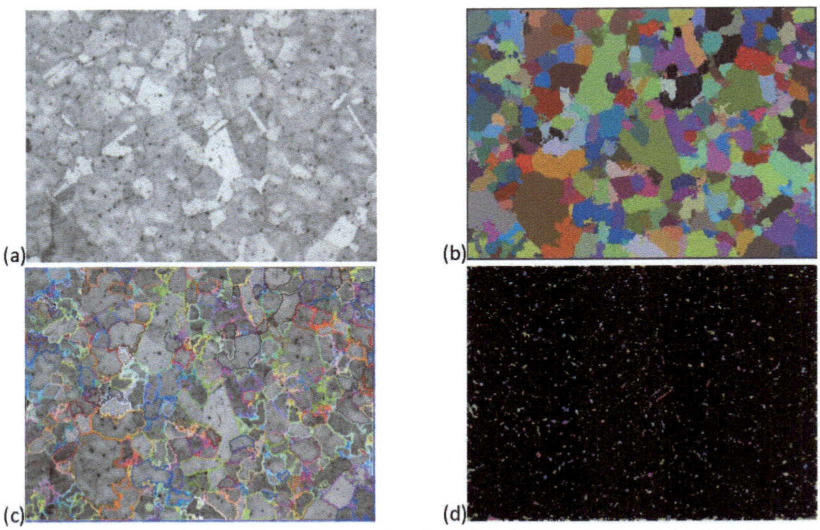

Figure 2: (a) practical quality micrograph of a Nickel alloy captured by the RoboMet.3D technology. The dark spots are the precipitates (b) Segmented grain mask (c) grain mask boundary overlaid on gray images (d) Segmented precipitates. Color pictures can be seen in electronic copy.

It has been argued that the size, shape and structural features distribution of the grains can be directly related to the material properties (1-6). For example, the Hall-Petch effect quantifies the trend of increasing strength and toughness with decreasing grain size i.e. fine grain size strengthens the material. Creep rates (Coble creep) increase with increasing grain boundary area (per unit volume), hence decreasing grain size. Grain size has the opposite effect at high temperatures where fine grain size weakens the material. To some extent, intensity of these effects is material dependent. Thus, it is essential to accurately mark grain boundaries and measure grain size for the quantitative analysis of the microstructures. Since the grains have irregular shapes and are 3D objects, it is imperative that we conduct all measurements in 3D space for a reliable data analysis.

A closer look at the available literature on image analytics for metallography shows that there not many software applications available for analysis of microstructures. Most of the available software are directly copied from biomedical applications and hence are either too difficult to use or just not right for analysis of metallographic data. The RobMet.3D technology commercially available through UES Inc. has generally relied on Avizo Fire software from FEI (formerly, VSG group). It is extremely difficult to automate microstructure analysis processes using this software and more often it require every step to be done manually and look for outside algorithms for segmentation and feature measurement. To rectify this, UES Inc. is launching a new suite of software tools, "R3Danalytics", specifically designed for the analysis of microstructures. Although the tools are developed with RoboMet.3D technology in mind, it is not specific to images collected by RoboMet.3D and can be used for analysis of microstructural image data acquired by any other similar device. This paper shows the application of this tool set in the analysis of a large set of 3D data of Nickel alloy collected using RoboMet.3D technology.

2. Image Analysis

Image analysis of 3D data obtained in the form of a sequence of micrographs of the material using RoboMet.3D technology consists of image alignment, image pre-processing, grain segmentation and basic feature measurement (7-10).

Image Alignment: The RoboMet.3D imaging technology consist of chemical etching and polishing process to remove material layer by layer for the 3D imaging purposes. Also, many mechanical movements of the specimen between each image capture cycle result in slight misalignment of the field of image captured. It is very difficult to control precision of the surface removal and hence the voxel lattice is generally non-linear and anisotropic. Accuracy of the voxel lattice can be improved by setting the etching and polishing process to remove minimum amount of material from the surface. This would result in slow process and a large amount of data to sift through. Thus the first step in the image analysis is to align the images in the stack of optical sections that form 3D data such that the grain boundary/surface is smooth and continuous across the spatial directions. In cases where the distance between consecutive image-slices is too high to establish the spatial continuity between grain boundaries, one should use appropriate interpolation techniques to create virtual image slices to establish grain surface continuity or insert a blank image to accept the discontinuity as is and avoid the error in segmentation. The R3Danalytics software suite provides three different algorithms for registering consecutive images based on similarity measures such as Pearson correlation, Statistical Regression and Euclidean distance. Graphical interface based interactive alignment and alignment correction tools are also provided. The parameters for alignment such as translational and rotational search size, acceptable level of correlation, etc., can be set prior to the alignment start. RoboMet.3D imaging technology has a good precision in its mechanical movements of the material and hence not rotational alignment is necessary. Figure 3 shows the result of aligning a block of five hundred optical sections of a Nickel alloy specimen aligned using statistical regression as the similarity measure between voxels of consecutive image-slices.

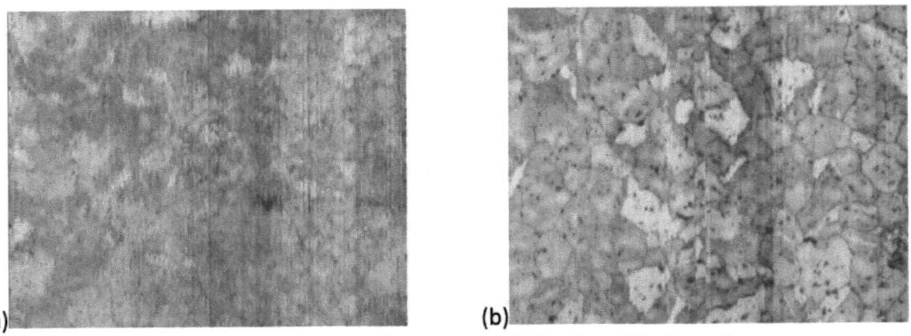

(a) (b)

Figure 3: A YZ section through a stack of 500 images (a) before (b) after alignment

Grain Segmentation: Segmentation of the grains is the most important and difficult part of the automated image analytics. Segmentation of 2D micrographs is relatively easy due to isotropic nature of the pixels and generally consistent gray scale statistics all through the image. But, 2D image analysis is nearly not as useful due to its partial representation of the grains and hence erroneous feature measurement and statistical analysis. The 3D segmentation of microstructures acquired using RoboMet.3D technology is complex. The complexity stems from nonlinear and anisotropic voxel lattice, inconsistent voxel statistics and histogram features from one optical section to another, large discontinuity in grain surfaces in the axial dimension leading to holes in grain surfaces and hence leakage of grain regions to its neighborhood during volumetric region based segmentation. The variation in the concentration and purity of the chemicals used for etching when imaged over a long time (required to collect 3D stack) and wear & tear in the polishing cloth result in inconsistent texture and precipitate deposits in the material surface.

The R3Danalytics suit provides a few automated image segmentation techniques based on amalgamation of boundary and region growing algorithms such as hybrid volume growing, gradient cue controlled regions, seeded volume growing technique, etc. Thresholding and clustering based segmentation techniques to segment the grains in the microstructures and establish grain boundary/surface are also provided. In this paper we are presenting the segmentation result obtained using hybrid volume growing technique. Properly aligned image stack is subject to several steps of preprocessing where the noise and artifacts are reduced and grain surface is fortified using non-linear adaptive diffusion filters that readjust its parameters based on the voxel anisotropicity as well as underlying voxel statistics. Continuous homogenous regions in the pre-processed data volume are considered as initial grain seeds and grown into neighborhood space under several constraints dictated by the neighborhood pixel statistics calculated in multiple scale space. In the final analysis the empty regions that are not represented in the mask area are provided unique labels to establish every voxel in the image is accounted and belongs to a particular grain. Figure 4 shows the result of volumetric segmentation on a Nickel alloy data stack consisting of 1000 image-sections.

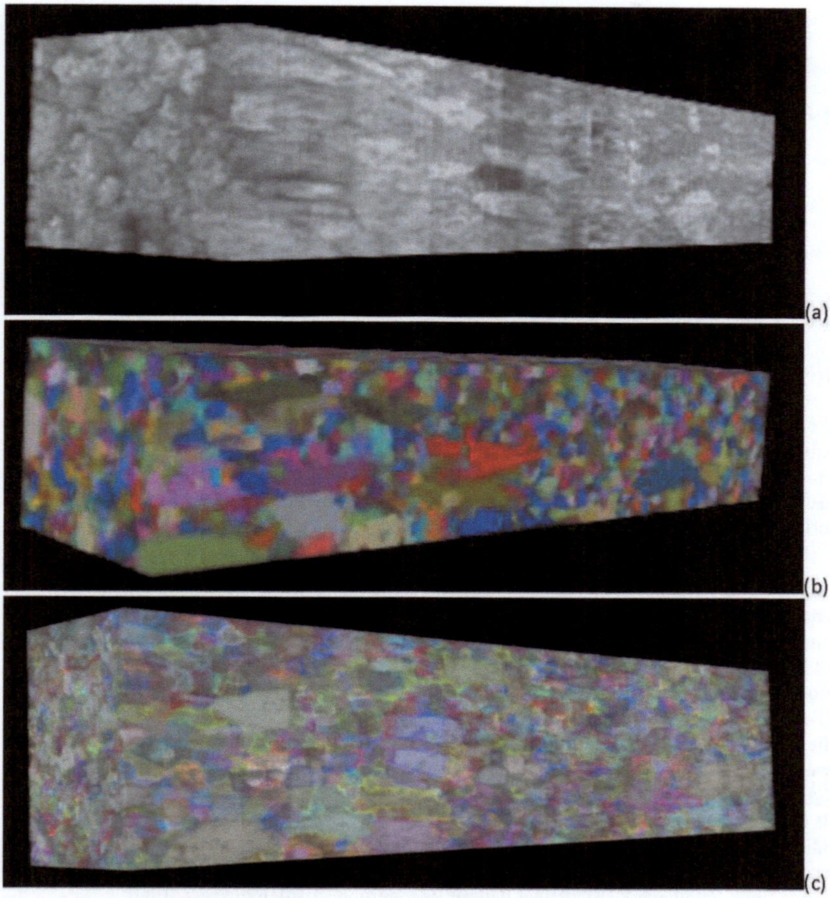

Figure 4: Segmentation of 3D data of Nickel alloy (a) Original aligned data block (b) Segmented masks of grains (c) Segmented mask boundary overlaid on original aligned data. The pictures are not in the same orientation. Color pictures can be seen in electronic copy

3. Experimental Results and Discussions

We have tested the algorithms on different metal, alloy and composite materials and analyzed the algorithms ability to segment the grains accurately. We then calculated the spatial centeroid position for each grain, grain size, basic object shape features, statistical texture features of each grain and simple histogram and amplitude features to analyze the inter-relation between grains in the sample tested. Figure 5, shows the graph of grain siize measured in terms of number of voxels per segmented grain in a large Nickel alloy sample.

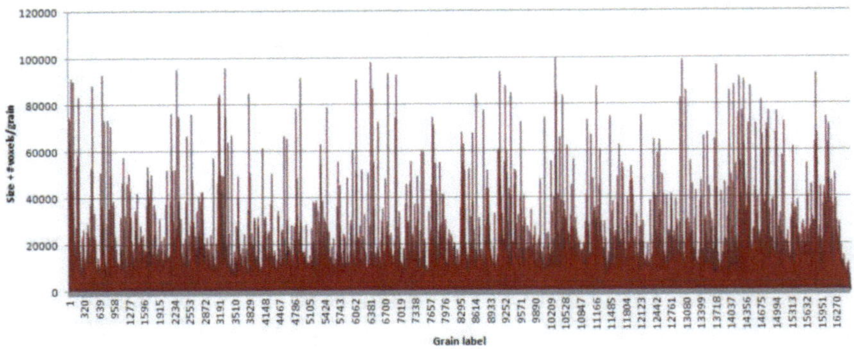

Figure 5: Distribution of grain size in a Nickel alloy sample

In the future studies, we plan to improve the accuracy of segmentation and compare it with possible manual annotation of the grain boundary/surface. Also we plan to build a data base of the material data processed by the RoboMet.3D and the analytics software along with the measured image features and investigate possible relation between image features and the material properties.

References

Mechanical Behavior of Materials (1966), F. McClintock and A. S. Argon, Addison Wesley.

Mechanical Behavior of Materials, T.H. Courtney, McGraw-Hill, ISBN 0-07-013265-8, 620.11292,C86M

Microstructure and Properties of Materials, J.C.M. Li, editor, World Scientific, ISBN 981-02-2403-6

Leslie, WC, The Physical Metallurgy of Steels, Hemisphere Press, McGraw-Hill

Llewellyn, DT & Hudd, RC, Steels, Metallurgy and Applications, Butterworths-Heinemann http://www.steeluniversity.org/content/html

Duda O, Hart P, Stork D. Pattern Classification: Wiley and Sons Press; 2001.

Haralick R. Statistical and structural approaches to texture. Proc IEEE. 1979;67:786-804.

Jahne B. Digital Image Processing: Concepts, Algorithms and Scientific Applications: Springer-Verland; 1997. Fakunaga K. Statistical Pattern Recognition: Academic Press, 1990.

Jain A. K. and Dubes R. C. Algorithms for clustering data, Prentice Hall, NJ 1988

2nd World Congress on Integrated Computational Materials Engineering
Edited by: Mei Li, Carelyn Campbell, Katsuyo Thornton, Elizabeth Holm, and Peter Gumbsch
TMS (The Minerals, Metals & Materials Society), 2013

A BRIEF REVIEW OF PRECIPITATION HARDENING MODELS FOR ALUMINUM ALLOYS

Guannan Guo[1], Qigui Wang [2], Gang Wang[3], Yiming Rong[1,3]

[1] Worcester Polytechnic Institute, Manufacturing Engineering, Worcester, MA 01609 USA
[2] GM Powertrain Materials Technology, Pontiac, MI, USA
[3] Tsinghua University Precision, Instruments and Mechanology, Beijing China

Keywords: A356, aging, yield strength, modeling, precipitate hardening

Abstract

This paper briefly reviews the precipitation hardening models for aluminum alloys. Several well-known precipitation and strengthening models are compared with our experimental data of aluminum A356 alloy. The differences among various models are presented and further improvement of precipitation hardening models is discussed.

1 Introduction

Aluminum alloys are increasingly used in structural applications because of their lightweight, relatively low manufacturing cost, and high strength to weight ratio particularly after heat treatment. Most aluminum alloys, like A356 used for critical structures are usually subjected to aging hardening. The aging strengthening is through the formation of precipitates, which act as point obstacles to inhibit the motion of dislocations. The early period of aging (i.e. under-aging) is governed by the dislocation mechanism of shearing, while the dislocation mechanism of bypassing dominates the later period of aging (over-aged). The type, size and volume fraction of precipitates depend upon the alloy compositions and heat treatment conditions. In Al-Mg-Si system, like A356 alloy, Mg/Si precipitates are the dominated strengthening phases after aging.

Modeling of precipitation hardening has been extensively studied in past years [1-7]. Several well-known strengthening models for aluminum alloys are reproduced in this paper. The model predictions are compared with our experimental data of A356 aluminum alloy. The differences among various precipitation and hardening models are presented and further improvement of hardening models is proposed.

2 Models for Precipitate Evolution

Mean value and discrete value approaches are two types of models in the literature to predict the size and volume fraction of the precipitate particles during aging. The mean value method does not consider the particle size distribution, [1, 3, 5, 6], taking all the particles as the same size. The discrete value approach, however, considers the particle size distribution based on selected radius classes. [9]

2.1 Mean value approach

In the mean value approach, the modeling of precipitates follows the classical nucleation and growth theory.[4] The basic principle for the growth of precipitate particles is diffusion mechanism of solution element. In each period of time during aging, volume fraction and the mean radius of particles follow different growth kinetics. Table 1 gives the experiment data for mean value approach method. C0, Ce, and Cp are the initial solute concentration, equilibrium solute concentration and solute concentration in precipitate, respectively. The result is given by Figure 1.

Table 1 Input Data for Figure 1

C0	Ce	Cp	D(Diffusion coefficient m^2/s)	γ(surface energy J/m^2)	V(volume per atom m^3)	a(lattice parameter nm)	T(K)
0.06	0.01	1	$5.0*10^{-20}$	0.13	$1.6*10^{-29}$	0.404	433

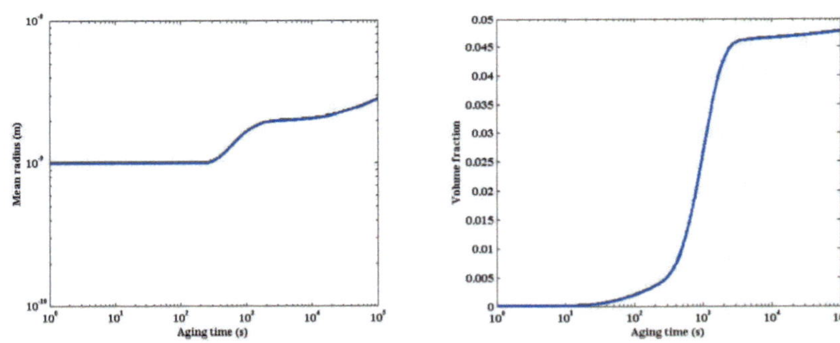

Figure 1 The evolution of mean particle size and volume fraction of precipitates in 6061 alloy predicted by Deschamps's model.

The nuclei remain at the critical radius at the nucleation stage. The nucleation rate drops to zero when no extra solute element remains in the solid solution. Growth period is corresponded with the dramatically increased mean radius of precipitates, meanwhile the volume fraction increases. In coarsening period, the radius slowly increases and the volume fraction almost remains the same at peak value.

2.2 Discrete value approach

In discrete value method, the newly formed nuclei at each time step are grouped and the size evolution of each group is tracked. The following plots (Figure 2) showing the changes of mean radius, volume fraction and particle size distribution during aging are based on this approach.[10]

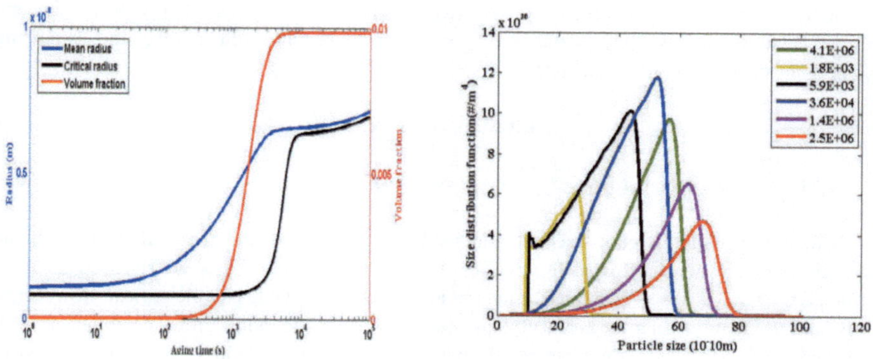

(a) (b)

Figure 2 The evolution of mean radius, critical radius, density distribution and volume fraction of precipitates in aluminum alloy A356 aged at 443K, predicted by Myhr's model.

With the density distribution at each radius group, the total density at each time step and the mean radius can be calculated by summing up all groups. The volume fraction can be then derived based on the assumption of spherical precipitate particles.

3 Models for Strengthening

3.1 Only shearing or bypassing mechanism considered

There are two types of dislocation hardening mechanisms: shearing and bypassing. Both mechanisms follow the similar strengthening prediction, which is given by [10]:

$$\sigma = \frac{MF}{bL} = Cr^m f^n \tag{1}$$

where M is the Taylor factor, F represents the average obstacle force, b is the burgers vector and L is the average space of particles. C is the coefficient decided by material and aging conditions, r and f represent the mean radius and volume fraction of the precipitates, respectively. m and n are different for shearing and bypassing mechanism. In Liu's model, only bypassing hardening mechanism was assumed.

3.2 Combined shearing and bypassing mechanism

Ashby and Shercliff combined both shearing and bypassing mechanisms using the harmonic value of shearing and bypassing strength. [1] But, Deschamps' model and Myhr's model separate shearing and bypassing mechanisms with critical radius, applying corresponding equations in different aging periods. [4] At the beginning of the aging process, the particles are small and coherent with matrix; the dislocation can shear these particles. [14] At peak aging and over-aged conditions, precipitates are large and incoherent with matrix and bypassing mechanism dominates deformation. [13]

4 Results and Discussion

4.1 Modeling of precipitate evolution during aging

Difficulty in building good precipitation model is in how to quantify and distinguish the nucleation, growth and coarsening periods. Critical radius approach is widely used in many models. Critical radius is defined as the minimum radius for stable particle growth. However, it is hard to identify when the coarsening period begins since the radius grows continuously with no obvious change in a short period of time. Deschamps *et al.* defines coarsening portion to calculate the growth and coarsening particles fraction: [4]

$$f_{coarse} = 1 - \mathrm{erf}\left(4\left(\frac{R}{R_0}\log\left(\frac{C}{C_e}\right) - 1\right)\right) \tag{2}$$

Volume fraction can be another way to build the precipitation model without considering radius. Fig. 3 gives the volume fraction evolution and yield strength change curve predicted by Ashby's model. Lloyd also predicted volume fraction using JMAK model which was calibrated by TEM.

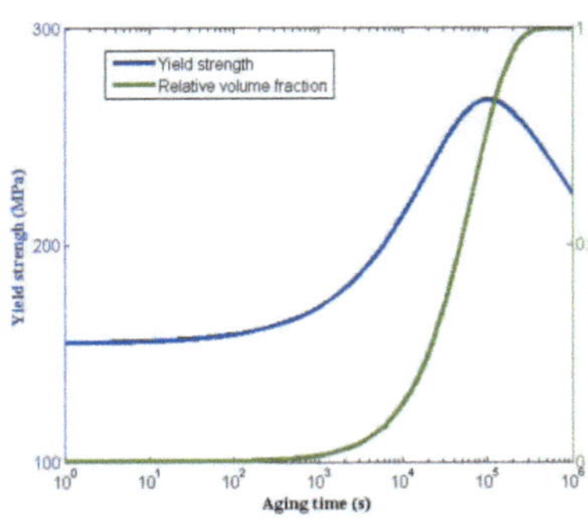

Figure 3 Yield strength and volume fraction for Al-Mg-Si at 443K.

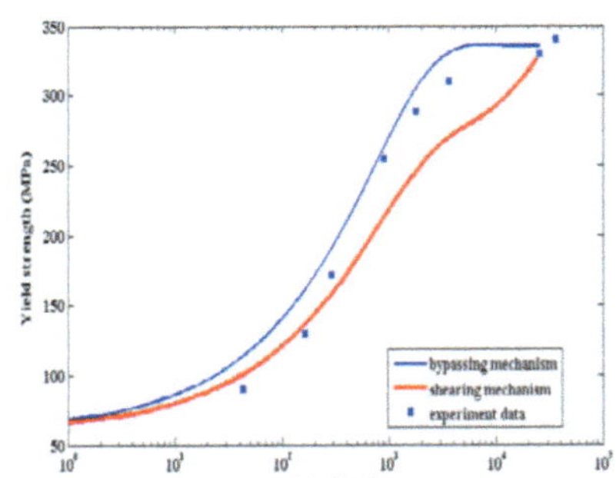

Figure 4 Shearing and bypassing mechanism for A6111 aging at 453K.

4.2 Modeling of yield strength

As mentioned above, the shearing and bypassing mechanisms are strongly related with the radius of particles. Lloyd made a comparison when considering only shearing mechanism or bypassing mechanism. [12] In Figure 4, our experimental data lie between the two prediction lines, which indicates that there should be a better method to combine two dislocation hardening mechanisms in order to make the prediction more reliable. Ashby's model takes the harmonic value of shearing strength and bypassing strength which matches well with experimental data before peak-aging, but poorly in overaging period.

The orientation and the shape of precipitates also affect the yield strength. [13] Liu *et al.* considered this effect in their model when predicting Al-Mg-Si alloy aging behaviors, [7] following the method given by Zhu *et al.* to evaluate the yield strength based on bypassing mechanism. [8]

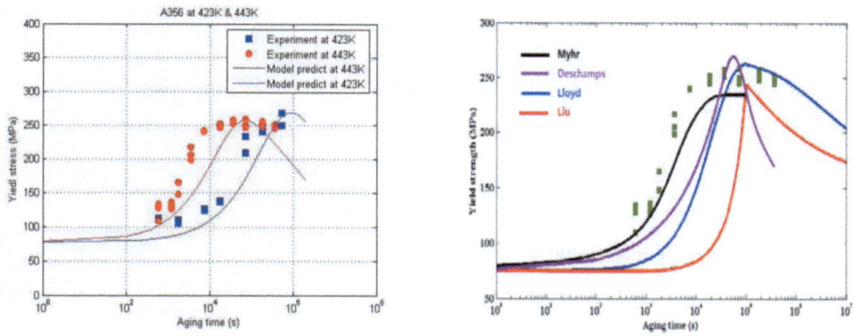

(a) (b)

Figure 5 (a) Ashby's model; (b) Lloyd, Liu, Deschamps & Myhr's models for A356 at aging temperature 443K, the green square data points are our experiment data.

Figure 5 compares the yield strength predictions from various models including Ashby, Loyld, Liu, Deschamps and Myhr's models with our experimental data of A356 alloy aged at 443K. It can be seen that Liu's model has the largest deviation from the experimental data and Ashby and Myhr's models match well with the experimental data.

5 Future works

Precipitate evolution model is critical to the aging hardening prediction. Mean or discrete value approaches are developed well for the spherical shape precipitates. However, the actual shape of the precipitate particles in aluminum alloys is not spherical. Following the principle of the discrete value approach, the length and the radius of the precipitate particles can be considered as two-axis coordinate to classify the group of particles. A better method is still needed to distinguish the growth and coarsening periods. In discrete value approach, each group can be considered as a unity to analyze its contribution to the yield strength by comparing with the critical radius.

6. Acknowledgement

This work is financially supported by Materials Technology Department of GM Global Powertrain Engineering.

7. Reference

1 H.R Shercliff, M.F. Ashby, *Acta mater,* Vol 38, No. 10, pp. 1789-1802, 1990

2 M. Perez, M. Dumont and D. Acevedo-Reyes, *Acta mater* 56(2008) 2119-2132

3 A. Deschamps, *Acta mater*, Vol 47, No. 1, pp 281-292, 1999

4 A. Deschamps, *Acta mater*, Vol 47, No. 1, pp 293-305, 1999

5 S. Esmaeili, *Acta mater* 51 (2003) 2245-2257,

6 A. Bahrami, *Metallurgical and Materials Transactions A,* Vol 43, Issue 11, pp 4445-4453, Nov 2012

7 G. Liu, G.J. Zhang, X.D. Ding, J. Sun, K.H. Chen,. *Materials Science and Engineering A*, Vol 344, Issues 1-2, pp 113-124, 15 March 2003

8 A.W Zhu, E.A Starke Jr, "Stress aging of Al-Cu Alloys: Computer Model", *Acta mater.* 49(2001)3063-3069

9 S. Esmaeili, D.J. Lloyd, *Metallurgical and Materials Transactions A*, Vol 34A, March 2003.

10 O.Myhr, *Acta mater.* 48(2000)1605-1615,

11 S. Esmaeili, D.J. Lloyd, *Acta mater* 51(2003) 2243-2257,

12 M. Ferrante, R.D. Doherty, *Acta mater* Vol. 27 pp. 1603-1614

13 O.R.Myhr, S.J. Andersen, *Acta mater*.49(2001) 65-75

14 M. Song, *Metallurgical and Materials Transactions A*, 443(2007) 172-177

2nd World Congress on Integrated Computational Materials Engineering
Edited by: Mei Li, Carelyn Campbell, Katsuyo Thornton, Elizabeth Holm, and Peter Gumbsch
TMS (The Minerals, Metals & Materials Society), 2013

CRYSTAL PLASTICITY FINITE ELEMENT MODELING OF SINGLE CRYSTAL NIOBIUM TENSILE TESTS WITH WEIGHTED DYNAMIC HARDENING RULE

Aboozar Mapar[1], Thomas R. Bieler[2], Farhang Pourboghrat[1], Christopher C. Compton[3]

[1]Department of Mechanical Engineering, Michigan State University
[2]Department of Chemical Engineering and Materials Science, Michigan State University
[3]National Superconducting Cyclotron Lab, Michigan State University
East Lansing, Michigan 48824, USA

Keywords: Crystal Plasticity, Hardening Rule, Niobium

Abstract

Fine grain Niobium (Nb) sheets are used to make superconducting cavities, which are the fundamental building blocks of particle accelerators. Slices of large grain Nb cut from as cast ingots are cheaper to produce than fine grain sheets. However, they are more prone to anisotropic deformation. Reproducible manufacturing of a cavity with anisotropic large grain Nb slices may be less expensive, but it needs greater care in anticipating the complex strain paths required to plastically transform the Nb slice into an accelerator cavity.

An alternative to the trial-and-error method of manufacturing, that also reduces cost, is to use Integrated Computational Materials Engineering (ICME) techniques. ICME requires the use of a constitutive material model to predict the behavior of Nb during the manufacturing operation. This constitutive model can then be implemented into a finite element code and used to design and optimize the manufacturing process. The premise of ICME for Nb slices is to successfully model the behavior of a single crystal of Nb. If successful, the single crystal theory can be extended to model the behavior of polycrystals.

In this research, a crystal plasticity code was developed for a single crystal and implemented into a commercial FEM solver. The classical crystal plasticity phenomenological hardening model is able to accurately predict the shape change of a Nb single crystal, although it is not as successful in predicting the stress history. To address this problem, the classical phenomenological hardening rule for a single crystal has been modified to dynamically account for conditions where single slip leads to a lack of hardening, which has improved stress predictions.

Introduction

Superconducting Radio Frequency (SRF) cavities are made from pure Niobium (Nb), which has the highest critical temperature (T=9.2 K) of pure metals [1]. Pure Nb is also very ductile, which makes Nb the preferred material for fabricating complex shaped superconducting cavities.

The mechanical behavior of Nb is anisotropic and shows a great directionality. The topology arising from differential strains on the surface degrade the SRF performance of Nb [2]. The lack of grain boundaries in large grain Nb slices lead to higher SRF performance with reduced fabrication cost, since the rolling process is not required. Fine grain sheets are more often used since they deform more uniformly [3].

Changing the grain size requires designing a new manufacturing process. Nb, however, is very expensive (in the range of $40K/ton) which makes the trial-and-error design process costly.

One way to overcome these difficulties and reduce the costs is to use a constitutive model to predict the deformation behavior of single crystal Nb. Considering that each polycrystal consists of single crystals, a model capable of predicting behavior of a single crystal can also be used to account for polycrystal deformation.

Developing the crystal plasticity model

Crystal Plasticity modeling constrains plastic deformation of materials to occur by shear on specific slip planes within a crystal [4]. The current crystal plasticity code is based on the model of Zamiri and Pourboghrat [5]. In this model they used an optimization method to define the yield surface. They also used a classical hardening rule, equation (1) to predict the evolution of shear stress.

$$\dot{\tau}^{\alpha}_{Class} = \sum_{\beta=1}^{N} h^{\alpha\beta}\left|\dot{\gamma}^{\beta}\right| \qquad (1)$$

Calibrating the model

Nine tensile specimens with different orientations were cut from a large grain Nb disk. The disk itself was cut from an as-cast ingot. The tensile samples were alphabetically named from “P” to “X”, and have tensile axis orientations indicated in Figure 1(a). Tensile test data was used to calibrate the model using a very simple FEM mesh consisting of 11 initially cube shaped hexahedral elements lined up in a row. The deformation was imposed with a specified displacement rate on one end of the model. The model was calibrated using data from tensile sample P, and an image of the deformed mesh for orientation P is illustrated in Figure 1(b). The model was then used to predict the stress-strain curve for the remaining eight orientations. Figure 2 compares the measured and predicted stress-strain curves for P, which has an orientation near the edge of the stereographic triangle, and Q, which has an orientation near the center of the triangle.

As can be seen in this figure, the predicted stress-strain curve for the Q orientation is much harder than the experiment. This occurs for other orientations that are not close to the edge of the triangle as well, which means the hardening rate in Nb is less than what classical hardening models can predict. Therefore, the hardening rate in the model should be lowered.

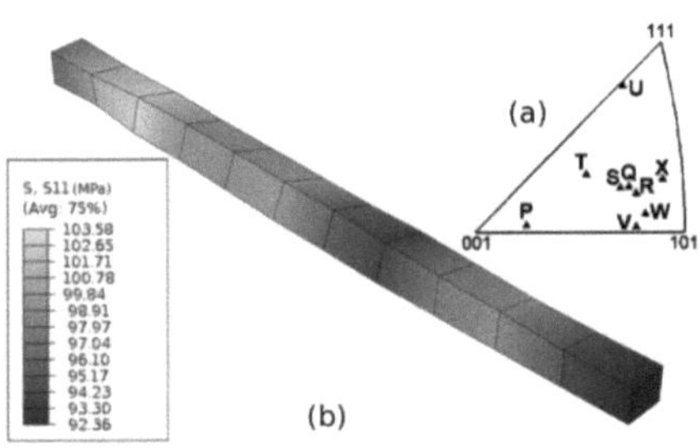

Figure 1. (a) The tensile axis for single crystal tensile samples P-X. (b) Simulated deformation of orientation P.

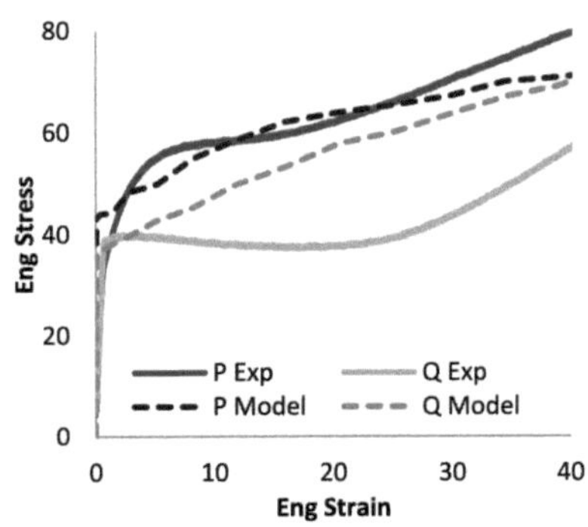

Figure 2. Comparison of flow curves for experimental and classical hardening model for orientations P and Q.

Proposed Dynamic Hardening Rule

BCC metals exhibit slip in four <111> directions, which can occur on as many as 12 different planes for each direction, leading to 48 slip systems. Hardening is the result of entanglement of slip in different directions. The classical hardening rule assumes that all slip systems harden (at different rates) as strain progresses, regardless of the slip activity details. While this is probably accurate for polycrystals, where the constraint of neighboring grains forces arbitrary shape changes that require operation of multiple slip systems, it is not always true for a single crystal.

Plastic deformation in a single crystal starts when the first slip system is activated, and dislocations can move through the material and exit from the opposite face without experiencing resistance. Single slip occurs when the tensile orientation is not at or near the edge of the stereographic triangle shown in Figure 1. During deformation, the crystal orientation rotates toward one of the edges of the triangle and other slip systems are activated, which interact and harden each other. To model the early stage of single slip deformation where there is no entanglement, the classical hardening rule in equation (1) is multiplied by the ratio of shear strain on the second to the first most active slip direction ($\frac{\gamma^{2nd}}{\gamma^{1st}}$). When only one slip system is active, this ratio is zero, so there is no increase in the shear stress for the next increment of deformation. When the second slip system is activated, the ratio will gradually increase and the ratio will approach one and model hardens. Hence, the rate of hardening depends on the rate of the second most slip direction activity, and equation (1) is rewritten as:

$$\dot{\tau}^{\alpha}_{\mathrm{Dynamic}} = \sum_{\beta=1}^{N} h^{\alpha\beta} \left|\dot{\gamma}^{\beta}\right| \left(\frac{\gamma^{2nd}}{\gamma^{1st}}\right) \quad (2)$$

This dynamic hardening rule improves the accuracy of the model, as expected; however, it predicts softer behavior than observed in the experiments, implying that there is some hardening that probably arises from dislocation interactions with pre-existing dislocations.

Weighted Hardening Rule

The stress-strain curves of many orientations fall between the predictions of these two models. Consequently, a combination of these two models can be used to more accurately model the behavior of Nb. Hence, a weighted hardening rule can be defined as follows:

$$\tau = (1 - w)\tau_{Class} + (w)\tau_{Dynamic} \quad (3)$$

When $w = 0$, equation (3) represents classical hardening rule, and when $w = 1$ it represents the dynamic hardening rule. For $0 < w < 1$ the prediction of the model is a combination of the classical and dynamic hardening rules. Figure 3 shows the results of the weighted hardening rule for orientation Q. As w goes from 0 to 1 the simulated curves move toward lower stresses.

The weighted hardening rule agrees well with experiments for most of the orientations when $0.6 < w < 0.9$. Figure 4 compares the weighted hardening rule for P and Q samples when the weight coefficient is $w = 0.7$. With the lower degree of hardening, the stability of the code suffers, as shown by terminated flow curves at lower strain for some weight values.

Conclusion

The classical hardening rule introduces too much hardening into the model. The proposed dynamic hardening ratio ($\frac{\gamma^{2nd}}{\gamma^{1st}}$) considerably improves the results of the model, however, it is softer than needed. A weighted hardening rule is used to adjust the rate of hardening. This works considerably better than the classical form when the weight is $0.6 < w < 0.9$.

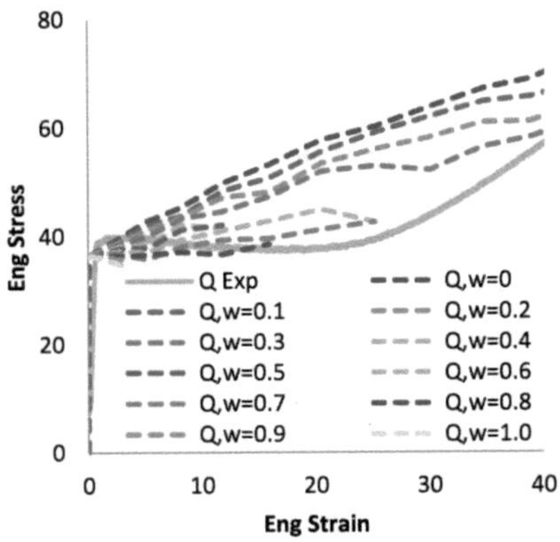

Figure 3. Results of Weighted Hardening Rule for orientation Q tensile sample.

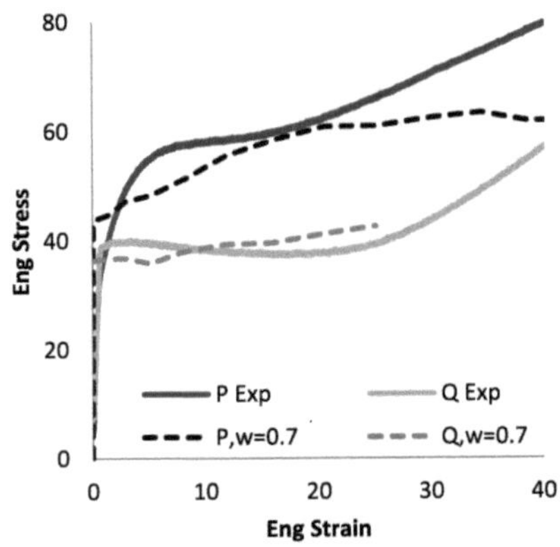

Figure 4. Comparing measured stress-strain results with the weighted hardening model for P and Q tensile samples when weight coefficient is $w = 0.7$.

References

[1] H. P. M. Peiniger, "A superconducting Nb3Sn coated multicell accelerating cavity," *IEEE Transactions on Nuclear Science*, vol. NS-32, no. 5, pp. 3610–3612, 1985.

[2] P. J. Lee, a. a. Polyanskii, a. Gurevich, a. a. Squitieri, D. C. Larbalestier, P. C. Bauer, C. Boffo, and H. T. Edwards, "Grain boundary flux penetration and resistivity in large grain niobium sheet," *Physica C: Superconductivity*, vol. 441, no. 1–2, pp. 126–129, Jul. 2006.

[3] L. Lilje, E. Kako, D. Kostin, A. Matheisen, W.-D. Möller, D. Proch, D. Reschke, K. Saito, P. Schmüser, S. Simrock, T. Suzuki, and K. Twarowski, "Achievement of 35 MV/m in the superconducting nine-cell cavities for TESLA," *Nuclear Instruments and Methods in Physics Research Section A: Accelerators, Spectrometers, Detectors and Associated Equipment*, vol. 524, no. 1–3, pp. 1–12, May 2004.

[4] G. I. Taylor, "Plastic strain in metals," *Twenty-Eighth May Lecture to the Institute of Metals*, 1938.

[5] A. R. Zamiri and F. Pourboghrat, "A novel yield function for single crystals based on combined constraints optimization," *International Journal of Plasticity*, vol. 26, no. 5, pp. 731–746, May 2010.

2nd World Congress on Integrated Computational Materials Engineering
Edited by: Mei Li, Carelyn Campbell, Katsuyo Thornton, Elizabeth Holm, and Peter Gumbsch
TMS (The Minerals, Metals & Materials Society), 2013

Three dimensional X-ray Diffraction Contrast Tomography Reconstruction of Polycrystalline Strontium Titanate during Sintering and Electron Backscatter Diffraction Validation

M. Syha[1], W. Rheinheimer[1], B. Loedermann[1], A. Graff[2], A. Trenkle[1], M. Baeurer[1], D. Weygand[1], W. Ludwig[3], P. Gumbsch[1,2]

[1]Karlsruhe Institute of Technology, IAM, Kaiserstr. 12, 76128 Karlsruhe, Germany
[2]Fraunhofer IWM, Walter-Hülse-Str. 1, 06120 Halle, Germany
[3]European Synchrotron Radiation Facility, 6 Rue Jules Horowitz, 38000 Grenoble, France

Keywords: Diffraction Contrast Tomography, Sintering, EBSD, 3D Materials Characterization, Microstructure Evolution

Abstract

The microstructural evolution of polycrystalline strontium titanate was investigated in three dimensions (3D) using X-ray diffraction contrast tomography (DCT) before and after ex-situ annealing at 1600°C. Post-annealing, the specimen was additionally subjected to phase contrast tomography (PCT) in order to finely resolve the porosities. The resulting microstructure reconstructions were studied with special emphasis on morphology and interface orientation during microstructure evolution. Subsequently, cross-sections of the specimen were studied using electron backscatter diffraction (EBSD). Corresponding cross-sections through the 3D reconstruction were identified and the quality of the reconstruction is validated with special emphasis on the spatial resolution at the grain boundaries, the size and location of pores contained in the material and the accuracy of the orientation determination.

Introduction

Only recently, time resolved 3D characterization of polycrystalline bulk materials during microstructure evolution became feasible through non-destructive imaging techniques such as 3D X-ray diffraction microscopy [1, 2, 3], X-ray DCT [4, 5] and differential aperture X-ray microscopy [6]. Being of invaluable use for investigations of sintering and microstructure evolution in ceramics, the DCT technique is applied here to the investigation of the annealing behavior of strontium titanate. Since the applied procedure for the analysis of diffraction data makes simplifying assumptions concerning the optics and the stress/strain state of the material [5] one focus of the present work lies on the validation of the resulting microstructure reconstructions by means of EBSD measurements taken in a subvolume of special interest. This region contains several big grains exhibiting a cube like shape and a preferred <100> interface orientation [7]. Moreover, statistics on orientation distributions and morphology evolution are investigated in the context of interface property anisotropies.

Methods

Sample Preparation

The polycrystalline DCT specimen was prepared from bulk strontium titanate made from $SrTiO_3$ powders processed from $SrCO_3$ and TiO_2 (both 99.9+%, Sigma Aldrich Chemie, Taufkirchen, Germany). The raw material was processed by the mixed oxide route using a molar Sr/Ti ratio of 0.996. After milling, calcining and isostatical pressing, the green bodies were sintered for 1h at 1600°C in oxygen atmosphere yielding a material with an average grain radius of 14.1±1.5μm. The DCT specimen was fabricated manually using a turning lathe and abrasive paper. The final sample dimensions are cylindric with $\sim$ 280μm diameter and a height of $\sim$ 350μm. In between the two DCT scans, the specimen was annealed for 1h at 1600°C in air. Detailed information on fabrication and annealing of the specimens are provided elsewhere [8, 9, 7].

Diffraction Contrast Tomography

DCT scans of the specimen before and after heating were performed using monochromatic synchrotron X-rays, set-up and technical details as described in [4, 7]. Diffraction as well as absorption information was acquired in 360° scans with an angular stepping of 0.05°. Using the full 360° rotation allows the exploitation of pairs of diffraction spots separated by 180°(Friedel pairs), which in turn allow to extract position and crystallographic orientation for each grain with high accuracy. Applying algebraic reconstruction to sets of Friedel pairs identified for a particular grain yields a 3D voxelated reconstruction of its shape. Placing the grains at their correct position inside the sample volume yields the microstructure reconstruction, which is complemented by a set of Rodriguez vectors, defining the crystallographic orientations of the individual grains. A detailed description of the data analysis procedure can be found in [5, 7]. For the post-annealing stage, these informations were complemented by PCT [10] data. The acquisition at a larger sample-detector distance allows a free space propagation leading to edge enhancement (Fresnel diffraction) which increases the visibility of small porosities. All synchrotron experiments were performed at beamline ID11 of the European Synchrotron Radiation Facility (ESRF).

Electron Backscatter Diffraction

For the EBSD measurements, the specimen was sectioned perpendicular to the cylinder axis starting from the cylinder top. Therefore, the specimen and four silicon wafer pieces were embedded in epoxin resin. The silicon ensures the electrical conductivity and the mechanical stability of the embedding and allows depth control during abrasion and alignment. Since automated focused ion beam ablation is not feasible considering the large surface area and the hardness of the material, the sectioning was done by mechanical grinding and polishing. Prior to each EBSD scan, the specimen was polished by Argon ion beam and covered with a thin carbon film. Mounted with 70° tilt to the EBSD detector in a working distance of 11mm, the whole cross section was scanned in a hexagonal grid with 1μm step size. Diffraction patterns were indexed using the cubic Perovskite structure with 3.905Å lattice constant and spacegroup 221. Neighboring pixels with

Figure 1: 3D reconstructions before (a) and after annealing (b).

orientation deviations below 3° were grouped as grains. A scanning electron microscope (Zeiss; Supra 55 VP) equipped with an EBSD system (EDAX TSL) was used at 15kV. Resulting images were used to generate grain boundary networks to be matched with corresponding cross-sections of the reconstructed microstructure. A detailed description of the identification of corresponding cross-sections can be found in [11].

Results

3D microstructure reconstructions of the specimen at both stages in microstructure evolution are presented in figure 1. The overall shape is identical and surface grains can easily be reidentified. The growth of the outer grains is reduced due to surface grooving effects, hindering the free motion of the grain boundaries. In order to determine the growth dynamics and porosity evolution, both reconstructions were aligned and identical subvolumes identified. The in-depth comparison of 25μm thick layer of the reconstructed volumes at both stages shown in figure 2(a) reveals the grain coarsening process. The image shows an overlay of several sections of the specimen prior to (magenta) and post (green) annealing, reflecting microstructural changes in a quasi 3D view. Figure 2(b) shows an overlay of the pores at both stages. Although pore clusters that stayed can be reidentified in the second stage, a decrease of the volume fraction of porosity was observed. From the tomography measurement the total pore volume for both stages is calculated. The volume fraction of porosity decreased from 2.6 vol-% in the initial stage to 1.2 vol-% in the post-annealing stage. During annealing, the number of grains changed from 849 to 797 resulting in an average volume growth of 5% per grain, under consideration of the varying porosity. Overall, the grains grow from an average grain radius of 14.7±2μm before annealing to 15.0±2μm after annealing.
Distributions of the local interface orientations in the pre- and post-annealed stage are given in multiples of the random distribution in figure 3. Orientation information has been extracted from Laplace smoothend surface tessellations of the grains that conserve the physically relevant microstructure elements. The distributions show a preference for certain interface orientations, reflected in a cumulation of 15% and 20%, respectively with respect to the random distribution. An overlay of grain boundary networks as obtained by EBSD and DCT for one corresponding cross-section is given in figure 4, alongside with an euclidean distance map [12] for these corresponding cross-sections pro-

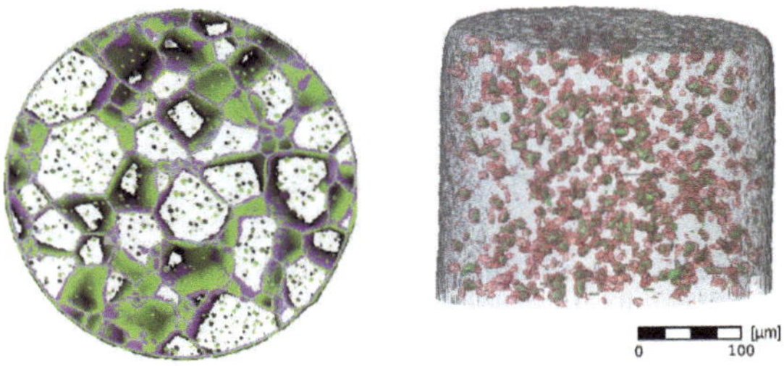

(a) (b)

Figure 2: (a) 25μm thick layer of the reconstructed structure before (magenta) and after (green) annealing, (b) collective pore ensembles before (red) and after (green) annealing.

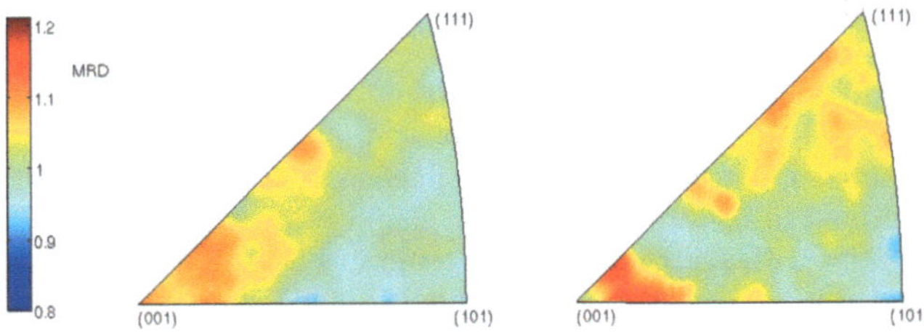

Figure 3: Distribution of local interface orientations for all bulk grains in the (a) pre- and (b) post-annealing state given in multiples of the random distribution.

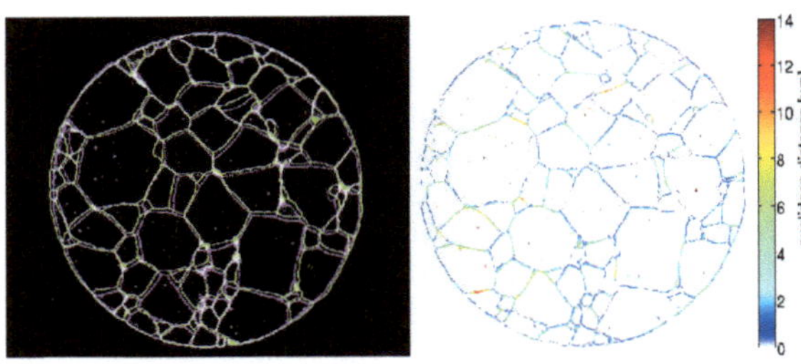

(a) (b)

Figure 4: (a) Overlay of grain boundary networks obtained by DCT (green) and EBSD (magenta), (b) DCT network colored according to euclidean distance from EBSD network.

jected on the DCT network. Clearly, the EBSD network shows a higher number of small intragranular pores. Apart from the disagreements resulting from these pores, deviations are mainly observed along the grain boundaries and not on triple points. The average euclidean distance between the corresponding grain boundary networks was found to be 2.26µm for the cross-section shown in 4. The number of grains found in both networks is fairly identical, as is the average grain size [11].

Discussion

The presented results prove the feasibility of 4D microstructure characterization in perovskites using X-ray DCT. High resolution grain orientations allow the reidentification of crystallites after heating so that the two microstructure reconstructions could be aligned for further investigations of volume, orientation and topology characteristics at the different stages. The reconstructed microstructures exhibit moderate grain coarsening and densification, the latter being most likely underestimated due to the fact that the porosity was measured with higher accuracy for the post annealing stage. The error of 2µm in the average grain size is the estimated error introduced by volumetric measurement of the average grain size in the finished reconstruction. However, additional uncertainty might be introduced by the reconstruction algorithm, which contains a dilation step for filling gaps in the initial voxel information [5]. Considering the resolution limit for small grains, which is about 300 voxels and using a greedy algorithm to fill the gaps in the undilated structure with smaller grains yields a more robust error estimate: Allowing for grains down to 200 voxels volume yields a change of 5.5µm in mean radius. However, the good agreement between EBSD and DCT characterization gives reason to assume the error introduced by the dilation procedure to be much smaller. Reported deviations between EBSD and DCT might occur due to uneven cutting plane resulting from the manual preparation of the cross sections for EBSD measurements.
Interface orientation distributions reveal a preference for orientations lying within a 5° range of the <100> orientation, which is in excellent agreement with the results from orientation imaging microscopy [13]. This orientation is the low energy orientation in the $SrTiO_3$ system [14]. The accuracy of the orientation determintaion is affected by the smoothing step. That said the momentary achievable accuracy is not yet sufficient to identify general anisotropies in grain boundary properties except for pronounced cases of faceting. DCT with higher resolution and/or a refined reconstruction technique might improve the spatial resolution at the grain boundaries. A significantly improved resolution of the porosity (30-50 nm) might be achieved by the application of optimized optical instruments in combination with Zoom-tomography [15].

References

[1] E. Lauridsen et al. "Growth kinetics of individual cube grains as studied by the 3D X-ray diffraction microscope". *Recrystallizationd and Grain Growth. Proceedings*, 1, (2001), 589–594.

[2] H. Poulsen. *Three-dimensional X-Ray diffraction microscopy. Mapping polycrystals and their dynamics* (Springer, 2004).

[3] R. Suter et al. "Forward modeling method for microstructure reconstruction using x-ray diffraction microscopy: Single-crystal verification". *Review of Scientific Instruments*, 77, (2006), 123905.

[4] G. Johnson et al. "X-ray diffraction contrast tomography: a novel technique for three-dimensional grain mapping of polycrystals. I. The combined case". *Journal of Applied Crystallography*, 41, (2008), 310–318.

[5] W. Ludwig et al. "Three-dimensional grain mapping by x-ray diffraction contrast tomography and the use of Friedel pairs in diffraction data analysis". *Review of Scientific Instruments*, 033905.

[6] B. Larson et al. "Three-dimensional X-ray structural microscopy with submicrometre resolution". *Nature*, 415, (2002), 887–890.

[7] M. Syha et al. "Three-dimensional grain structure of sintered bulk strontium titanate from X-ray diffraction contrast tomography". *Scripta Materialia*, 66, (2012), 1–4.

[8] M. Bäurer et al. "Grain growth anomaly in strontium titanate". *Scripta Materialia*, 61 (6), (2009), 584 - 587.

[9] M. Bäurer, H. Kungl, and M. Hoffmann. "Influence of Sr/Ti Soiciometry on the Densification Behaviour of Strontium Titanate". *Journal of the American Ceramical Society*, 92 (3), (2009), 601–606.

[10] P. Cloetens et al. "Observation of microstructure and damage in materials by phase sensitive radiography and tomography". *Journal of Applied Physics*, 81, (1997), 5878.

[11] M. Syha et al. "Validation of 3D diffraction contrast tomography reconstructions by means of EBSD characterization". *Journal of Applied Crystallography, subm.*

[12] P. Danielsson. "Euclidean Distance Mapping". *Computer Graphics and Image Processing*, 14, (1980), 227–248.

[13] D. Saylor et al. "Distribution of Grain Boundaries in SrTiO3 as a Function of Five Macroscopic Parameters". *Journal of the American Ceramical Society*, 84 (4), (2004), 670–676.

[14] T. Sano, D. Saylor, and G. Rohrer. "Surface Energy Anisotropy of SrTiO3 at 1400 °C in Air". *Journal of the American Ceramical Society*, 86 (11), (2003), 1933–1939.

[15] R. Mokso et al. "Nanoscale zoom tomography with hard X-rays using Kirkpatrick-Baez optics". *Applied Physics Letters*, 90, (2007), 144107.

2nd World Congress on Integrated Computational Materials Engineering
Edited by: Mei Li, Carelyn Campbell, Katsuyo Thornton, Elizabeth Holm, and Peter Gumbsch
TMS (The Minerals, Metals & Materials Society), 2013

TOWARDS THE INTERFCAE LEVEL UNDERSTANDING OF INTERNALLY OXIDIZED METAL-OXIDE COMPOSITES: Cu-Al_2O_3

Yong Jiang[1,2*], Guoqiang Lan[1], Canhui Xu[1]

1 Materials School, Central South University, Changsha, China
2 State Key Laboratory for Powder Metallurgy, Central South University, Changsha, China

Keywords: Internal oxidation, Interfacial thermodynamics, Interfacial kinetics, First principles

Abstract

We report a systematic strategy for characterizing the interface-level microstructure evolution in metal-oxide composites during fabrication from first principles interfacial thermodynamics and classic kinetics analyses, with a sample study of the Cu-Al_2O_3 composite fabricated by in-situ internal oxidation. First, equilibrium atomic structures and corresponding properties of Cu-Al_2O_3 interfaces were determined, as a function of temperature (T) and the ambient oxygen partial pressure (p_{O2}). The results were then used to construct the interface phase stability diagrams. Thermodynamic equilibrium solubilities and diffusion kinetics of oxygen in Cu matrix were further analyzed, to establish the connection between p_{O2} and the local oxygen activity inside the composite. Finally we correlated the interfacial microstructures and adhesion properties with processing parameters during the composite fabrication.

1. Introduction

To optimally design and fabricate metal-oxide composites, one requires a fundamental understanding of the thermodynamics and kinetics of the relevant interfaces and how they correlate with fabrication processing parameters. In recent decades, considerable experimental efforts [1-9] have been devoted to characterizing atomic interface microstructures based on high resolution transmission electron microscopy (HRTEM). However, HRTEM cannot provide information such as the chemical interaction at interface and the associated energetics. Here we propose a theoretical strategy of combining the first principles thermodynamics with classical diffusion kinetics, to systematically investigate the processing-microstructure-property interactions related to internal interfaces. This is demonstrated for the *in-situ* internal oxidation fabrication of the Cu-Al_2O_3 composite, one of the best candidate material for high temperature electrical devices and high heat exchanger components. During the internal oxidation, oxygen dissolves into the Cu(Al) matrix and *in-situ* oxidizes Al to form strengthening Al_2O_3 precipitates. Although thermodynamics and diffusion kinetics for internal oxidation have been previously studied [10-12], they have never been correlated and clarified at the interface level. This work allows us to develop a knowledge-based design strategy for oxidation processing parameters.

2. Theoretical Methodology

Our first-principles calculations were performed on periodic supercells using the semi-commercial code VASP [13]. Blöchl's projector augmented wave method (PAW) was used for the electron-core interaction. The exchange correlation functional of PAW-PW91 yields the best prediction of bulk Cu and Al_2O_3, in well agreement with experiments [14,15]. The interface calculations employed a $Cu/Al_2O_3/Cu$ sandwich configuration with a vacuum thickness of at least 12 Å and a 3×3×1 M-P k-mesh. In the sandwich model, each Cu block consists of four atomic layers and the alumina of four O and eight Al-layers. Full relaxation uses a high energy cutoff of 550 eV, until the total force on each ion converged to within 0.02 eV/Å.

3. Results and Discussion

3.1. Interface structures and properties

To model a heterogeneous interface, the interface orientation of Cu(111)($\sqrt{3}\times\sqrt{3}$) /α-Al_2O_3 (0001)(1×1) is employed as suggested by HRTEM observations [16]. The work of separation, W_{sep}, as a measurement of adhesion strength, is sensitive to interfacial stoichiometry (reflecting the deviation of the Al-to-O ratio on the interface from the nominal value of 2/3). Upon different oxygen exposure, the interface termination may change from stoichiometric to nonstoichiometric, and the associated W_{sep} can vary by several folds [17]. In this work, we consider both stoichiomtric and non-stoichiometric terminations, with at least three different coordination types. Total energy calculations of the resulted interface ensembles can determine the most thermodynamically favored structures in Fig. 1, with the corresponding adhesion strength, W_{sep}.

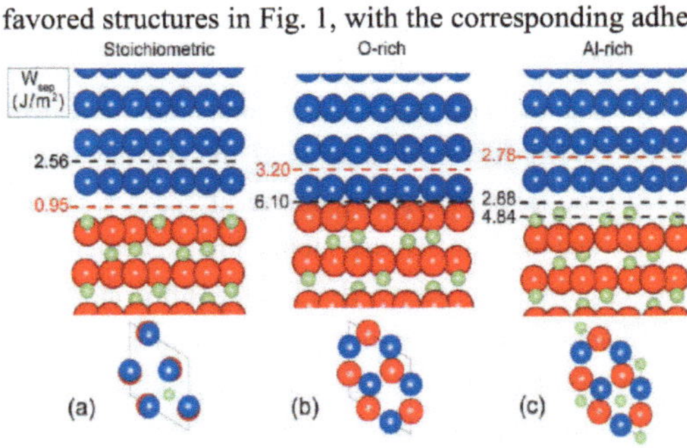

Fig. 1 The most stable interface structures with different terminations: (a) the stoichiometric, (b) the O-rich, and (c) the Al-rich. Blue, green and red balls are for Cu, Al and O, respectively.

3.2 Interfacial thermodynamics and phase stabilities

Under the thermodynamic equilibrium, the interface energy can be expressed as [18]

$$\gamma_I = \frac{1}{2A}\left[G_0 - \frac{1}{3}N_O\mu^0_{Al2O3} - N_{Cu}\mu^0_{Cu} - (N_{Al} - \frac{2}{3}N_O)\times(\mu^0_{Al} + kT\ln a_{Al})\right] \quad (1)$$

where A and G_0 are the area and the Gibbs free energy of the interface. N_i and μ_i are, respectively, the atom number and chemical potential of species i. According to the oxidation of Al in Cu(Al),

$$\text{Reaction 1:}\quad 2[Al]_{Cu}+\frac{3}{2}O_2=Al_2O_3,\qquad \Delta G_1^0=320.59\text{T-}1677000 \tag{2}$$

the Al activity a_{Al} and the oxygen partial pressure p_{O2} can be related as

$$p_{O_2}^{(1)}=\left(a_{Al}\right)^{-\frac{4}{3}}\exp\left(2\Delta G_1^o/3RT\right)\quad (atm). \tag{3}$$

Using Eq. (1-3), we calculate the interface energy w.r.t. a_{Al} or p_{O2} for T = 1223 K in Fig. 2. The Al-rich interface dominates for $p_{O2}<10^{-27}$ *atm*. At $p_{O2}>10^{-11.7}$ *atm*, the interface may change to O-rich. The formation of the weakest stoichimetric interface can be avoided by keeping $p_{O2}<10^{-27}$ or $>10^{-11.7}$ *atm*. Those critical a_{Al} that dictate the interfacial termination transitions are, respectively, $p_{O_2}^{(2)}=a_{Al}^{-1}a_{Cu}^{-1}\exp(\Delta G_2^o/RT)$ and $p_{O_2}^{(3)}=a_{Cu}^{-4}\exp(2\Delta G_3^o/RT)$.

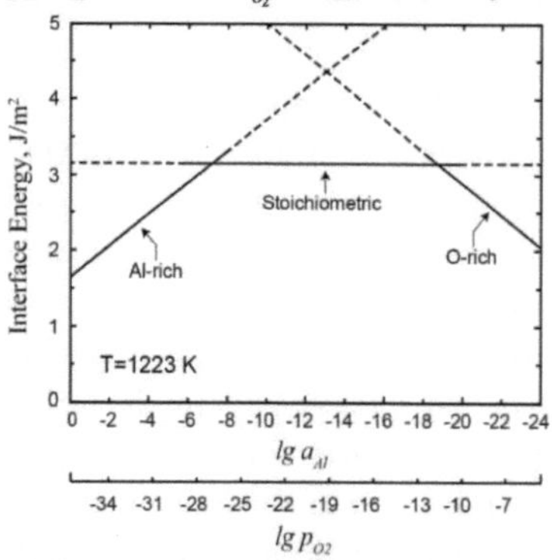

Fig. 2 Calculated interface energies vs. Al activity and oxygen partial pressure at 1223 K

Based on thermodynamic data in Refs. [19, 20], we further deduce the Gibbs free energies of reactions, ΔG, and the corresponding critical p_{O2} for other possible oxides.

$$\text{Reaction 2:}\quad Cu+[Al]_{Cu}+O_2=CuAlO_2,\quad \Delta G_2^0=\text{-}934219\text{+}201.30\text{T} \tag{4a}$$

$$\text{Reaction 3:}\quad 2Cu+\frac{1}{2}O_2=Cu_2O,\quad \Delta G_3^0=\text{-}169702\text{+}74.12\text{T} \tag{4b}$$

$$a_{Al\text{-}rich\_Stoi}=\exp[(G_{Al\text{-}rich}^o-G_{Stoi}^o-2\mu_{Al}^o)/2kT],\quad a_{Stoi\_O\text{-}rich}=\exp[(G_{Stoi}^o-G_{O\text{-}rich}^o-2\mu_{Al}^o)/2kT] \tag{5}$$

Using Eqs. (2-5), we can calculate the phase stability diagram for T=1223K in Fig. 3 (left). At the initial a_{Al} range (point e, predicted as $\sim10^{-6.5}<a_{Al}<10^{-4.5}$ and marked as a grey bar), reaction (1) dominates, resulting in the formation of Al-rich Al_2O_3. As the oxidation proceeds, a_{Al} gradually decreases and p_{O2} at the interface gradually increases along line (1). The Al_2O_3 interface termination may change from Al-rich to stoichiometric (point c), and further to O-rich (point d). When point b or a is reached, the Al_2O_3 might convert to $CuAlO_2$ or even Cu_2O. The exclusive formation of Al_2O_3 in Cu can only be achieved for p_{O2} below point b.

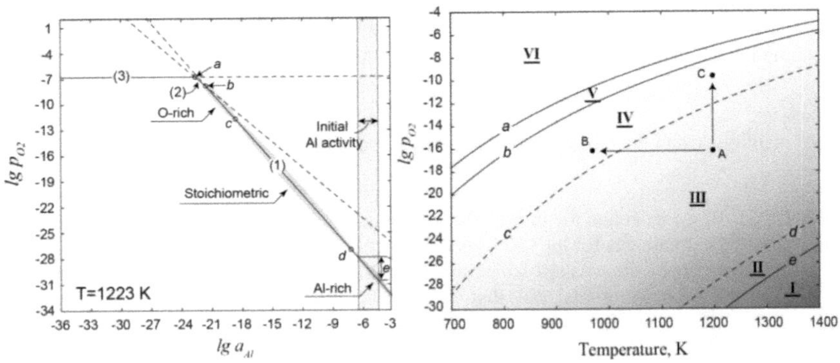

Fig. 3 (Left) Calculated phase stability diagram at T=1223 K. Lines (1-3) indicate the equilibrium of reactions (1-3). (Right) Calculated phase diagram for internal oxidation of Cu(Al). Zones I: Cu(Al), II: Cu(Al)+Al_2O_3(Al-rich), III: Cu(Al)+Al_2O_3(Stoichiometric), IV: Cu(Al)+ Al_2O_3(O-rich), V: Cu(Al)+Al_2O_3(O-rich)+$CuAlO_2$, VI: Cu_2O+Al_2O_3(O-rich)+$CuAlO_2$.

Tracking all these phase transitions (points *a* to *e*) at various temperatures enables the construction of the full phase diagram in Fig. 3 (right). The phase boundaries that are constituted by those transition points divide the phase diagram into six phase stability zones (I-VI). Let us consider an oxidation condition, say point A. To favor the formation of O-rich interfaces, one could either reduce the temperature (from A to B), or increase p_{O2} (from A to C). The former results in retarded oxidation kinetics. The latter is more favorable, and can be regarded as the foundation mechanism for the so-called "high pressure internal oxidation techniques" [21]. Further increasing the temperature at the same p_{O2} level (= e^{-16}), might facilitate the oxidation kinetics but cannot improve the (stoichiometric) interface microstructure.

3.3 Interfacial kinetics during internal oxidation

The Wagner's model [22] predicts that during oxidation, the position of the oxidation front (ξ) moves inwards with time by following the parabolic rate law. The oxygen concentration profile in the oxidation region ($x \le \xi$) and the Al concentration profile in the unoxidized region ($x \ge \xi$) can be derived separately [22, 23]

$$N_{[O]}(x,t) = N_{[O]}^{S}\left\{1 - erf\left(\frac{x}{2\sqrt{D_O t}}\right) / erf\gamma\right\},\quad N_{Al}(x,t) = N_{Al}^{0}\left\{1 - erfc\left(\frac{x}{2\sqrt{D_{Al} t}}\right) / erfc\left(\theta^{\frac{1}{2}}\gamma\right)\right\} \quad (7)$$

where $N_{[O]}^{S}$ is the equilibrium solubility of O in Cu. N_{Al}^{0} and D_{Al} are the initial atomic concentration and the diffusion coefficient of Al in Cu, respectively. $\theta = D_O/D_{Al}$. Assuming oxygen diffusivity is insensitive to the presence of oxide phases, one can correlate the internal oxygen concentration $N_{[O]}$ with p_{O2}, as

$$\lg p_{O2} = 2\lg N_{[O]}(x,t) - 3720/T + 4.1743 \quad (T = 1023 \sim 1303 K), \quad (8)$$

3.4 Predictions on microstructure evolution

We demonstrate our prediction strategy for the internal oxidation of dilute Cu(Al) using oxidizer Cu_2O, where the ambient p_{O2} equals to the dissociation pressure of Cu_2O. Fig. 4(a) presents the calculated interface energies for T=1223 K. Fig. 4(b) presents the calculated oxygen concentration profiles using oxidizer Cu_2O. Fig. 4(c) presents the predicted microstructure after oxidation for t=32 hr, based on the important insights provided in Fig. 4(a) and (b).

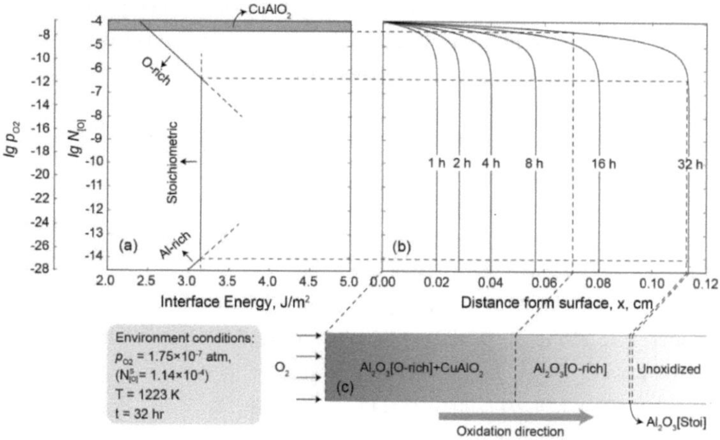

Fig. 4 The use of combined diagram to predict the microstructure evolution during internal oxidation of the Cu(Al) alloy with oxidizer Cu_2O at T=1223 K. (a) the isothermal section of the interface phase diagram at 1223 K. (b) the calculated depth profiles of oxygen concentration in the Cu matrix. (c) The predicted microstructure after oxidation for 32 hours.

It is predicted that a gradient microstructure would be resulted along the depth. Al_2O_3+$CuAlO_2$ precipitates may dominate from the surface down to a depth of ~700 μm, followed by the Al_2O_3 precipitation region of ~400 μm thickness. These Al_2O_3 precipitates are favorably O-rich with high adhesion strength. At the oxidation front, a thin layer (only 10~20 μm thickness) dominated with the weakest stoichiometric Al_2O_3 precipitates may present, due to the sharply decreasing p_{O2} at the oxidation front in Fig. 4(b). The oxidation front is thus most vulnerable to fracture. To avoid this vulnerable thin layer, one might choose to either restrict the maximum radius size of the Cu(Al) alloy powders to within ~1100 μm (=~700+~400 μm), or alternatively, to increase the oxidation time till the full oxidation of the largest alloy powder, so as to ensure the full oxidation over all different alloy powder radii.

4. Concluding Remarks

We have thoroughly described the thermodynamics and kinetics of Cu(Al)/α-Al_2O_3

interfaces during the internal oxidation process. It is found that interface structures and adhesion strongly depend on interfacial stoichiometry. The adhesion of a non-stoichiometric interface is about three times higher than that of the stoichiometric counterpart. The first principles based thermodynamics and diffusion kinetics can be correlated at the interface level through the local oxygen concentration. A gradient microstructure is predicted for oxidation using oxidizer Cu_2O at 1223 K for 32 hours. The oxidation front is dominated by the weakest stoichiometric Cu/Al_2O_3 interfaces and thus is most vulnerable to fracture. The optimal processing parameters can be suggested accordingly, to ensure the microstructure (and thus property) homogeneity.

References

[1] W.P. Vellinga, J.T.M.D. Hosson, Acta Mater, 45 (1997), 933.

[2] B.J.Kooi, H.B. Groen, J.T.M.D. Hosson, Acta Mater, 45 (1997), 3587.

[3] G. Renaud, P. Guenard, A. Barbier. Phys Rev B, 58 (1998), 7310.

[4] D.A. Shashkov, M.F. Chisholm, D.N. Seidman, Acta Mater, 47 (1999), 3939.

[5] D.A. Shashkov, D.A. Muller, D.N. Seidman, Acta Mater, 47 (1999), 3953.

[6] B.J. Kooi, J.T.M.D. Hosson, Acta Mater, 48 (2000), 3687.

[7] S. Mogck, B.J. Kooi, J.T.M.D. Hosson, Acta Mater, 52 (2004), 5845.

[8] S. Cazottes, Z.L. Zhang, R. Daniel, J.S. Chawla, D. Gall, G. Dehm, Thin Solid Films, 519 (2010), 1662.

[9] G. Zhou, Appl Phys Lett, 94 (2009), 233115.

[10] K. Song, J. Xing, Q. Dong, P. Liu, B. Tian, X. Cao, Mater Sci Eng: A, 380 (2004), 117.

[11] S. Liang, Z. Fan, L. Xu, L. Fang, Composites Part A: Appl Sci Manuf, 35 (2004), 1441.

[12] S.G. Tian, L.T. Zhang, H.M. Shao, L.J. Zou. Acta Metall SINICA, 9 (1996), 387.

[13] G.Kresse, J. Furthmüller, Phys Rev B, 54 (1996), 11169.

[14] HM Otte. J Appl Phys, 32 (1961), 1536.

[15] LW Finger, RM Hazen. J Appl Phys 1978, 49, 5823.

[16] G. Dehm, M. Rühle, G. Ding, R. Raj. Philos Mag Part B 1995, 71, 1111.

[17] J.R. Smith, Y. Jiang, A.G. Evans. Inter J Mater Res 2007, 98, 1214.

[18] W. Zhang, J.R. Smith, A.G. Evans. Acta Mater 2002, 50, 3803.

[19] J.M.W. Chase. NIST-JANAF Themochemical Tables, Fourth Edition; J. Phys. Chem. Ref. Data, 1998; Vol. Monograph 9.

[20] K.T. Jacob, C.B. Alcock, J Am Ceram Soc 1975, 58, 192.

[21] S. Tanaka, T. Hirata, M. Yida, Ag-SnO2 electrical contact materials by high pressure internal oxidation, US Patent 5078810, Publ. date 1992-01-07.

[22] C. Wagner. Zeitschrift für Elektrochemie, Berichte der Bunsengesellschaft für physikalische Chemie 1959, 63, 772.

[23] R.A. Rapp. Corrosion, 21 (1965), 382.

2nd World Congress on Integrated Computational Materials Engineering
Edited by: Mei Li, Carelyn Campbell, Katsuyo Thornton, Elizabeth Holm, and Peter Gumbsch
TMS (The Minerals, Metals & Materials Society), 2013

DISLOCATION DENSITY BASED CRYSTAL PLASTICITY FINITE ELEMENT MODEL OF POLYCRYSTALS WITH GRAIN BOUNDARY EFFECT

Zhe Leng[1], Alankar Alankar[2], David P. Field[1], Nathalie Allain-Bonasso[3], Francis Wagner[3]

[1]School of Mechanical and Materials Engineering, Washington State University
[2]Los Alamos National Laboratory, Los Alamos 87544, NM
[3]LEM3, Univ. of Metz, Metz, France.

Keywords: Grain boundary, Geometrically necessary dislocation, Crystal plasticity, Finite element simulation

Abstract:

Grain boundaries play an important role in determining the mechanical properties of metallic materials. The impedance of dislocation motion at the boundary results in a strengthening mechanism. In addition, dislocations can pile-up, be transmitted or be absorbed by the grain boundaries based on the local stress state and grain boundary character. In this study, a dislocation density based crystal plasticity finite element model is applied to incorporate the interaction between the dislocations and the grain boundaries, and a simulation is conducted on polycrystalline alpha iron deformed to 12% in uniaxial tension. The results indicate that the geometrically necessary dislocation density is generally higher near the grain boundary than within the grain interior. Taylor factor mismatch sometimes reveals strong localization effects near the grain boundaries.

Introduction:

Plastic deformation in metallic materials is controlled by dislocations. The dislocation movement on a slip plane and along a given slip direction under the influence of the local stress give rise to the permanent deformation, and the interaction between the dislocations results in forest hardening [1][2]. The dislocations also interact with the grain boundaries and can be absorbed, transmitted, or piled up at the boundaries, based on their character [3][4].

In order to investigate the effects of grain boundaries in metallic materials and the relationship between the microstructure and the mechanical properties, a uniaxial tension test has been applied to a polycrystalline alpha iron specimen, and the corresponding crystal plasticity finite element modeling is conducted based on the microstructure of the tested material.

Crystal Kinematics:

The kinematics of crystal plasticity formulation is based on the developments of Asaro and Rice [5], where the total deformation gradient F is decomposed into an elastic component F^e and a plastic component F^p:

$F = F^e F^p$.

The plastic deformation gradient is the part due to slip, it corresponds to a stress free intermediate configuration, where the crystal lattice is undeformed and unrotated. The elastic

deformation gradient involves the stretching and rotation of the crystal lattice.
The rate of plastic deformation gradient is given as:

$$\dot{F}^p = \left\{ \sum_{\alpha=1}^{n} \dot{\gamma}^\alpha \, m_0^\alpha \otimes n_0^\alpha \right\} F^p$$

m^α and n^α are the slip direction and slip plane normal of slip system α respectively. $m^\alpha \otimes n^\alpha$ is known as the Schmid tensor. In the current study, only 12 <111>(110) slip systems are considered, and compared with the 24 and 48 slip system, it is more cost saving and follows the full model prediction in spite of a relatively higher local error.

Constituitive Law:

For a single crystal, the second Piola-Kirchoff stress is defined in the intermediate configuration:

$$\tau = \det(F^e) F^{e-1} \sigma F^{e-T}$$

σ is the Cauchy stress, the Piola-Kirchhoff stress is related to the work conjugate elastic Green strain E^e through:

$$\tau = \mathcal{L} : E^e \qquad E^e = \frac{1}{2}(F^{eT} F^e - I)$$

Dislocation Density Based Framework:

The dislocation density based formulation proposed by Arsenlis and Parks [6][7] is adopted, and as is briefly stated here. The dislocation evolves in the form of a square loop and each dislocation segment has pure edge or pure screw character with different polarities. The plastic strain rate from the Orowan equation is given as:

$$\dot{\gamma}^\alpha = (\rho_{e+}^\alpha \bar{v}_{e+}^\alpha + \rho_{e-}^\alpha \bar{v}_{e-}^\alpha + \rho_{s+}^\alpha \bar{v}_{s+}^\alpha + \rho_{s-}^\alpha \bar{v}_{s-}^\alpha) |b^\alpha|$$

where b is the Burgers vector and $\bar{v}$ is the average velocity of the dislocation.

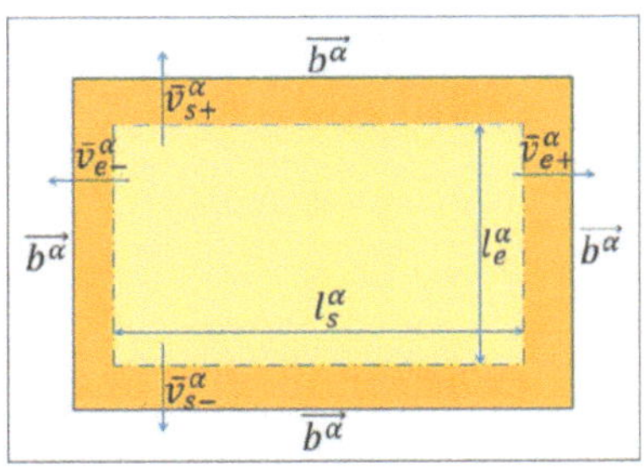

Fig.1 Illustration of the square dislocation loop assumed in the model framework. [7]

The evolution of dislocation includes generation, annihilation and dislocation flux terms:

$$\dot{\rho}^\alpha = \dot{\rho}_{gen}^\alpha + \dot{\rho}_{ann}^\alpha + \dot{\rho}_{flux}^\alpha$$

It is assumed that the generation of dislocation density mainly results from the expansion of the dislocation loop as shown in Fig.1, so the dislocation generation rate depends on the current dislocation state, such as density, velocity and average segment length.

$$\dot{\rho}_{e+(gen)}^\alpha = \dot{\rho}_{e-(gen)}^\alpha = \frac{\rho_{s+}^\alpha |\bar{v}_{s+}^\alpha|}{l_{s+}^\alpha} + \frac{\rho_{s-}^\alpha |\bar{v}_{s-}^\alpha|}{l_{s-}^\alpha}$$

$$\dot{\rho}^{\alpha}_{s+(gen)} = \dot{\rho}^{\alpha}_{s-(gen)} = \frac{\rho^{\alpha}_{e+}|\bar{v}^{\alpha}_{e+}|}{\bar{l}^{\alpha}_{e+}} + \frac{\rho^{\alpha}_{e-}|\bar{v}^{\alpha}_{e-}|}{\bar{l}^{\alpha}_{e-}}$$

Dislocation annihilation results from the dipole interactions:

$$\dot{\rho}^{\alpha}_{e+(ann)} = \dot{\rho}^{\alpha}_{e-(ann)} = -\rho^{\alpha}_{e+}\rho^{\alpha}_{e-}R_e(|\bar{v}^{\alpha}_{e+}| + |\bar{v}^{\alpha}_{e-}|)$$

$$\dot{\rho}^{\alpha}_{s+(ann)} = \dot{\rho}^{\alpha}_{s-(ann)} = -\rho^{\alpha}_{s+}\rho^{\alpha}_{s-}R_s(|\bar{v}^{\alpha}_{s+}| + |\bar{v}^{\alpha}_{s-}|)$$

R_e and R_s are the critical radii of interactions for edge and screw dislocations, respectively. In a discrete dislocation basis, the dislocation tensor is calculated as:

$$\mathbf{A} = \sum_{\alpha}(\rho^{\alpha}_{e+} - \rho^{\alpha}_{e-})\, b^{\alpha}_0 \otimes P^{\alpha}_0 + (\rho^{\alpha}_{s+} - \rho^{\alpha}_{s-}) b^{\alpha}_0 \otimes m^{\alpha}_0$$

and it is related to plastic deformation by:

$$\mathbf{A} = \mathrm{Curl}(\mathbf{F}^{p})\mathbf{F}^{p^{T}}$$

by taking the time rate form of the above two equations and making some physical considerations, for example, accumulation of a certain amount of ρ^{α}_{e+} is equal to loss of the same amount of ρ^{α}_{e-}, the dislocation flux divergence of each type of dislocation can be calculated [7]. In the current study, it is assumed that the GND is only a subset of total dislocation density, and the GND density of a specified type of dislocation can be quantified by the net polarity of that type of dislocation. In other words, both the dislocation flux and the GND density mainly depend on the local plastic deformation history.

Dislocation resistance is mainly controlled by forest dislocations, and is represented in a Taylor type equation of hardening:

$$S^{\alpha}_{d} = \mu b \sqrt{\sum_{\beta=1}^{N} (G^{\alpha\beta} \rho^{\beta})}$$

where G is the dislocation strength interaction matrix. Dislocation velocity v^{α} is calculated as:

$$\bar{v}^{\alpha}_{e} = v_e \exp\{-\frac{F_e}{kT}(1 - (\frac{|\tau^{\alpha}|}{S_{de}})^{pe})^{qe}\}$$

$$\bar{v}^{\alpha}_{s} = \frac{Lb^2}{l_c^2} v_d \exp\left\{-\frac{F_s}{kT}\left(1 - \left(\frac{|\tau^{\alpha}|}{\xi}\right)^{ps}\right)^{qs}\right\}$$

where T is the absolute temperature, k is Boltzmann's constant, L is segment length, l_c is the critical length for double kink nucleation, v_d is the Debye frequency. ξ is the Peierls stress on screw dislocations. τ^{α} is the Schmid stress. It should emphasized here that although the Schmid rule is often a good approximation for simple metals, the violation of Schmid law in BCC metals is inevitable.[8].

Grain Boundary Effect:

The grain boundaries act as obstacles for dislocation motion. When they encounter a grain boundary, mobile dislocations will accumulate at the grain boundary in the form of pile ups and give rise to the stress concentration there. Among all the types of interactions between dislocations and grain boundaries, slip transmission makes important contribution to the polycrystalline deformation, and the grain boundary effect is included in the model by introducing an thermally activated slip transmission concept [9]. The activation enthalpy quantifies the transmission probability for the mobile dislocations and it is considered as an energy barrier for those grain boundary elements. This energy barrier E^{α}_{GB} is equal to the elastic energy of forming the misfit dislocation at the interface [9][10].

$$l^{\alpha'} = l^{\alpha}(\boldsymbol{n}_{gb} \times \boldsymbol{n}^{\alpha})$$

$$l^{\beta'} = l^{\beta}(\boldsymbol{n}_{gb} \times \boldsymbol{n}^{\beta})$$

$$E_{GB}^{\alpha} = \frac{1}{2} * (l_{GB}^{\alpha} G {b_{GB}^{\alpha}}^{2})$$

$l^{\alpha'}$is incoming dislocation segment, and $l^{\beta'}$is outgoing dislocation segment,:

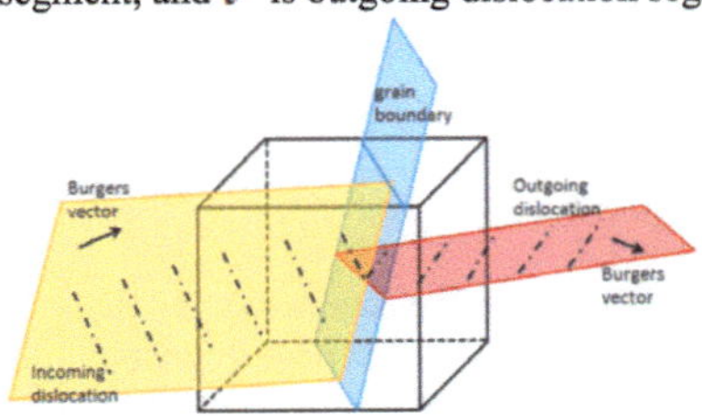

Fig.2 Illustration of slip transmission behavior.[9]

It is likely that the slip transmission will occur at the minimum energy expenses, thus for the grain boundary elements, for each incoming slip system, all the possible outgoing slip system will be investigated, the one with minimum energy barrier will be considered as the corresponding outgoing slip system.

$$E_{GB}^{\alpha} = \frac{1}{2} \min_{\beta} G {b_{GB}^{\alpha}}^{2} l_{GB}^{\alpha}$$

The grain boundaries in the model are represented by the bi-crystal volume elements [11], each having the crystallographic lattice orientations of its adjacent crystals. The grain boundary effects are considered on the near boundary bi-crystal volume elements. The F_{eff}^{α} will be used instead of F_e or F_s compared with the ingrain elements.

$$F_{eff}^{\alpha} = F_{e/s} + E_{GB}^{\alpha}$$

Uniaxial Tensile Test:

The uniaxial tension test was conducted on polycrystalline ferrite. The material was subjected to 12% strain, and the simulation was conducted only on a small region with about 6 grains. The crystal orientation map of the small region was taken before and after the experiment in order to investigate the microstructure evolution.

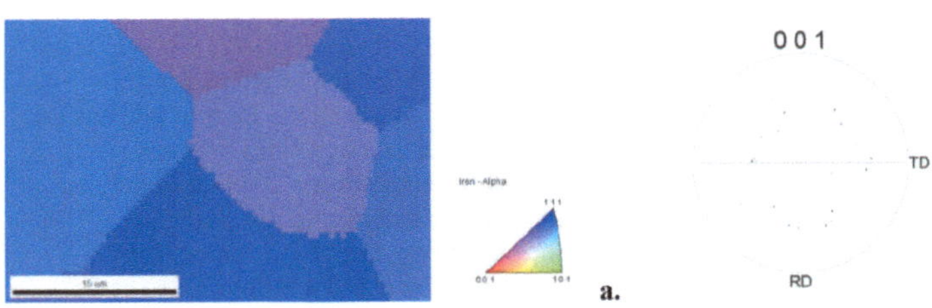

a. b.

Fig.3 a) Orientation image map and b) Pole figure of alpha-iron before deformation.

Simulation Results and Concusions:

The effective stress and the statistical dislocation density distribution is shown in the following Fig.4a and Fig.4b, respectively. Statistical dislocation density is another subset of the total dislocation density and it is about the same as the total dislocation density, since most of the dislocation in the total dislocation density do not contribute to geometric effect such as lattice curvature. A high stress state is likely to be located at grain boundaries. As a result of the stress concentration at these boundaries, a high dislocation density state is more favored there.

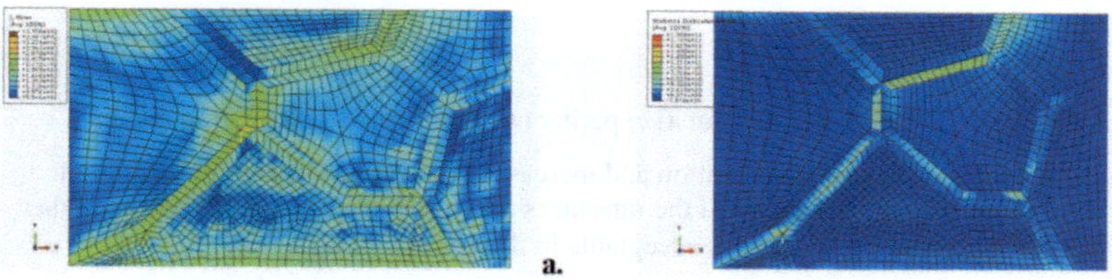

a. b.

Fig.4 The simulation result of a) Effective stress and b) statistical dislocation distribution of alpha-iron after tension

Fig.5a is micromechanical Taylor factor [12] determined from the simulation, it is calculated from the local stress and strain state. Most of the region has a Taylor factor between 1 to 4, and the Taylor factor near the grain boundaries is higher than the grain interior, indicating a higher plastic effect near these regions.

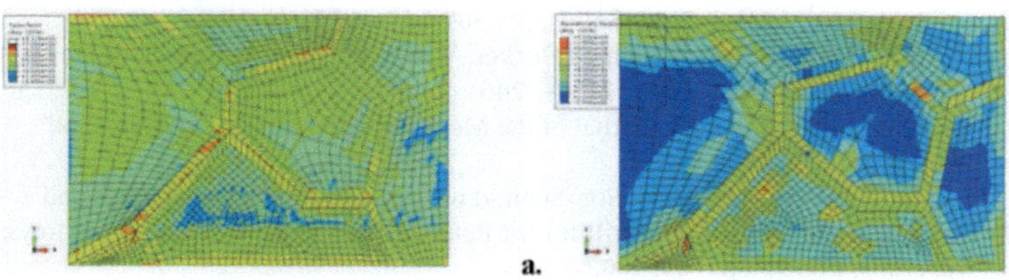

a. b.

Fig.5 a) Taylor factor and b) Geometrically necessary dislocation determined from the model

One important role of geometrically necessary dislocations is to maintain the lattice curvature [13][14], thus the geometrically necessary dislocation here is derived from the plastic strain gradient [15]. Fig.5b is the GND density determined from the model, it is about 1% of the total dislocation density, and higher GND density is localized near the grain boundaries and triple junctions.

Fig.6b is the predicted texture of the polycrystal after deformation, compared with the experimental result in Fig.6a. The results are similar, and both of them indicate orientation spread during deformation.

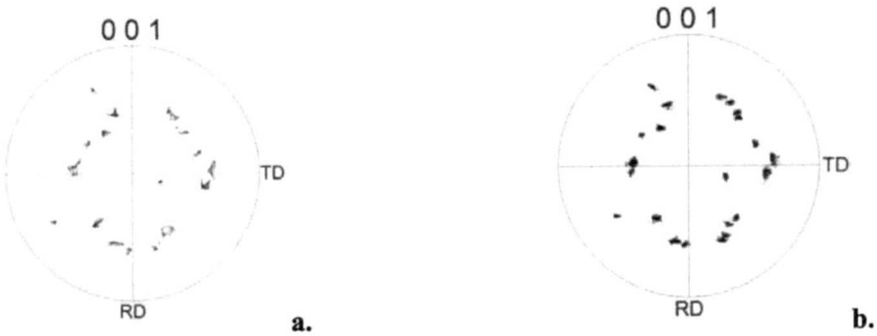

Fig.6 The 001 pole figure from a) experimental result and b) simulation result.

The simulation shows the stress localization and increased dislocation concentration at grain boundaries, and the high Taylor factor at the interfaces and triple junctions indicates the higher hardness there. The texture prediction is acceptable in this model.

Reference:

[1] M. Tang, M. Fivel, L. P. Kubin, Materials Science and Engineering A309–310 (2001) 256–260.

[2] B. Devincre, L. Kubin, T. Hoc, Scripta Materialia 54 (2006) 741–746.

[3] A. Acharya, Philosophical Magazine Vol. 87, Nos. 8–9, 11–21 March 2007, 1349–1359.

[4] H. Lim, M.G. Lee, J.H. Kim, B.L. Adam, R.H. Wagoner, International Journal of Plasticity 27 (2011) 1328–1354.

[5] Asaro, R.J. and Rice, J.R. , J.Mech.Phys. Solids, 25, 309-338, 1977.

[6] Athanasios Arsenlis, David M. Parks, Richard Becker, Vasily V. Bulatov, Journal of the Mechanics and Physics of Solids, 52 (2004) 1213 – 1246.

[7] Athanasios Arsenlis, David M. Parks, Journal of the Mechanics and Physics of Solids, 50 (2002) 1979 – 2009.

[8] Duesbery. M.S., Vitek, V. and Cserti, J. :Non-Schmid plastic behaviour in bcc metals and alloys in Understanding Materials: A Festschrift for Sir Peter Hirsch, edited by C. J. Humphreys, The Institute of Materials, London, p. 165, 2002.

[9] A. Ma, F. Roters, D. Raabe, Acta Materialia 54 (2006) 2181–2194.

[10] A. Ma, F. Roters, D. Raabe, International Journal of Solids and Structures 43 (2006) 7287–7303.

[11] L.P. Evers, D.M. Parks, W.A.M. Brekelmans, M.G.D. Geers, Journal of the Mechanics and Physics of Solids, 50 (2002) 2403 – 2424.

[12] D. Raabe, M. Sachtleber, Z. Zhao, F. Roters and S. Zaefferer, Acta mater. 49 (2001) 3433–3441.

[13] M. F. Ashby, (1970) Philosophical Magazine, 21: 170, 399 — 424.

[14] J. F. Nye, (1953) Acta Metallurgica, 1, 153–162.

[15] Tetsuya Ohashi, Philosophical Magazine Letters, 1997, Vol. 75, NO. 2, 51- 57.

2nd World Congress on Integrated Computational Materials Engineering (ICME)

ICME Challenges and Education

2nd World Congress on Integrated Computational Materials Engineering
Edited by: Mei Li, Carelyn Campbell, Katsuyo Thornton, Elizabeth Holm, and Peter Gumbsch
TMS (The Minerals, Metals & Materials Society), 2013

INTEGRATED REALIZATION OF ENGINEERED MATERIALS AND PRODUCTS: A FOUNDATIONAL PROBLEM

Janet K Allen[1], Farrokh Mistree[1], Jitesh Panchal[2], BP Gautham[3], Amarendra Singh[3], Sreedhar Reddy[3], Nagesh Kulkarni[3], and Prabhash Kumar[3]

[1]The Systems Realization Laboratory@ OU
The University of Oklahoma, Norman, OK, USA
[2]Department of Mechanical Engineering
Purdue University, West Lafayette, Indiana, USA
[3]Tata Research Development and Design Centre
Tata Consultancy Services, 54B Hadapsar Industrial Estate, Pune-411013, INDIA

Keywords: Integrated design; Uncertainty management; Complexity management.

Abstract

In the last several years, considerable ICME related research has been directed towards the study of multiscale materials design. However, relatively little effort has gone into the study of the integrated design of engineered materials and products. In this paper, we describe a foundational problem that is being used to validate and demonstrate the ICME construct and enable design approaches and a computational platform. The foundational problem involves the integrated design of steel and gears, traversing across the chain of steel making, mill production, component fabrication and performance testing tracking the evolution of the material during these processes and linking this to the mechanical design of the component. The design approaches are focused on addressing the following key challenges in the integrated design of products and materials: uncertainty management, complexity management, and holistic verification and validation. The management of uncertainty and complexity is critical to understanding the relationship between computational costs and the value of the resulting information in the design decision making process.

Frame of Reference

The vision of Integrated Computational Materials Engineering (ICME) is to "enable the optimization of the materials, manufacturing processes, and component design long before components are fabricated, by integrating the computational processes involved into a holistic system" [1]. Further, "until materials engineering, component design, and manufacturing engineering are integrated, designers will not attempt to optimize a product's properties through processing and one route to improving the competitiveness of US manufacturers will be closed off" [1]. Pollock and co-authors [1] highlight a number of issues within ICME, such as modeling, scale bridging, modeling uncertainty, data management, and integration with product design efforts.

One of the key aspects of ICME is the use of modeling and simulation tools to study the evolution of material state through a series of manufacturing process steps for production of a component and its performance under service conditions. This requires integration across manufacturing process and performance analysis tools as well as integration across multiple length scales of the material. While major challenges exist in the ability to model and simulate material performance characteristics across multiple length scales, even where reasonable tools exist, one needs to make judicious choices to balance the computational cost and the value delivered through the information provided by the simulation tools. Another major difficulty is the inherent uncertainties that exist in the simulation models in design and uncertainties introduced in manufacturing as well as the stochastic nature of the material itself. These uncertainties are propagated through the chain of processes the material undergoes and are manifest in diverse ways. While it is possible to reduce some of these uncertainties, it is not always possible to achieve this satisfactorily within the constraints of available processes and costs. Then one must find ways to manage the uncertainty in such a way that reduces the final impact on material performance leading to a robust product. Above all, many aspects of the chain of manufacturing processes, material performance and product design play against each other and a designer must make judicious selections and compromises.

According to the NRC research study, "identifying and then beginning to solve some *foundational engineering problems* could be the first steps in developing and demonstrating an ICME framework" [1]. In this paper, we discuss a foundational problem that we are using to demonstrate and validate new design approaches [2, 3] and a computational platform. The foundational problem involves the evolution of a steel component from the alloy selection/development through its manufacturing steps in a steel mill to an intermediate product and subsequent component manufacture, its impact on the component performance and the impact of this information on component design (see Figure 1). The problem is discussed next.

A Foundational Problem for ICME Methodologies and Framework

One of the largest consumers of steel is the automotive industry and an automotive gear is a key transmission element that uses special grades of carburizing steels. Within the Gear Vision document [4], the American Gear Manufacturers Association (AGMA) articulates some of the challenges faced by the gear industry that will be impacted by materials. Some of these challenges include the need to enhance the power density, need to achieve 100% reliability for 20 years for large gears, etc., and it anticipates the use of steels that heat treat to RC70+ and higher with cleaner steel to enable greater power density; new steels to reduce/eliminate intergranular oxidation; shot peening and other innovative manufacturing processes for improved performance; and high quality and stable production processes. Such requirements can be addressed well by looking at the complete production route starting from steel-making to product design and performance testing, encompassing operations that spread across steel mill, automotive ancillary companies and OEMs [5].

A steel mill typically produces a variety of intermediate products such as slabs, billets, blooms and finished flat and long products such as sheets, bars, etc. One common product from a steel mill is a round bar, which is produced through a long product production line and is the raw material for the production of a transmission gear. The evolving structure and the resulting properties of the bar in the mill are dictated by the chemistry and cleanliness of molten steel, casting process and subsequent deformations and thermal processes. Some of the key influencing factors of a bar product include, besides the steel composition, the segregation of alloying elements, distribution of inclusions and the microstructure of the product. These can be predicted to a reasonable accuracy in a simulation framework [6].

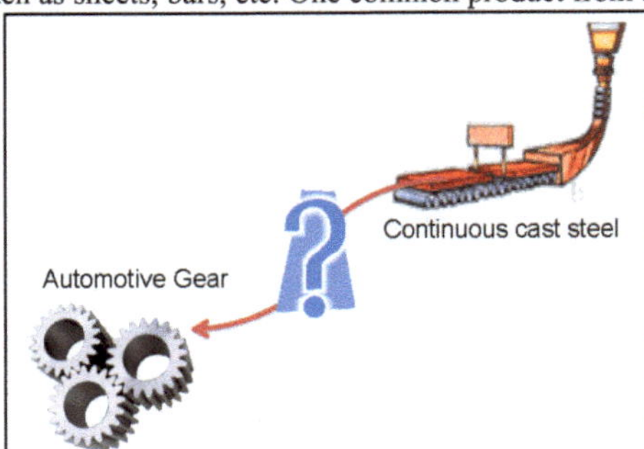

Figure 1 – The foundational problem: Integrated design of steel, manufacturing processes and automotive gears

Armed with the above information, the subsequent component manufacturing processes such as hot-forming, machining, heat treatment can be studied with appropriate simulation tool integration horizontally across the processes and vertically across the material length scales. This finally extends to the component performance evaluation in a simulation environment, and its fatigue life and wear can be determined.

Modeling and simulating the evolution of the structure of the material and determining the influence of this evolving structure on various properties is a challenging task. Apart from the modeling challenges, this requires an enabling platform to integrate interconnected models of various unit operations horizontally, shown in Figure 2, across processes and vertically across multiple length scales. These must have access to appropriate knowledge and databases to obtain the optimized design set points for the various unit operations along the entire production path.

Enabling such a study would help in decision making regarding the specifications required for the mill product. For example, a higher degree of segregation would lead to greater distortion of the gear blanks during forging and heat treatment [7], leading to more machining. Similarly, the extent of inclusions and their distribution influences the fatigue life of the component. However, making stringent requirements on segregation and inclusions can increase the cost of mill production significantly. A combined study would help make judicious decisions and using this information, significant reductions in "total cost" with enhanced performance could be obtained.

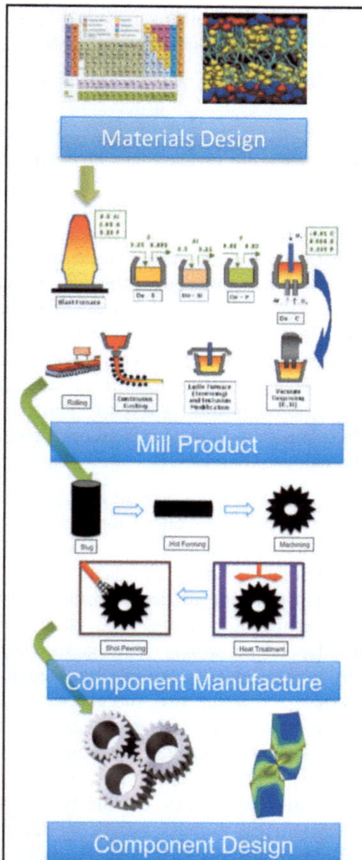

Figure 2 – Schematic of various stages of the foundational problem under consideration

Key Requirements for an ICME Methodology

We view ICME as a systems design process where the goal is to explore the design space, compare design alternatives, and to make informed decisions both at the materials level and at the product level. Exploring the design space is fundamentally different and broader than design optimization. Design optimization implies that the design problem has been parameterized, the ranges of design variables are known, and the design objectives have been quantitatively well defined. While this may be true for components during the detailed design phase, such formulations are generally unavailable for systems in the early stages of design. During the early stages, the design space is typically large, requirements are available at the system level only, the amount of information available is low, and the uncertainty is high. The ICME approach and the platform must adapt to such evolving design contexts. In the following, we discuss a few core issues in ICME from an integrated design perspective.

Managing Uncertainty: Uncertainty within ICME can be from a variety of sources such as: a) imprecise control of process parameters such as temperature in the ladle and the uncertainty in heat transfer rates in the tundish; b) incomplete knowledge of physical phenomena (eg. bainitic transformations during quenching, if present); c) inability to account for all aspects of physics in a model; d) errors during aggregation of information at lower-scale models to link to upper-scale models, etc., for example, bringing the influence of inclusions at microscopic length scale to flow curves used in higher length scale simulations. Approaches and tools to account holisitically for uncertainty in the integrated design of products and materials must address four aspects: uncertainty quantification, propagation, mitigation and management [8]. While uncertainty quantification and propagation aspects have to be addressed by analysts or model developers, uncertainty mitigation and management are activities performed by system designers during the design process. Management of uncertainty involves making meta-design decisions such as i) which aspects of uncertainty to reduce, e.g., by performing experiments or executing higher fidelity models and ii) when to generate simplified metamodels for design exploration [9-11]. Uncertainty management is an open challenge, and has received little attention from the materials design community.

Complexity management in design processes: The increasing complexity of systems being designed results in a corresponding increase in the complexity of their design processes, which is a function of the number of activities in the process and the number and strengths of interactions between the activities. For example, studying the influence of segregation during casting on the gear performance in detail would entail complex and expensive simulations. It is clear that considering all interactions leads to better designs. However, accounting for all interdependencies may result in simulation processes that are so complex or costly to run that they prohibit design exploration. Hence, simpler design processes where some interactions are ignored are faster and more resource efficient. To determine the right level of resource allocation, designers must evaluate tradeoffs between efficiency and complexity of design processes. This involves evaluating the appropriateness of different design process alternatives [12, 13]. Panchal, et al. [11, 13] have used concepts such as value of information from information economics to quantify the tradeoff between the cost and the benefit of simplifying a design process to make satisficing design decisions about design processes. Further research in complexity management is necessary to realize the ICME potential.

Verification and Validation: Panchal and co-authors [8] discuss five different levels of verification and validation (V&V) within ICME. These include a) *individual model V&V*, where the emphasis is on a single model, typically at a single length and time scales, b) *multiscale model V&V*, which involves a set of linked models at multiple length/time scales, c) *multi-physics model V&V*, in which models of the material and product for multiple physical phenomena are validated, d) *design process validation*, where the focus is on ensuring that the design process will yield a solution that satisfies design requirements efficiently, and e) *design outcome validation* that involves comparing design outcomes to system-level requirements. Typically, the discussion on V&V within the ICME community is centered on model V&V through experiments, which includes the first three levels listed above. However from a design standpoint, model V&V is necessary but not sufficient for achieving good design outcomes. Identifying the best process of achieving a design, which involves how decisions are formulated and executed, the sequence in which simulation and experimental activities are carried out, and the way in which models are validated significantly influence the design outcome and the resources needed. The authors are establishing methodologies to perform holistic verification and validation within ICME.

Knowledge Engineering: Design space can be explored much more efficiently if knowledge guided assistance can be provided at various decision points in the design process. For example, knowledge of material compositions and their properties can help shortlist the materials for detailed simulations. Similarly knowledge of structure-property relationships and process-structure relationships can help select the right processing steps for the material. Knowledge engineering plays a critical role in enabling this. Learning is an integral part of knowledge engineering. Results obtained from previous design problems can be systematically mined to gain knowledge that can make future design exploration more efficient.

Requirements for an ICME Platform

The research group at Tata Consultancy Services, Pune, India is developing an ICME platform, PREMΛP, to support the integrated design of products and materials. In order to support the ICME methodology discussed above, the platform should have the following capabilities:

1. Ability to define and provide support for different types of products, materials, manufacturing processes, design processes and so on. Ability to simulate these processes.
2. Ability to integrate simulation tools spanning across processes and length scales.
3. A comprehensive database on materials, products and processes. Ability to connect to external databases.
4. Ability to provide knowledge-guided assistance to users in design-related decision making.
5. Knowledge engineering with ability to capture knowledge from a number of different sources, and the ability to learn from previous design activities and results.
6. Support for systems engineering approach to set up problems
7. Support for multidisciplinary design optimization and decision support
8. Support visualization of aggregated simulation results to enable detailed analysis by users

Summary

It has been well documented that the integration of materials design in product development processes has the potential to provide greater flexibility in achieving system performance and reducing costs [1,7]. However, existing efforts are primarily limited to developing computational approaches for specific problems. There is a long way to go before product development enterprises can adopt integrated products and materials design approaches in their design processes. Existing approaches are *material-property mediated,* i.e., the material properties become the interface between the two domains. The primary advantage of this procedure is that it decouples the system design problem and reduces its complexity. However, using materials properties as the interface between products and materials design may be limiting, particularly given the emerging manufacturing processes where it is possible to tailor geometry and material properties at each point. This is because (a) it restricts the joint exploration of the components and material structure and some preferred designs may be left out, (b) it is difficult to account for manufacturing constraints, and (c) it may not be feasible to control the microstructure independently from the component geometry. The ideal scenario involves integrating the structure of both the components and the material to perform joint design exploration. A comprehensive ICME strategy must address the management of uncertainty and complexity, in addition to modeling and simulation aspects.

/

Acknowledgments

The authors acknowledge the support and encouragement given by the management of Tata Consultancy Services Ltd., India. F. Mistree and J. K. Allen gratefully acknowledge support from the NSF Grant CMMI 1258439 and the L.A. Comp and John and Mary Moore Chairs, respectively. J.H. Panchal acknowledges support from NSF Grant CMMI 1042350.

References

1. *Pollock, T.M., Allison, J. E., Backman, D.G., Boyce, M. C., Gersh, M., Holm, E.A., LeSar, R., Long, M., Powell IV, A.C., Schirra, J.J., Whitis, D.D., and Woodward, C., 2008, "Integrated Computational Materials Engineering: A Transformational Discipline for Improved Competitiveness and National Security", National Materials Advisory Board, NAE, National Academies Press, Report Number: ISBN-10: 0-309-11999-5.*
2. *Kumar, P., Goyal, S., Singh, A.K., Allen, J.K., Panchal, J.H., and Mistree, F., 2013, "PREMAP - Exploring the Design Space for Continuous Casting of Steel", in ICoRD 2013, Lecture Notes in Mechanical Engineering, IIT Madras, Chennai, India, Springer, India.*
3. *Kulkarni, N., Zagade, P.R., Gautham, B.P., Panchal, J.H., Allen, J.K., and Mistree, F., 2013, "PREMAP - Exploring the Design and Materials Space for Gears", in ICoRD'13 - Lecture Notes in Mechanical Engineering, IIT Madras, Chennai, India, Springer, India.*
4. *AGMA, 2004, Gear Industry Vision: A vision for the gear industry in 2025, [cited 2013 March 06]; Web Link: http://agma.server294.com/images/uploads/gearvision.pdf.*
5. *Gautham, B.P., Singh, A.K., Ghaisas, S.S., Reddy, S.S., and Mistree, F., 2013, "PREMAP: A Platform for the Realization of Engineered Materials and Products", in ICoRD'13, Lecture Notes in Mechanical Engineering, Springer India, pp. 1301-1313.*
6. *Singh AK, Pardeshi R, Goyal S., 2011, Integrated modeling of tundish and continuous caster to meet quality requirements of cast steels. In: 1st world congress on integrated computational materials engineering, vol 1, pp. 81–85.*
7. *Şimşir, C., Hunkel, M., Lütjens, J., and Rentsch, R., 2012, "Process-chain simulation for prediction of the distortion of case-hardened gear blanks," Materialwissenschaft und Werkstofftechnik, **43**, pp. 163.*
8. *Panchal, J. H., Kalidindi, S. R., and McDowell, D. L., 2012, "Key Computational Modeling Issues in Integrated Computational Materials Engineering," J. Computer Aided Design, **45**(1), pp. 4-25.*
9. *Panchal, J., Paredis, C.J.J., Allen, J.K., and Mistree, F., 2007, "Design Process Simplification Via Scale and Decision Decoupling – A Value-of-information Based Approach", in ASME Design Automation Conference, Las Vegas. Paper: ASME-DETC2007-35686.*
10. *Panchal, J.H., 2005, "A Framework for Simulation-Based Integrated Design of Multiscale Products and Design Processes", PhD Dissertation, The GW Woodruff School of Mechanical Engineering, Georgia Institute of Technology, Atlanta, GA.*
11. *Panchal, J.H., Choi, H.-J., Allen, J.K., McDowell, D.L., and Mistree, F., 2007, "A Systems Based Approach for Integrated Design of Materials, Products, and Design Process Chains," J. Computer-Aided Materials Design, **14**(Supplement 1), pp. 265-293.*
12. *Panchal, J.H., Choi, H.-J., Allen, J.K., McDowell, D.L., and Mistree, F., 2006, "Designing Design Processes for Integrated Materials and Products Realization: A Multifunctional Energetic Structural Materials Example", in Design Automation Conference, Philadelphia, PA. Paper Number: DETC2006-99449.*
13. *Panchal, J.H., Paredis, C.J.J., Allen, J.K., and Mistree, F., 2009, "Design Process Simplification Via Scale and Decision Decoupling – A Value-of-information Based Approach," J. Computing and Information Sciences in Engineering, **9**(2), pp. (021005)1-12.*

2nd World Congress on Integrated Computational Materials Engineering
Edited by: Mei Li, Carelyn Campbell, Katsuyo Thornton, Elizabeth Holm, and Peter Gumbsch
TMS (The Minerals, Metals & Materials Society), 2013

ICME – A MERE COUPLING OF MODELS OR A DISCIPLINE OF ITS OWN?

Markus Bambach[1], Georg J. Schmitz[2], Ulrich Prahl[3]

[1]Institute of Metal Forming, RWTH Aachen University
[2]ACCESS e. V., RWTH Aachen University
[3]Institute of Ferrous Metallurgy, RWTH Aachen University

Keywords: ICME, standardization, economy, education, viability, platform

Abstract

Technically, ICME - Integrated computational materials engineering - is an approach for solving advanced engineering problems related to the design of new materials and processes by combining individual materials and process models. The combination of models by now is mainly achieved by manual transformation of the output of a simulation to form the input to a subsequent one. This subsequent simulation is either performed at a different length scale or constitutes a subsequent step along the process chain. Is ICME thus just a synonym for the coupling of simulations? In fact, most ICME publications up to now are examples of the joint application of selected models and software codes to a specific problem. However, from a systems point of view, the coupling of individual models and/or software codes across length scales and along material processing chains leads to highly complex meta-models. Their viability has to be ensured by joint efforts from science, industry, software developers and independent organizations. This paper identifies some developments that seem necessary to make future ICME simulations viable, sustainable and broadly accessible and accepted. The main conclusion is that ICME is more than a multi-disciplinary subject but a discipline of its own, for which a generic structural framework has to be elaborated and established.

Introduction

A number of success stories e.g. [1-5] suggest that Integrated Computational Materials Engineering (ICME) [6] is a versatile, ready-to-use solution for the development of new materials, products and the design of the corresponding manufacturing process chains. The present paper challenges this view by critically examining the viability of ICME simulations. The main question is: Is ICME just a toolbox for creating ICME meta-models by combining different black-box type materials and process models? Or does the coupling of models and simulation codes lead to new 'meta-models' and 'meta-codes' rising dedicated new scientific and technologic challenges? Can these challenges then be met by a joint effort of different disciplines or do they turn ICME into a discipline of its own whose viability has to be ensured by dedicated efforts?

A study on "Integrated Computational Materials Engineering – a transformational discipline" [6] already uses the term "discipline", however several factors seem to determine the definition and viability of ICME as a 'discipline' in terms of scientific/technological, economic and even sociologic aspects, Figure 1. These factors will be discussed in the following, identifying both the current status and possible future developments.

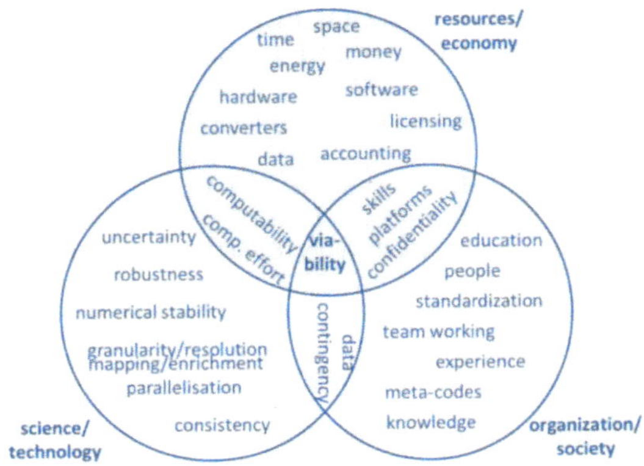

Figure 1: Factors determining the viability of ICME. Each of these factors comprises a technological, an economic and a sociologic dimension.

Resources and Economic Aspects of ICME

ICME has emerged from economically driven questions and requirements for the optimization of materials and products. The discussion of its economic dimension – perhaps in contrast to purely scientific disciplines – thus seems justified and adequate. First ICME examples have been based on combinations of available resources:

- *Software:* simulation codes covering almost all manufacturing processes and all the relevant scales in the material/component,
- *Hardware:* Powerful hardware and storage capacities – whether centralized or distributed across the grid – to perform such simulation chains and to store respective results,
- *Data:* for properties of materials and components and for process conditions,
- *Initial and boundary conditions:* on all relevant length scales.

Performing an ICME type simulation by its nature comprises the integrated use of multiple simulation tools, respective models and data – i.e. "software applications" – disregarding whether these are academic and/or commercial codes. All these tools have to be programmed or to be procured, to be installed, to be regularly up-dated and maintained and eventually to be run in a suitable hardware environment. Also, the computer hardware and network infrastructure has to be procured, to be maintained and to be operated including non-negligible related costs.

The most significant costs will however be evoked by adequately skilled people operating an ICME environment. Additionally, the different software applications will require specialists with a dedicated expertise.

In particular, small and medium sizes enterprises (SMEs) and academia, although being highly interested in ICME solutions to their problems, are not in the position to maintain own simulation expertise and infrastructure. In the long term area of "Calculation on Demand" will thus become important for ICME, where users only license software and lease hardware for a specific simulation task in a concept of "Software as a Service" (SaaS). The configurability of a modular and adaptable ICME setting will allow providing tailored individual solutions for different users. Simulation chains meeting the topics of interest of an individual SME will then be designed and created by a skilled team of ICME engineering experts as a service. Once the simulation chain is operational it is up to the SME user to modify individual parameters of this

ICME simulation and to exploit their effect on his products and their production process in the frame of parameter variations.
Especially meta-data being collected along the entire simulation chain like e.g. cycle times for individual processes, cost estimates for raw materials, for energy consumption, pollution, cycling material, required investments and others are most valuable for decision making processes.
In the frame of "SaaS" respective accounting and licensing schemes for commercial codes have to be implemented into a future ICME concept. Especially confidential transfer of sensitive data via internet channels has to be guaranteed requiring crypting schemes and secure data storage.
In summary, aspects of economic viability of ICME encompass a number of other disciplines like telecommunication, economy, internet technology/distributed simulations and cryptology.

Organizational and Societal Aspects of ICME

Besides the availability of resources and their economic aspects, there are still other aspects that determine the future success of ICME. Potential obstacles concern organizational aspects of configuring complex ICME tasks and societal aspects in terms of education, skills and acceptance for performing ICME tasks.
As a generic condition for performing ICME, scientists are needed being able to formulate and implement models in tools, to accelerate tools by parallelization or model reduction technologies, to control IT resources via internet, and to design interfaces to control ICME requests and to visualize and to evaluate ICME results. These experts have to be educated in a multidisciplinary framework and along with their technological skills they need to have additional soft skills, e.g. team working in multi-cultural conditions between different scientific disciplines like engineering, natural science, IT and economics. The education and training of such ICME experts thus requires a detailed definition of a multi-disciplinary teaching course agenda. Actually emerging efforts spent in the education of highly-skilled material scientists to become familiar with modeling and numerical methods represent a first step into this direction and a number of Computational Engineering Science or PhD programs have been established recently. Still much more effort has to be spent into this kind of interdisciplinary education in future.
Even if all of the above mentioned resources in terms of tools and experts are available, the individual simulations along the simulation path have to be configured, executed and evaluated. These steps can be performed manually, requiring a significant effort to compose even a single ICME simulation chain. To reduce these efforts in future and to make ICME useable for a broad range of non-ICME experts, an ICME-specific cyber-infrastructure has to be created. As a generic solution a standardized, modular and expandable platform for ICME computations has been proposed and is available now [7]. Recently, a European activity has been launched aiming at formulating a global standard for information exchange between a variety of commercial and academic simulation codes operating on different lengths scales [8].
Performing ICME simulations in practice is still cumbersome as no 'master' pre- and postprocessor for the generation and evaluation of ICME meta-models and results is available by now. Efficient ICME pre- and postprocessing has to be based on modern, user-friendly Graphical User Interfaces (GUI) allowing both to control the simulation chain and to visualize the results for non-(ICME-)experts from materials and production engineering practice.
Eventually, industrial reluctance to accept ICME must be overcome. This reluctance is due to the slow conversion of science-based tools to engineering tools, a lack of awareness and investment, and a shortage of trained computational materials engineers.
Furthermore, a general mistrust against virtual solutions has to be overcome by a continuous publication of relevant success stories as a most effective confidence-building measure. Actually, companies providing ICME infrastructure usually have own concepts to integrate their

proprietary models and tools and tend to protect respective own platform solutions. This substantially hinders an open, easy-to-use exchange of results with tools from other providers or from academia.

Scientific and Technological Aspects of ICME:

When ICME is applied to a specific problem like the design and optimization of a product manufacturing sequence, individual software codes along this chain and across the length scales have to be coupled to form a specific 'ICME simulation path', as indicated schematically for the example of a new steel grade for gearing components [9], Figure 2.

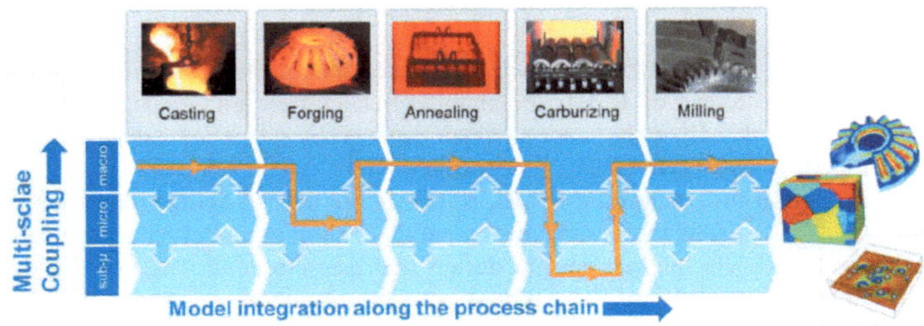

Figure 2: ICME simulation path for simultaneous materials and process design for a gear wheel [9]. For this example especially the finite element code DEFORM [10] has been used for the forging simulation, the multi-phase field code MICRESS [11] in combination with Thermo-Calc and respective databases [12] for the description of microstructure evolution, MatCalc [13] to describe the evolution of precipitate size distributions. Note that the "path" actually will consist of a number of mutually coupled, individual paths at the different scales.

This *simulation path* constitutes a meta-model, which requires several resources to be combined. The development of the individual resources themselves remains a task of the different individual scientific disciplines. The essential topic of ICME is their mutual combination and particularly all scientific and technological questions arising from this combination(s), e.g.:

- Global optimum instead of optima of individual models
- Specification of a global frame of reference and handling of different time scales,
- Data contingency along the simulation chain and across the length scales,
- Meshing and remeshing procedures and respective algorithms,
- Self-consistent coupling of scales,
- Data reduction/effective properties,
- Robustness and stability of ICME type simulation chains.

Finding a global optimum instead of optimizing the individual processes represents one of the generic key benefits of the ICME approach. Some of the other topics can be easily addressed like e.g. the specification of a common frame of reference and a common timeline for all models in the simulation path. Other topics may require modifications/extensions of individual models in the chain, e.g. to guarantee data contingency. An example could be segregation patterns of alloy elements being channeled through a deformation simulation although not being important for this simulation itself, but being again important for simulations of a downstream manufacturing process like e.g. heat treatment or welding. Meshing and remeshing procedures and respective algorithms have to be provided to allow for seamless conversion between different types of meshes, mesh refinement and coarsening and other operations like e.g. modular combinations of meshes e.g. for simulating joining of different components. An important issue will be a self-

consistent coupling between different scales to secure e.g. mass and energy conservation [14]. Another important task is data reduction, resp. the determination of effective values being fed back to the larger scale. Besides simple volume averaging approaches, especially mathematical homogenization schemes and virtual testing and allow evaluating even anisotropic values [15].
The *robustness* of ICME meta-models evoked by the increased complexity of an ICME simulation chain as compared to an individual simulation is affected by the following issues:
Propagation of Uncertainty: An ICME meta-model may comprise a large number of individual simulations. All input data for these individual models are subject to uncertainty and errors. Each simulation itself contributes additional uncertainty due to unknown influences of the numerical solution like round-off errors, discretization errors, errors produced by mapping the results between codes, and others. Thus the ICME meta-model may be strongly affected by uncertainties, their propagation and their accumulation. Once the overall uncertainty exceeds a critical threshold, the entire ICME simulation becomes meaningless. Hence, methods for the quantification of uncertainty in ICME model predictions are urgently required.
Ambiguity/stability of ICME models: A number of phenomena related to the evolution of a microstructure are discontinuous, i.e. certain thresholds have to be crossed to trigger phenomena such as solidification, recrystallization or phase transformations. In an ICME meta-model, where a process model invokes a certain material model once a certain critical condition is crossed (e.g. temperature or degree of deformation) it can be easily imagined that in the same process model – when being executed with only slightly different initial conditions or settings – the material model might be invoked earlier, later or even not at all leading to drastically different results. Hence, methods are needed to assess the stability of the predictions made using ICME meta-models against minor variations in the initial and boundary conditions as well as parameter related to the numerical solution.
Run-time of ICME meta-models: A strong coupling between process models and material models with full spatial resolution at the scale of the microstructure and along the entire process chain would be highly desired in the field of ICME. The computational efforts are still tremendous and largely exceed present hardware capabilities. Hence, ICME applications in the mid and long term demand for speeding up individual simulations, e.g. by parallelization or model reduction.
Among others, these factors determine the viability of ICME from a technological / scientific point of view. Especially resolving the questions on stability and uncertainty requires joint efforts from materials science, mathematics, computer science and production engineering.

Conclusions

In 2008 it was stated that for ICME to succeed, one of three things to happen is that "ICME must be embraced as a discipline in the materials science and engineering community, leading to changes in education, research, and information sharing." [6]. The term "discipline" (e.g. according to Wikipedia) is not unambiguously defined. ICME may probably to be considered an interdisciplinary field – like biochemistry or biophysics, which both seem to emerge as own disciplines – or even as a multidisciplinary field with following disciplines contributing: (i) 1st level disciplines like physics, chemistry, mathematics, computer science, materials science, production engineering and (ii) 2nd level disciplines like economics, telecommunications, cryptology, education/didactics, psychology and others.
The scope of future ICME research and education has to be on a holistic view on materials and their processing into components and products. The major topics to be addressed in the frame of dedicated ICME research all should all be related to the *combination of models* into meta-models, while the individual models will be further developed within the individual disciplines. The present paper has outlined a number of aspects, where the *combination of models* as a focus

for future ICME research provides new benefits, leads to drawbacks, rises new scientific and technological questions, opens new opportunities all being beyond of what is tackled by the individual disciplines contributing to ICME. Along with several other features characterizing a discipline like e.g. an own research community with own dedicated journals (e.g. IMMI) and dedicated workshops and conferences (like the 1st and 2nd ICME Conferences on ICME in 2011 and 2013), as well as already existing dedicated curricula in academic education we come to the conclusion that ICME indeed may be considered an own discipline.

Acknowledgements

We gratefully acknowledge the financial support of the Deutsche Forschungsgemeinschaft (DFG) within the Cluster of Excellence "Integrative Production for High Wage Countries".

References

[1] Allison, JE; Liu, BC; Boyle, KP; Hector, L; McCune, R: ín Agnew, SR; Neelameggham, NR; Nyberg, EA; Sillekens, WH (eds.) "Integrated Computational Materials Engineering for Magnesium in Automotive Body Applications", Proc. Magnesium Technology (2010) Seattle, WA, pp. 35-40

[2] Hirsch, J (eds): "Virtual Fabrication of Aluminium Products", Wiley VCH, Weinheim, ISBN: 352731363X

[3] Hirsch, J; Karhausen, KF: "History of ICME in the European Aluminium Industry" in Allison, J; Collins, P.; Spanos, G. (eds):"Proc. of the 1st World Congress in Integrated Computational Materials Engineering (ICME)" (2011) p.203-210, John Wiley&Sons VCH, Weinheim, ISBN: 352731363X

[4] Schmitz, GJ; Benke, S; Laschet, G; Apel, M; Prahl, U; Fayek, P; Konovalov, S; Rudnizki, J; Quade, H; Freyberger, S; Henke, T; Bambach, M; Rossiter, EA; Jansen, U; Eppelt U: Production Engineering & Research (2011) DOI: 10.1007/s11740-011-0322-1

[5] Backman, DG; Wei, DY; Whitis, DD; Buczek, MB; Finnigan, PM; Gao, DM: "JOM 58 11 (2006) 36

[6] National Research Council: Integrated Computational Materials Engineering: A Transformational Discipline for Improved Competitiveness and National Security; National Academic Press, Washington, D. C. (2008), ISBN: 0-309-12000-4

[7] Schmitz, GJ; Prahl, U (eds): Integrative Computational Materials Engineering - Concepts and Applications of a Modular Simulation Platform, Wiley VCH Verlag Weinheim, ISBN 978-3-527-33081-2 (2012)

[8] Integrated Computational Materials Engineering expert group – a coordination and support action of the European Commission; www.icmeg.eu (2013)

[9] Prahl, U; Schmitz, GJ; Benke S; Konovalov, S; Freyberger, S; Henke, T; Bambach, M: in Zoch HW, Lübben T (eds.): 3rd Int. Conf. on Distortion Engineering 2011, IDE 2011, 14.-16.09.2011, Bremen, pp. 429-436

[10] www.deform.com

[11] MICRESS- The MICRostructure Evolution Simulation Software; www.micress.de

[12] Thermo-Calc Software and databases: www.thermocalc.se

[13] http://matcalc.tuwien.ac.at/

[14] Böttger, B; Eiken, J; Apel, M: J. Comput. Phys. (2009), 6784-6795

[15] Laschet G, Kashko T, Benke S, Öte M, Bobzin K: Chapter 5 in [7] (2013)

2nd World Congress on Integrated Computational Materials Engineering
Edited by: Mei Li, Carelyn Campbell, Katsuyo Thornton, Elizabeth Holm, and Peter Gumbsch
TMS (The Minerals, Metals & Materials Society), 2013

KNOWLEDGE ASSISTED INTEGRATED DESIGN OF A COMPONENT AND ITS MANUFACTURING PROCESS

BP Gautham, Nagesh Kulkarni, Danish Khan, Pramod Zagade, Sreedhar Reddy, Rohith Uppaluri

Tata Research Development and Design Centre
Tata Consultancy Services, 54B Hadapsar Industrial Estate, Pune – 411013, INDIA

Keywords: Integrated design, Knowledge base, User interface

Abstract

Integrated design of a product and its manufacturing processes would significantly reduce the total cost of the products as well as the cost of its development. However this would only be possible if we have a platform that allows us to link together simulations tools used for product design, performance evaluation and its manufacturing processes in a closed loop. In addition to that having a comprehensive knowledgebase that provides systematic knowledge guided assistance to product or process designers who may not possess in-depth design knowledge or in-depth knowledge of the simulation tools, would significantly speed up the end-to-end design process. In this paper, we propose a process and illustrate a case for achieving an integrated product and manufacturing process design assisted by knowledge support for the user to make decisions at various stages. We take transmission component design as an example. The example illustrates the design of a gear for its geometry, material selection and its manufacturing processes, particularly, carburizing-quenching and tempering, and feeding the material properties predicted during heat treatment into performance estimation in a closed loop. It also identifies and illustrates various decision stages in the integrated life cycle and discusses the use of knowledge engineering tools such as rule-based guidance, to assist the designer make informed decisions. Simulation tools developed on various commercial, open-source platforms as well as in-house tools along with knowledge engineering tools are linked to build a framework with appropriate navigation through user-friendly interfaces. This is illustrated through examples in this paper.

Introduction

Recent developments in computational materials science and engineering have raised significant hope of utilizing computational techniques for development of new materials, observing their evolution through the manufacturing processes and finally developing materials and manufacturing processes in a closed loop with product development. This paradigm is expected to lead to faster development of materials and products with lower cost and reduce the experimental techniques used in development significantly. The US National Academy of Sciences report on Integrated Computational Materials Engineering (ICME) [1] has articulated this in detail. A large part of research pertaining to ICME addresses various aspects of materials science across length scales and bridging them. However, as articulated in [1], there is also need to look at other aspects such as databases, platforms for tool integration, standards and knowledge engineering methods. This would be required to make ICME available to different stakeholders from researchers to product designers [2]. A strong knowledge enriched platform would be significantly helpful to make ICME paradigm reach the engineers and designers in the industry [3]. The focus of this paper is to explore in a closed loop, product design with material selection and manufacturing process design to enable decision making across these two activities, breaking the barriers between product designers and manufacturing process experts.

Simulation tools for evolution of material state during the manufacturing process and product performance evaluation along with design methodologies are all incorporated in a single platform. In order to make the platform accessible to mechanical component designer who may not be an expert in selection of relevant manufacturing processes, various types of user assistance based on domain knowledge are provided to allow the designer to study the impact of manufacturing processes on performance and thus make informed decisions on the manufacturing processes. This process is illustrated through the steps of mechanical design, heat treatment operations simulation and performance evaluation in a single platform supporting various stages of decision making for a steel gear used for load transfer applications. The aim of this work is to eventually develop methods and templates for such an integrated platform. This paper restricts itself to the illustration of the application studied and lays emphasis on the process than the results.

Description of the Case Study

Design process of an automobile transmission component, gear, is chosen for the case study. Designing a gear that meets user specified functional and performance requirements is a challenging task. The designer has to deal with conflicting goals such as high reliability and low cost and compactness. High reliability demands gears of larger size and high-strength material, which conflicts with the compactness and low cost requirements. The cost to size ratio is also not linear as smaller size gears can be made using expensive heat treatment and other manufacturing processes. There are a number of possible routes to manufacture a gear, such as (a) Blanking → Gear Hobbing → Carburization-Quenching → Tempering → Fine Grinding or (b) Hot Forging → Form Grinding → Carburization-Quenching → Tempering → Fine Grinding →Shot Peening. In the current study, we limit the study to AGMA based gear design, material selection, carburization-quenching and tempering processes (here in referred to as CQT process) and their impact on the gear performance. Figure 1 shows the schematic of scope of the current study. It illustrates key steps of a typical process route to be followed while carrying out industrial gear design and manufacture.

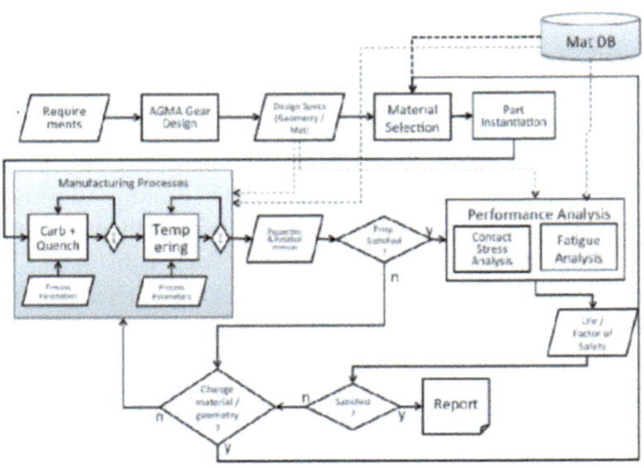

Figure 1: Flowchart for gear design and manufacture

The process starts with capturing functional requirements and progresses through geometric design, material selection, and prediction of effect of manufacturing processes on part and finally the evaluation of fatigue performance under applied loads. This sequence of processes is explained in detail in the next section along with the knowledge guided assistance the platform provides to the designers. Rules are used at several places in the process chain to decide process parameters to fetch appropriate data from the database and to perform informed corrective actions to achieve the specified targets. Integration of various steps of the design process is achieved through seamless data interchange using appropriate process and product templates.

Details of Implementation

The case study described in above section is carried out using various tools for design, and simulation of manufacturing processes and performance. These tools are linked through a platform. The platform also provides appropriate knowledge support through user interaction facilities in the GUI. It provides interfaces to the user where he/she can perform tasks such as

- Providing functional and performance requirements,
- Setting up design space,
- Setting up process set points and simulation controls with knowledge support
- Making GO/NOGO decisions

Fig. 2 (a) shows the User Interface (UI) screen for capturing the functional and performance requirements. As we are dealing with gear design here, appropriate functional and performance requirements such as design torque, gear ratio, input speed (RPM), minimum expected reliability, fatigue life, factor of safety in bending and contact etc., are captured. Except for information on key requirements, default values are provided for all other parameters based on prior knowledge/rules used in designing a gear. The user can change these default values where required and carry out the preliminary design of the gear. Our objective here is to generate initial guesses for gear geometry and material specifications so as to narrow down the vast design material space. User can tweak the geometry, material and/or manufacturing process to explore the synergy between these using the platform.

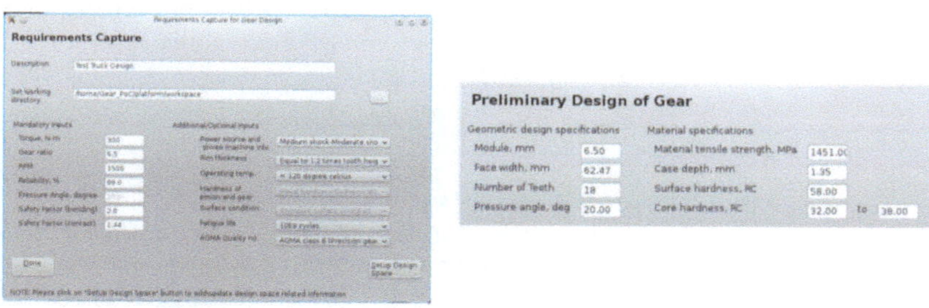

(a) (b)

Figure 2: Preliminary gear design (a) Requirements capturing, (b) Geometrical design and material specs

The platform uses tool adapters to generate simulation tool specific input files, executes the simulation tools and post-processes their results. Fig. 2 (b) shows typical output of the preliminary gear design, for which the geometry is defined in terms of module, face-width, number of teeth and pressure angle. The platform provides a knowledge service for selecting materials from the database based on strength and hardness requirements estimated during the design as shown in Figure 2 (b). The database used in this work consists of carburizing and through hardenable grades of steels taken from publically available National Institute for Material Science (NIMS) database. The material strength and hardness values used for selection based on the Cartesian distance between the required values and material properties in the database after normalization. Normalization is carried out by centering with respect to the mean followed by division by variance for each property.

The next stage allows the user to create a customized finite element mesh required to carry out further manufacturing and virtual testing simulations. Here the user need not construct the mesh in any meshing tool, but merely needs to give a few details such as the number of segments required in pre-specified critical regions. The mesh is automatically generated through a script that sets up a process for mesh generation in a commercial FE pre-processor with appropriate mesh controls and generates various node and element sets required for boundary condition specification and post processing. A set of common terms are used to help the platform interpret and pass the information across the tools.

Subsequent to mesh generation, the carburization-quenching and tempering processes are simulated. All these three stages are simulated together internally in a number of simulation steps with a single external input screen and output report screen. Figure 3 shows the main input screen for this process. The inputs represent the heating step followed by two carburization boost steps, two carbon diffusion steps, quenching and tempering. User can set up individual process step set points such as temperature, time, carbon potential, mass transfer coefficient etc.

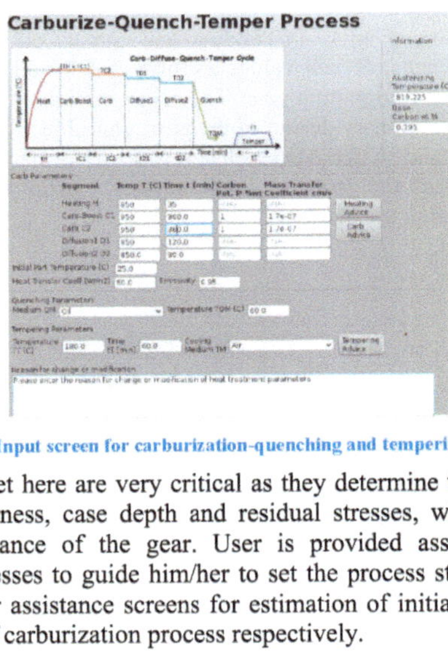

Figure 3: Input screen for carburization-quenching and tempering processes

The process parameters set here are very critical as they determine the final material properties of the gear such as hardness, case depth and residual stresses, which play a crucial role in determining the performance of the gear. User is provided assistance based on standard knowledge of these processes to guide him/her to set the process step set points appropriately. Figure 4 shows such user assistance screens for estimation of initial set points for heating and estimation of set points of carburization process respectively.

In the screen shown in Figure 4(a), estimated time for heating is computed using the time required for attaining a temperature within a small gap of the set-point temperature based on the volume of the material. A feedback on the adequacy of this will be provided after the simulation is completed as will be discussed later. Figure 4(b) shows a screen to provide initial guidance on total time for carburization in the boost phase based on the hardness requirement, given the carbon potential as an input. This uses well-known empirical equations. A number of similar guidance screens are incorporated across the process flow at several places.

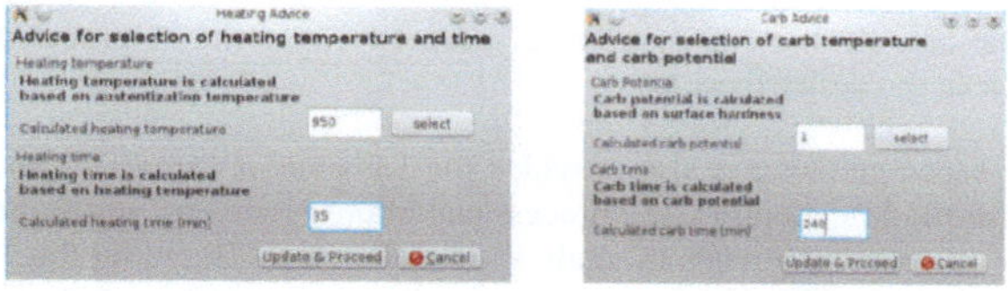

(a) (b)

Figure 4: Knowledge assisted advice for (a) Heating time, (b) Carburization time

Apart from the knowledge assistance in setting up a process step, platform also provides assistance to enable the user to study the impact of manufacturing processes on performance and thus make informed decisions. This is illustrated in Figure 5, which shows the output of CQT simulations. The first two rows show whether the desired temperature is reached in the set time and the actual time needed to achieve the temperature. The time to heat can be corrected in the next round of simulation. The achieved values of material properties are compared with the material property requirements derived from the design and appropriate flags are set based on preset rules. These rules can range from simple bounds checks to ones involving complex nested conditions. A green flag indicates satisfactory achievement of the target specification and a red flag indicates that the user's attention is needed. User can modify the process step set points based on the rules/guidance provided by the platform and carry out the simulations again to observe the impact of the modifications on property achievement. Based on the outcome, the user can either proceed for the next step or go back to change the process parameters for CQT stage as can be seen from the buttons at the bottom of Figure 5.

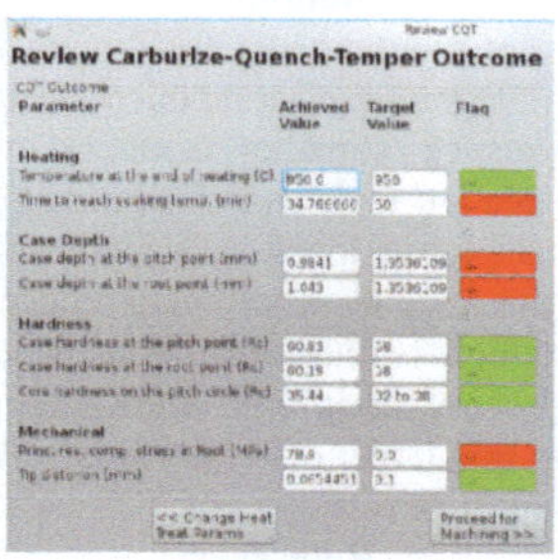

Figure 5: Output of CQT

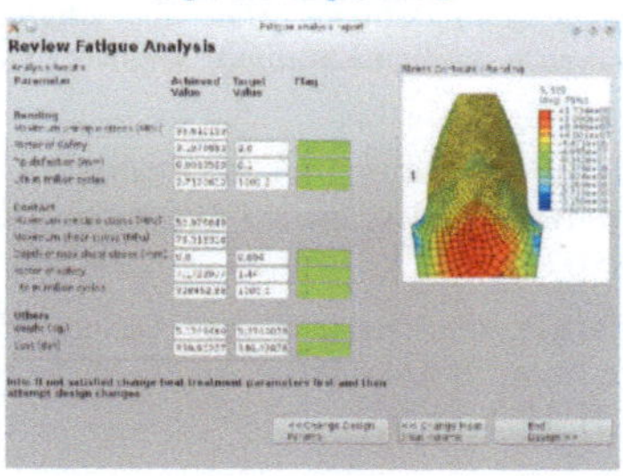

Figure 6: Output of fatigue analysis

A similar knowledge based assistance system is available for virtual testing simulations. Figure 6 shows an example of this for contact and bending fatigue analysis simulations. Based on the outcome, the user can change one or more of geometry, material or processing conditions as necessary to achieve the desired results.

Rule/knowledge-based guidance is also provided for decisions at various stages. Figure 7 shows sample screens for modification of CQT process conditions, geometry and material respectively, with information of previous rounds of study shown alongside. Based on this information from earlier rounds and back-end knowledge support, user can make appropriate decisions for the next round of simulation and continue until satisfied.

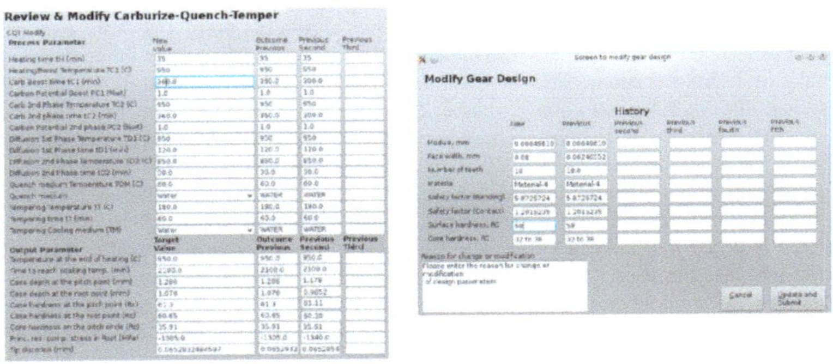

Figure 7: GUI for modification of (a) CQT parameters, (b) Gear geometry and material

Summary

Increased competition and stringent requirements are forcing industries to look for innovative methods for design of products. Integration of component design, material design/selection and manufacturing, along with performance estimation is known to be a promising way of reducing the design trials. Linking the simulation tools used for product design, manufacturing processes and performance evaluation in a closed loop on a single platform, along with appropriate knowledge guided assistance to the designer would help reduce the total cost of product as well as its cost of development. An example is illustrated in this paper for achieving integrated design of product and manufacturing process along with knowledge support for decision making at various stages. This work is a part of a larger effort towards building an ICME enabling platform at the authors' organization.

References

1. Pollock T., ed., *Integrated Computational Materials Engineering: A Transformational Discipline for Improved Competitiveness and National Security* (The National Academies Press, 2008).

2. B P Gautham et al. "PREMAP: A Platform for the Realization of Engineered Materials and Products", *ICoRD'13, Lecture Notes in Mechanical Engineering*, ed. A. Chakrabarti and R. V. Prakash , (Springer India 2013), 1301-1313.

3. Manoj Bhat et al. "PREMAP: Knowledge Driven Design of Materials and Engineering Process", *ICoRD'13, Lecture Notes in Mechanical Engineering*, ed. A. Chakrabarti and R. V. Prakash , (Springer India 2013), 1315-1329.

AUTHOR INDEX
2nd World Congress on Integrated Computational Materials Engineering

T

U

V

W

X

Y

Z

SUBJECT INDEX
2nd World Congress on Integrated Computational Materials Engineering

β

2

3

7

A

B

C

D

E

F

G

H

MIX
Papier aus verantwortungsvollen Quellen
Paper from responsible sources
FSC® C105338

If you have any concerns about our products,
you can contact us on
ProductSafety@springernature.com

In case Publisher is established outside the EU,
the EU authorized representative is:
Springer Nature Customer Service Center GmbH
Europaplatz 3, 69115 Heidelberg, Germany

Printed by Libri Plureos GmbH
in Hamburg, Germany